Process Systems Engineering for Environmental Protection

Process Systems Engineering for Environmental Protection

Guest Editor

Javier Martínez-Gómez

MDPI

Basel • Beijing • Wuhan • Barcelona • Belgrade • Novi Sad • Cluj • Manchester

Guest Editor
Javier Martínez-Gómez
Departamento de Teoría de la
Señal y Comunicación
Universidad de Alcalá
Alcalá de Henares
Spain

Editorial Office
MDPI AG
Grosspeteranlage 5
4052 Basel, Switzerland

This is a reprint of the Special Issue, published open access by the journal *Processes* (ISSN 2227-9717), freely accessible at: https://www.mdpi.com/journal/processes/special_issues/P18D7L3P06.

For citation purposes, cite each article independently as indicated on the article page online and as indicated below:

Lastname, A.A.; Lastname, B.B. Article Title. *Journal Name* **Year**, *Volume Number*, Page Range.

ISBN 978-3-7258-6246-7 (Hbk)
ISBN 978-3-7258-6247-4 (PDF)
https://doi.org/10.3390/books978-3-7258-6247-4

Contents

About the Editor

Javier Martínez Gómez

Javier Martínez Gómez is a Professor in the Department of Signal Theory and Communications (Mechanical Engineering Area) at the Polytechnic School of the Universidad de Alcalá, Spain. He holds a Ph.D. in Materials Science and Engineering from Universidad Carlos III de Madrid and a European Fusion Doctorate from FUSENET. His research focuses on thermal design, characterization of materials, and decision analysis applied to energy systems and sustainable technologies. Dr. Martínez Gómez has authored more than 50 indexed publications and contributed to over 90 scientific works, with an h-index of 24. He has participated in more than 20 research projects, serving as principal investigator in four, and collaborated in international initiatives such as Horizon Europe programs. His work spans advanced materials for energy, phase change materials for thermal storage, and multi-criteria decision making for energy planning. In addition to his research, he has extensive teaching experience in mechanical and materials engineering and has contributed to energy policy studies in Latin America.

Preface

This Reprint, "Process Systems Engineering for Environmental Protection", brings together cutting-edge research and innovative methodologies at the intersection of process systems engineering (PSE) and sustainability. The subject of this collection reflects the growing need to integrate advanced modeling, optimization, and control strategies to address environmental challenges in industrial processes.

The scope of the Reprint encompasses recent developments in process design, product design, and dynamic control, with a strong emphasis on reducing environmental impact and promoting sustainable development. The contributions included here explore multi-scale approaches, data-driven technologies, and integrated platforms that enable efficient resource utilization, energy savings, and enhanced safety in complex production systems.

The motivation for this Special Issue stems from the urgent global demand for cleaner technologies and sustainable industrial practices. By leveraging advances in artificial intelligence, big data analytics, and digitalization, PSE offers powerful tools to optimize operations and minimize ecological footprints. These efforts not only support environmental protection but also align with economic and societal objectives, ensuring a holistic approach to process innovation.

This Reprint is intended for researchers, engineers, and practitioners in chemical engineering, systems engineering, and related fields who seek practical solutions and theoretical insights for sustainable process development. We hope that the ideas presented here will inspire further research and foster collaboration toward a greener and more resilient future.

Javier Martínez-Gómez
Guest Editor

processes MDPI

Editorial

Process Systems Engineering for Environmental Protection: Overview on Methods, Models, and Applications

Javier Martínez-Gómez [1,2]

1 Departamento de Teoría de la Señal y Comunicación, (Área de Ingeniería Mecánica) Escuela Politécnica, Universidad de Alcalá, 28805 Alcalá de Henares, Madrid, Spain; javier.martinezgomez@uah.es
2 Facultad de Arquitectura e Ingenierías, Universidad Internacional SEK, Albert Einstein s/n and 5th, Quito 170302, Ecuador

Over the past decade, Process Systems Engineering (PSE) has undergone a significant transformation, evolving from a discipline primarily focused on process optimization and control to a key enabler of environmentally sustainable industrial development. This evolution has been driven by the urgent need to address global environmental challenges such as climate change, resource depletion, and pollution [1].

One of the most significant advancements in PSE for environmental protection lies in the shift from isolated process optimization to holistic process design and optimization. This paradigm emphasizes the integration of environmental objectives alongside traditional economic and technical criteria, enabling the development of processes that are not only efficient and cost-effective but also environmentally benign. At its core, holistic process design involves the simultaneous consideration of multiple interrelated factors—including energy and material efficiency, emissions reduction, waste minimization, and resource circularity—across the entire process life cycle. This approach contrasts with conventional sequential design methodologies, where environmental considerations are often addressed post hoc, leading to suboptimal or reactive solutions.

Hrithik P. M. et al. [1] present a time-series forecasting approach for CO_2 emissions in India using the ARIMA model, addressing the critical need for accurate environmental predictions to support climate policy. The study applies the Box–Jenkins methodology, incorporating stationarity checks, ACF/PACF analysis, and model selection via AIC to ensure robustness. Results demonstrate that ARIMA (0,2,4) effectively captures historical emission trends and provides reliable short- to medium-term forecasts, validated through error metrics such as RMSE and MAPE. By enabling evidence-based decision-making, this work exemplifies how data-driven predictive modeling within Process Systems Engineering can guide strategic interventions for carbon mitigation and sustainable development planning.

Recent methodological advances have enabled the incorporation of multi-objective optimization techniques that balance trade-offs between conflicting goals, such as minimizing operational costs while reducing carbon footprint or water usage. These techniques often employ mixed-integer nonlinear programming (MINLP), evolutionary algorithms, or multi-criteria decision analysis (MCDA) to navigate complex design spaces and identify Pareto-optimal solutions. Moreover, the integration of process simulation tools (e.g., Aspen Plus, gPROMS) with environmental assessment frameworks (e.g., Life Cycle Assessment, exergy analysis) has facilitated the evaluation of environmental impacts during the early stages of process development [2]. This integration allows for the identification of environmental hotspots and the design of mitigation strategies before capital-intensive decisions are made. Another key aspect of holistic optimization is the consideration of process integration opportunities, such as heat and mass integration, which can significantly enhance energy efficiency and reduce utility consumption. Techniques like pinch analysis, heat exchanger network synthesis, and water network optimization are increasingly being embedded within broader PSE frameworks to ensure systemic improvements.

Furthermore, the emergence of data-driven modeling and machine learning has opened new avenues for real-time optimization and adaptive process control, enabling systems to respond dynamically to changing environmental conditions, feedstock variability, and operational disturbances. These capabilities are particularly relevant in the context of integrating renewable energy sources and managing decentralized production systems.

Recent research on mechanoactivation-assisted nitric–sulfuric leaching of molybdenite concentrates demonstrates how Process Systems Engineering methodologies can significantly enhance the sustainability of metallurgical operations [3]. The study optimizes acid concentration, temperature, and leaching time using response surface methodology, achieving up to 72.6% molybdenum recovery under combined nitric–sulfuric acid conditions with oxygen sparging. By integrating experimental design, mechanochemical activation, and data-driven optimization, the approach reduces environmental burdens associated with traditional pyrometallurgical routes, such as CO_2 and SO_2 emissions, while improving process efficiency. This example highlights the role of multi-objective optimization and hybrid experimental–computational workflows in developing cleaner, more efficient extraction technologies aligned with circular economy principles.

A key contribution in this Special Issue is the work of Pastor et al. [4], which proposes the re-ISSUES model (Renewable Energy-Linked Interoperable Smart and Sustainable Urban Environmental Systems). This model integrates urban environmental management with renewable energy systems using semantic and technical interoperability approaches. It is based on citizen science, the use of low-cost IoT sensors, and collaboration with local renewable energy companies, all within a systems engineering methodology. The study presents a motivational case study focused on urban odor management, quantifying the costs and benefits of implementing environmental measurement and verification systems (EMVSs) and exploring opportunities to optimize their economic sustainability through digitalization, circular economy, and citizen participation strategies. Furthermore, the re-ISSUES model is validated using ontologies and an analysis of scientific literature, demonstrating its potential to improve urban air quality and foster energy resilience in smart cities.

Wentian Lu et al. [5] present an analytical framework for optimizing energy storage participation in primary frequency regulation within low-carbon power systems. The study introduces a reduced second-order aggregation model and compensation-based design to calculate virtual inertia and damping coefficients for distributed energy resources (DERs), ensuring grid stability under high renewable penetration. Additionally, an adaptive control strategy dynamically adjusts these parameters based on the state of charge (SOC), enhancing resource utilization and preventing operational constraints. By combining model simplification with SOC-aware optimization, the approach supports responsive and sustainable frequency control, exemplifying how Process Systems Engineering contributes to decarbonized and resilient energy infrastructures.

Guayanlema et al. [6] analyze the long-term benefits of promoting electric cooking in Ecuador as a strategy to replace liquefied petroleum gas (LPG) in the residential sector. Using the LEAP (Long-range Energy Alternative Planning) model, the study quantifies reductions in LPG demand, associated green-house gas emissions, and subsidy expenditures under different policy scenarios. Results indicate that widespread adoption of induction stoves, supported by targeted incentives, can significantly lower carbon emissions, enhance energy sovereignty, and contribute to national climate commitments. This work exemplifies how Process Systems Engineering integrates long-term energy planning with environmental objectives, aligning household-level interventions with system-wide sustainability goals.

In the context of PSE for environmental protection, Life Cycle Thinking (LCT) and environmental assessment (EA) have emerged as indispensable tools for guiding sustainable process development. While traditional process design often focuses on optimizing operational performance within plant boundaries, LCT expands the scope to encompass the entire life cycle of a product or process, from raw material extraction through production, use, and end-of-life disposal or recycling [7].

The integration of Life Cycle Assessment (LCA) into PSE frameworks enables a quantitative evaluation of environmental impacts across multiple categories, such as global warming potential, eutrophication, acidification, and resource depletion. This comprehensive perspective is essential for identifying environmental trade-offs that may not be apparent when focusing solely on direct emissions or energy use. For instance, a process modification that reduces energy consumption may inadvertently increase water usage or generate more hazardous waste—impacts that LCA can help uncover and mitigate [8–10].

From a methodological standpoint, recent advances have facilitated the coupling of LCA with process simulation and optimization tools, allowing for dynamic and iterative assessments during the design phase. This coupling supports eco-design strategies, where environmental performance indicators are treated as design objectives or constraints within multi-objective optimization problems. Techniques such as goal programming, ε-constraint methods, and evolutionary multi-objective algorithms are increasingly employed to navigate the trade-offs between economic and environmental criteria [11].

Moreover, LCT complements holistic process design and optimization by ensuring that sustainability is not only achieved at the unit or plant level but also across the broader value chain. This is particularly relevant in the design of biorefineries, waste valorization systems, and renewable energy integration, where upstream and downstream impacts can significantly influence the overall sustainability profile [12].

The adoption of hybrid LCA approaches, which combine process-based and input–output models, further enhances the robustness of environmental assessments, especially in complex systems with multiple interdependencies. Additionally, the emergence of dynamic LCA allows for time-dependent impact assessments, which are crucial for evaluating systems with variable performance over their operational lifespan, such as those incorporating intermittent renewable energy sources.

The ongoing digital transformation of the process industries—often referred to as Industry 4.0—has introduced a new paradigm in PSE, enabling the realization of smart manufacturing systems that are more adaptive, efficient, and environmentally sustainable. Digitalization provides the technological foundation for integrating real-time data, advanced analytics, and intelligent control into process design and operation, thereby enhancing the capacity of PSE to address environmental challenges. At the heart of this transformation is the digital twin, a virtual replica of a physical process or system that continuously receives data from sensors and control systems. Digital twins enable real-time monitoring, simulation, and optimization, allowing engineers to predict system behavior under varying conditions, detect anomalies, and implement corrective actions proactively. This capability is particularly valuable for minimizing energy consumption, reducing emissions, and ensuring compliance with environmental regulations.

Machine learning (ML) and artificial intelligence (AI) further augment the decision-making capabilities of PSE frameworks. By learning from historical and real-time data, ML models can uncover complex, nonlinear relationships between process variables and environmental performance indicators. These insights can be used to develop data-driven surrogate models that accelerate optimization routines, or to implement predictive control strategies that dynamically adjust operations to maintain optimal environmental performance under uncertainty [13].

Digitalization also enhances the implementation of Life Cycle Thinking (LCT) by enabling the integration of real-time environmental data into Life Cycle Assessment (LCA) models. This allows for dynamic LCA, where environmental impacts are assessed as a function of time and operational conditions, providing a more accurate and responsive basis for sustainability-oriented decision-making. For example, the environmental footprint of a process can be continuously updated based on actual energy mix variations, feedstock changes, or process disturbances [14].

Furthermore, smart manufacturing systems facilitate closed-loop optimization, where feedback from environmental performance metrics (e.g., emissions, energy use, and waste generation) is used to iteratively refine process operations. This aligns with the principles of holistic process optimization, ensuring that environmental objectives are not only considered during the design phase but are actively pursued throughout the operational life cycle. The integration of cloud computing, industrial Internet of Things (IIoT), and edge computing further supports distributed data acquisition and processing, enabling decentralized decision-making in complex, multi-unit, or multi-site systems. This is particularly relevant for renewable energy integration, waste valorization networks, and smart grids, where system dynamics are influenced by external factors such as weather variability, market conditions, and regulatory constraints [15].

The transition from a linear "take-make-dispose" industrial model to a circular economy (CE) paradigm represents a fundamental shift in how resources are managed across the life cycle of products and processes [16]. Within the framework of PSE, this transition is catalyzed by the development of advanced methodologies for resource recovery, process integration, and system-wide optimization, all aimed at minimizing waste and maximizing value retention. In a circular economy, the emphasis is placed on closing material and energy loops, thereby reducing the extraction of virgin resources and the generation of waste. PSE contributes to this goal by enabling the systematic design and optimization of closed-loop systems, such as waste valorization networks, industrial symbiosis, and integrated biorefineries. These systems are designed to recover valuable materials and energy from waste streams, transforming them into secondary raw materials or energy carriers that can be reintegrated into production cycles. A key enabler of circularity in PSE is the integration of process synthesis and optimization techniques that consider multiple pathways for resource recovery. For example, superstructure-based optimization models can be used to evaluate alternative configurations for waste treatment, recycling, and reuse, identifying the most sustainable and cost-effective options. These models often incorporate multi-objective criteria, balancing environmental benefits (e.g., reduced emissions, lower resource depletion) with economic performance [17].

The synergy with Life Cycle Thinking (LCT) is particularly important in this context. By embedding Life Cycle Assessment (LCA) into the design and evaluation of circular systems, engineers can ensure that resource recovery strategies do not lead to unintended environmental burdens elsewhere in the system. For instance, the energy required for recycling or reprocessing must be weighed against the environmental savings from avoided raw material extraction. Digitalization and smart manufacturing further enhance the implementation of circular economy principles by enabling real-time monitoring and control of resource flows. Through the use of sensor networks, blockchain-based traceability, and data analytics, it becomes possible to track material usage, detect inefficiencies, and optimize recovery processes dynamically. Digital twins of circular systems can simulate the impact of different operational strategies or policy interventions, supporting decision-making under uncertainty [8–10].

Moreover, process integration techniques such as heat and mass exchange networks, pinch analysis, and water reuse optimization are instrumental in designing systems that minimize utility consumption and internalize waste streams. These approaches align with the broader goals of holistic process optimization, ensuring that circularity is not an afterthought but a foundational design principle.

The pursuit of environmental sustainability in industrial systems necessitates a shift from isolated process-level optimization to a broader, systems-level perspective that encompasses entire supply chains and energy systems. In this context, PSE provides a powerful set of modeling, simulation, and optimization tools to design and manage sustainable supply chains and low-carbon energy systems, aligning operational decisions with environmental and economic objectives [18].

A sustainable supply chain integrates environmental considerations into every stage of the product life cycle—from raw material extraction and manufacturing to distribution, use, and end-of-life management. PSE contributes to this integration by enabling multi-scale modeling and optimization of supply chain networks, accounting for factors such as transportation emissions, resource availability, energy consumption, and waste generation. These models often employ mixed-integer linear/nonlinear programming (MILP/MINLP) and stochastic optimization to handle the inherent complexity and uncertainty in global supply chains [19–21].

The incorporation of LCA into supply chain modeling further enhances sustainability by quantifying the environmental impacts of different configurations and sourcing strategies. This allows decision-makers to evaluate trade-offs between cost, carbon footprint, and resource efficiency, and to identify hotspots where interventions can yield the greatest environmental benefits. For example, sourcing raw materials from geographically closer suppliers may reduce transportation emissions, while also improving supply chain resilience.

In parallel, energy systems modeling has become a critical area within PSE, particularly in the context of the global transition toward decarbonized and decentralized energy systems. PSE methodologies support the design and operation of integrated energy systems that combine renewable energy sources, energy storage, demand-side management, and sector coupling (e.g., power-to-heat, power-to-gas). These models often rely on multi-period optimization, agent-based modeling, and scenario analysis to evaluate the performance of energy systems under varying demand profiles, policy constraints, and technological developments.

The role of digitalization is particularly pronounced in this domain. The availability of high-resolution data from smart meters, IoT devices, and geospatial information systems (GISs) enables the development of data-driven models for real-time monitoring and optimization of supply chains and energy networks. Digital twins of supply chains and energy systems can simulate the impact of disruptions (e.g., supply shortages, price volatility, and extreme weather events) and support resilient planning and adaptive control strategies [22].

Moreover, multi-agent systems and blockchain technologies are being explored to enhance transparency, traceability, and coordination across decentralized supply chains and energy markets. These tools facilitate peer-to-peer energy trading, dynamic pricing, and collaborative logistics, all of which contribute to reducing environmental impacts while maintaining economic viability.

In synergy with holistic process design, Life Cycle Thinking, and smart manufacturing, sustainable supply chain and energy systems modeling enables a comprehensive approach to environmental protection. By extending the scope of optimization beyond the plant level to encompass entire networks and infrastructures, PSE empowers stakeholders to make informed decisions that align with the principles of circular economy, climate neutrality, and resource stewardship.

This Special Issue demonstrates the pivotal role of Process Systems Engineering (PSE) in addressing complex environmental challenges through advanced modeling, optimization, and integration strategies. The contributions collectively highlight how PSE enables multi-scale decision-making, coupling process-level improvements with system-wide sustainability objectives. Key technical insights include the integration of Life Cycle Assessment (LCA) with process simulation for dynamic eco-design, the deployment of digital twins for real-time optimization, and the application of multi-objective optimization frameworks to balance economic and environmental trade-offs.

Despite these advances, several research gaps and opportunities remain. Future investigations should focus on the following:

- Development of hybrid modeling approaches combining first-principles, data-driven, and AI-based methods to enhance predictive accuracy under uncertainty.
- Formalization of interoperability standards for environmental and energy systems, enabling semantic and technical integration across heterogeneous platforms.
- Expansion of dynamic LCA methodologies to incorporate real-time operational data streams, supporting adaptive sustainability assessments.
- Design of resilient supply chain and energy network models that integrate stochastic optimization and agent-based simulation for climate risk mitigation.
- Exploration of blockchain and distributed ledger technologies for transparent carbon accounting and traceability in circular economy frameworks.
- Advancement of digital twin architectures for multi-domain systems, including coupling with uncertainty quantification and scenario-based policy analysis.

By addressing these research directions, PSE can further strengthen its role as a cornerstone of sustainable industrial transformation, bridging the gap between theoretical innovation and practical implementation in pursuit of climate neutrality and resource efficiency.

Conflicts of Interest: The authors declare no conflicts of interest.

References

1. Hrithik, P.M.; Rehman, M.Z.; Dar, A.A.; Wangmo, A.T. Forecasting CO_2 Emissions in India: A Time Series Analysis Using ARIMA. *Processes* **2024**, *12*, 2699. [CrossRef]
2. Villacreses, G.; Jijón, D.; Nicolalde, J.F.; Martínez-Gómez, J.; Betancourt, F. Multicriteria Decision Analysis of Suitable Location for Wind and Photovoltaic Power Plants on the Galápagos Islands. *Energies* **2023**, *16*, 29. [CrossRef]
3. Kenzhaliyev, B.; Ultarakova, A.; Lokhova, N.; Mukangaliyeva, A.; Kassymzhanov, K. Optimized Hydrometallurgical Extraction of Molybdenum via Mechanoactivation and Nitric–Sulfuric Leaching. *Processes* **2025**, *13*, 1486. [CrossRef]
4. Pastor, R.; Lecuona, A.; Fraga, A. re-ISSUES—Renewable Energy-Linked Interoperable Smart and Sustainable Urban Environmental Systems. *Processes* **2024**, *12*, 1815. [CrossRef]
5. Lu, W.; Tan, E.; Cheng, L.; Zhang, K.; Liu, W. Optimizing Energy Storage Participation in Primary Frequency Regulation: A Novel Analytical Approach for Virtual Inertia and Damping Control in Low-Carbon Power Systems. *Processes* **2025**, *13*, 1146. [CrossRef]
6. Guayanlema, V.; Martínez-Gómez, J.; Fontalvo, J.; Espinoza, V.S. Modeling the Benefits of Electric Cooking in Ecuador: A Long-Term Perspective. *Processes* **2024**, *12*, 2400. [CrossRef]
7. Ali, Z.A.; Zain, M.; Hasan, R.; Pathan, M.S.; AlSalman, H.; Almisned, F.A. Digital twins: Cornerstone to circular economy and sustainability goals. *Environ. Dev. Sustain.* **2025**, *1*, 1–42. [CrossRef]
8. Sala, S.; Amadei, A.M.; Beylot, A.; Ardente, F. The evolution of life cycle assessment in European policies over three decades. *Int. J. Life Cycle Assess.* **2021**, *26*, 2295–2314. [CrossRef]
9. Iqbal, H.; Stevenson, A.; Sarwat, A.I. Impact Analysis and Optimal Placement of Distributed Energy Resources in Diverse Distribution Systems for Grid Congestion Mitigation and Performance Enhancement. *Electronics* **2025**, *14*, 1998. [CrossRef]
10. Kikuchi, Y. Application of PSE into social changes: Biomass-based production, recycling systems, and regional systems design and assessment. In *Computer Aided Chemical Engineering*; Elsevier: Amsterdam, The Netherlands, 2022; Volume 49, pp. 41–46.

11. Ferdous, J.; Bensebaa, F.; Hewage, K.; Bhowmik, P.; Pelletier, N. Use of process simulation to obtain life cycle inventory data for LCA: A systematic review. *Clean. Environ. Syst.* **2024**, *14*, 100215. [CrossRef]
12. Passoni, C.; Palumbo, E.; Pinho, R.; Marini, A. The LCT challenge: Defining new design objectives to increase the sustainability of building retrofit interventions. *Sustainability* **2022**, *14*, 8860. [CrossRef]
13. Soofastaei, A. Revolutionizing Chemical Engineering 4.0: Artificial Intelligence Innovations and Machine Learning. In *Advanced Analytics for Industry 4.0*; CRC Press: Boca Raton, FL, USA, 2025; pp. 271–288.
14. da Costa, T.P.; da Costa, D.M.B.; Murphy, F. A systematic review of real-time data monitoring and its potential application to support dynamic life cycle inventories. *Environ. Impact Assess. Rev.* **2024**, *105*, 107416. [CrossRef]
15. An, J.; Zou, Z.; Chen, G.; Sun, Y.; Liu, R.; Zheng, L. An IoT-based life cycle assessment platform of wind turbines. *Sensors* **2021**, *21*, 1233. [CrossRef] [PubMed]
16. Daga, S.; Yadav, K.; Singh, P.; Mishra, V. Beyond the Take-Make-Dispose Model—Unlocking the Power of Circular Economy for an Environmentally Resilient Future. In *Circular Economy and Environmental Resilience: Solutions for a Sustainable Tomorrow*; Springer Nature: Cham, Switzerland, 2025; Volume 1, pp. 1–11.
17. Rashid, S.; Malik, S.H. Transition from a Linear to a Circular Economy. In *Renewable Energy in Circular Economy*; Springer International Publishing: Cham, Switzerland, 2023; pp. 1–20.
18. Espinosa Gutiérrez, G.; Casas Ledón, Y. Life Cycle Thinking: An Essential Framework for Eco-Industrial Development. In *Eco-Industrial Development as an Industrial Strategy: Contributions from a German-Chilean Research Partnership*; Springer Nature: Cham, Switzerland, 2024; pp. 69–92.
19. Zhou, P.; Lv, Y.; Wen, W. The low-carbon transition of energy systems: A bibliometric review from an engineering management perspective. *Engineering* **2023**, *29*, 147–158. [CrossRef]
20. Berbiche, N.; Chakir, M.; Hlyal, M.; El Alami, J. An Integrated Inventory-Production-Distribution Model for Crisis Relief Supply Chain Optimization: A Systematic Review and Mixed Integer Programming Formulation. *J. Eur. Des. Syst. Autom.* **2024**, *57*, 899–920. [CrossRef]
21. Pamulapati, T.; Cavus, M.; Odigwe, I.; Allahham, A.; Walker, S.; Giaouris, D. A review of microgrid energy management strategies from the energy trilemma perspective. *Energies* **2022**, *16*, 289. [CrossRef]
22. Rind, Y.M.; Raza, M.H.; Zubair, M.; Mehmood, M.Q.; Massoud, Y. Smart energy meters for smart grids, an internet of things perspective. *Energies* **2023**, *16*, 1974. [CrossRef]

processes

MDPI

Article

Optimized Hydrometallurgical Extraction of Molybdenum via Mechanoactivation and Nitric–Sulfuric Leaching

Bagdaulet Kenzhaliyev, Almagul Ultarakova, Nina Lokhova, Arailym Mukangaliyeva * and Kaisar Kassymzhanov

The Institute of Metallurgy and Ore Beneficiation, Satbayev University, Almaty 050013, Kazakhstan; bagdaulet_k@satbayev.university (B.K.); a.ultarakova@satbayev.university (A.U.); n.lokhova@satbayev.university (N.L.); k.kassymzhanov@satbayev.university (K.K.)

* Correspondence: a.mukangaliyeva@satbayev.university; Tel.: +7-777-3934621

Abstract: This study explores the intensification of molybdenite concentrate processing through a synergistic hydrometallurgical approach using sulfuric acid, nitric acid, and their combination to enhance leaching efficiency while minimizing environmental impact. Molybdenum, a strategic metal widely used in advanced engineering and catalytic systems, presents extraction challenges due to the refractory nature of molybdenite (MoS_2). The experimental approach incorporated oxygen sparging and mechanoactivation to improve dissolution kinetics and molybdenum availability. A central composite design (CCD) of response surface methodology (RSM) was employed to develop a predictive model for optimizing the leaching parameters. Acid concentration, temperature, and leaching time were systematically varied, allowing for the identification of statistically significant factor interactions and optimal operating conditions. The model demonstrated strong predictive capability with high adjusted and predicted R^2 values, validating its suitability for process optimization. Optimal leaching conditions were identified as 50 g/dm^3 HNO_3 + 200 g/dm^3 H_2SO_4, a temperature of 95 °C, a leaching time of 240 min, and a solid-to-liquid ratio of 1:6, resulting in a maximum molybdenum extraction efficiency of 72.6%. This performance was attributed to enhanced oxidative decomposition and stable complexation of molybdenum species. This study provides a scalable and environmentally conscious framework for molybdenum extraction, with implications for sustainable metallurgy and industrial applications.

Keywords: molybdenite concentrate; acid leaching; mechanoactivation; oxygen sparging; response surface methodology (RSM); central composite design (CCD)

1. Introduction

Molybdenum is a strategically critical transition metal whose demand continues to rise due to its indispensable role in advanced material systems, high-performance alloys, catalysts, and corrosion-resistant components [1,2]. According to the USGS Mineral Commodity Summaries 2025, global mine production of molybdenum in 2024 was approximately 250,000 metric tons, with reserves estimated at 15 million metric tons, primarily in China, the United States, and Chile [3]. Globally, molybdenum consumption exceeded 300,000 metric tons in 2022, with annual growth projected at 4–5%, driven largely by its increasing applications in energy, aerospace, electronics, and green technologies, with uses such as a desulfurization catalyst and hydrogen storage material. The metal's unique combination of high melting point (2623 °C) [4,5], low thermal expansion, and superior mechanical and chemical stability under extreme conditions makes it particularly valuable for engineering solutions in high-temperature and high-pressure environments [6,7].

The primary mineralogical source of molybdenum is molybdenite, which is composed predominantly of molybdenum disulfide (MoS_2) [8] and accounts for over 85% of global molybdenum extraction [9]. Industrially, molybdenum is recovered through two distinct processing routes: pyrometallurgical and hydrometallurgical, which differ fundamentally in their mechanisms of oxidative transformation and resource utilization [10].

Pyrometallurgical processing involves high-temperature oxidative roasting of molybdenite, often exceeding 600–700 °C [7], and is typically governed by the nature of modifying reagents that drive the phase transformation of disulfide compounds [11]. Although historically dominant, this approach has raised serious environmental concerns [12]. Oxidative roasting releases significant quantities of CO_2 and SO_2 [13,14], with average emissions reaching 1.2–1.5 tons of CO_2 per ton of Mo produced, and SO_2 emissions exceeding 0.8 tons per ton in some facilities [15,16]. These emissions not only contribute to climate change and acid rain but also pose occupational health hazards. In energy-challenged regions such as Kazakhstan [17], where electricity shortages impact over 20% of industrial facilities during winter months, the energy-intensive nature of pyrometallurgy is increasingly viewed as unsustainable [18,19]. These constraints have prompted calls for alternative, lower-emission processes and the exploration of stable energy sources such as nuclear power to support metallurgical operations [20–23].

In response, hydrometallurgical technologies have gained prominence as a more environmentally benign and energetically efficient alternative. These processes operate at comparatively lower temperatures and emit significantly fewer greenhouse gases [24]. Hydrometallurgy is particularly suitable for processing complex ores and concentrates with variable compositions, offering higher selectivity for target metal phases and facilitating closed-loop recycling of reagents [25].

In recent years, mechanical activation has been increasingly adopted as a pretreatment step to enhance the reactivity of refractory sulfide minerals such as molybdenite [26]. This process, involving high-energy grinding in systems like planetary or vibratory mills, introduces structural defects, reduces crystallinity, and significantly increases surface area. Such physicochemical changes lead to accelerated leaching kinetics and improved extraction efficiency by facilitating the formation of soluble oxidation products [27,28]. The integration of mechanical activation with hydrometallurgical leaching thus presents a powerful strategy for intensifying metal recovery processes under milder conditions [29].

Among hydrometallurgical techniques for molybdenum extraction, leaching processes—especially in nitric acid media—have been widely investigated due to their adaptability and thermodynamic favorability [30–32].

Variants include pressure leaching in autoclaves at temperatures above 180 °C, oxidative dissolution with hypochlorite or dichromate oxidants, and electrochemically assisted leaching driven by redox-active electrode reactions [33,34]. These methods reduce reliance on thermal inputs and allow for the fine-tuned control of selectivity via process parameters such as pH, redox potential, and oxygen partial pressure [35].

Nevertheless, hydrometallurgical molybdenum extraction is not without its limitations. Industrial-scale implementation is hindered by the corrosive nature of the chemical media, necessitating expensive, corrosion-resistant equipment. Furthermore, maintaining optimal reaction kinetics often requires elevated pressures and continuous oxygen sparging, increasing capital and operational costs [6].

Several reagent-based pathways have been explored, including direct acid dissolution of molybdenite in concentrated nitric acid (25–50%), which is thermodynamically one of the most favorable routes for oxidative destruction of MoS_2 [36]. In such systems, the electron-acceptor capacity of the medium is intensified at higher temperatures, facilitating the liberation of soluble molybdenum oxide species [37]. However, when sulfuric acid is introduced into the system, stoichiometrically driven speciation changes occur: a portion of molybdenum transitions into stable anionic sulfate complexes, modifying the dissolution kinetics and altering the distribution of ionic forms in solution [38]. This dual-acid approach introduces both synergies and challenges, particularly in terms of optimizing conditions to prevent undesired precipitation or side reactions.

Moreover, despite promising laboratory results, many of these techniques face operational bottlenecks. These include the formation of colloidal and difficult-to-filter pulps, recovery rates as low as 65–70% in unoptimized systems, and high specific consumption of oxidative reagents, which may reach up to 8–10 kg per ton of concentrate, thereby affecting process economics. Electrochemical oxidation using sodium chloride electrolytes, while attractive for in situ generation of active chlorine species, has shown only moderate efficiency in practice, with recovery efficiency typically below 60%, limiting its industrial scalability [39].

Given these challenges, the present study aims to systematically investigate and optimize the leaching of molybdenite concentrates using a dual-acid system of nitric and sulfuric acids. Special emphasis is placed on quantifying the effect of oxygen sparging on molybdenum recovery efficiency with the goal of identifying conditions that enhance extraction performance while minimizing environmental and operational drawbacks. This work contributes to the development of robust, scalable, and sustainable hydrometallurgical strategies for future molybdenum production.

2. Materials and Methods

2.1. Materials

The object of the study was a molybdenite concentrate provided by the Aktogay Mining and Processing Plant, East Kazakhstan Region, Republic of Kazakhstan [40], with a chemical composition determined by X-ray fluorescence spectroscopy (XRF) as shown in Table 1. Nitric acid (65%, Sigma-Aldrich, St. Louis, MO, USA) and sulfuric acid (98%, Merck, Darmstadt, Germany) were used as leaching agents. High-purity oxygen (99.9%, Linde Gas, Almaty, Kazakhstan) was used for sparging. Deionized water was used for all experiments. Mechanoactivation was performed using an IV-1 pulverizer (Tekhnolit, Almaty, Kazakhstan) operating at 300 rpm for 2 h, reducing particle size and enhancing the reactivity of the molybdenite concentrate.

Table 1. The composition of the molybdenum concentrate.

Concentrate	Composition, wt.%
MoS_2	41.7
Cu_2S	1.14
FeS_2	7.5
SiO_2	3.2
SO_2	46.46

The chemical composition of the concentrate is presented in Table 1, detailing the product designation and its composition, which was determined using X-ray fluorescence spectroscopy (XRF) on a PANalytical Almelo, The Netherlands, Axios 1 kW wavelength-dispersive spectrometer.

2.2. Analytical Techniques

This study utilized advanced analytical techniques to investigate the molybdenum concentrate and to elucidate the mechanisms of phase transformations. The phase composition of the samples was analyzed using a Bruker D8 Advance X-ray diffractometer

(Bruker, Ettlingen, Germany). Elemental composition was determined using an Axios 1 kW wavelength-dispersive X-ray fluorescence spectrometer (PANalytical, Almelo, The Netherlands), with data processing and interpretation performed using SuperQ5 software (Omnian 37). The surface microstructure was examined using a JXA-8230 electron probe microanalyzer (JEOL, Tokyo, Japan) operated at an accelerating voltage of 20 kV, an electron beam current below 1 nA, and aperture diaphragm № 3. Energy-dispersive spectrometry (EDS) microanalysis (JEOL, Tokyo, Japan) was conducted with an electron beam current up to 6 nA and a dead time of up to 14%. Quantitative investigations of metal concentrations in solutions and solid samples were conducted using an Optima 8300DV inductively coupled plasma atomic emission spectrometer (PerkinElmer, Inc., Waltham, MA, USA) and an AA-7000 atomic absorption spectrometer (Shimadzu, Kyoto, Japan).

2.3. Optimal Experimental Design for Leaching Process

The leaching process from molybdenite concentrate was analyzed using response surface methodology (RSM) and the central composite design (CCD) approach to determine optimal conditions and enhance the accuracy and reproducibility of the results [41].

The solid-to-liquid (S/L) ratio was fixed at 1:6 g/dm^3 (60 g of molybdenite concentrate per liter of leachant) based on preliminary experiments and the literature, which suggest that this ratio ensures sufficient reagent availability and manageable slurry viscosity for laboratory-scale leaching. Including S/L as a variable was not prioritized, as acid concentration, leaching time, and temperature were deemed the primary drivers of reaction kinetics and molybdenum oxidation, and adding S/L would have increased experimental complexity. The independent variables considered included acid concentration (HNO_3 + H_2SO_4), leaching time, and temperature. The total acid concentration (HNO_3 + H_2SO_4) varied from 50 to 250 g/dm^3, with a fixed ratio of HNO_3:H_2SO_4 = 1:4, determined through preliminary experiments to balance the oxidative strength of the nitric acid and the ionic stability provided by sulfuric acid. Both acids were adjusted proportionally to maintain this ratio across all CCD levels. The application of the mathematical RSM approach enabled the development of a second-order model describing the relationship between the output response (molybdenum leaching recovery) and the key process parameters.

The general form of the model is expressed as follows:

$$y_0 = b_0 + \sum_{i=1}^{k} b_i x_i + \sum_{i=1}^{k} b_{ii} x_i^2 + \sum_{i=1}^{k-1} \sum_{j=i+1}^{k} b_{ij} x_i x_j, \quad (1)$$

where y denotes the predicted value of molybdenum leaching recovery, b_0 is the constant coefficient, b_i is the linear coefficient, b_{ii} is the quadratic coefficient, b_{ij} represents the interaction coefficient between variables, and k is the number of factors.

Experimental design was conducted using the Design Expert 7.0 software package (Stat-Ease, Inc., Minneapolis, MN, USA), which enabled the construction of a second-order regression model providing high predictive accuracy of the results.

The value ranges and levels shown in Table 2 represent the experimental settings used in the central composite design (CCD) to evaluate the effects of temperature, acid concentration (HNO_3 + H_2SO_4), and leaching time as independent variables (Table 2). The levels in Table 2 were selected based on a combination of the literature and preliminary experiments. Acid concentrations (50–250 g/L) were informed by studies on nitric acid leaching of molybdenite, which recommend 25–50% HNO_3 for effective oxidation. Preliminary tests confirmed this range's suitability for the dual-acid system. Leaching times (30–240 min) were chosen to cover rapid initial extraction to near-equilibrium conditions, as reported in the hydrometallurgical literature [42]. Temperatures (20–95 °C) were selected to maximize kinetics within laboratory equipment constraints, avoiding the need for high-pressure autoclaves typically used at 150–200 °C.

Table 2. Levels and codes of factors for CCD.

Factors	Symbol	Coding Level		
		−1	0	1
Concentration $HNO_3 + H_2SO_4$ (g/dm³)	A	50	150	250
Time (min)	B	30	135	240
Temperature (°C)	C	20	60	95

A rotatable central composite design (CCD) with $\alpha = 1.682$ was employed to ensure uniform precision across the design space. The axial points ($\pm\alpha$) for each factor were calculated as follows: acid concentration (65.9, 234.1 g/dm³), leaching time (46.6, 223.4 min), and temperature (28.5, 91.5 °C). Temperature was varied at five levels (20 °C, 28.5 °C, 60 °C, 91.5 °C, and 95 °C) in the CCD to capture its influence on molybdenum leaching recovery, ensuring a robust model within the experimental range of 20–95 °C, which was constrained by laboratory equipment to avoid high-pressure autoclave conditions. These levels, combined with factorial points (−1, 0, +1), enabled the fitting of a full quadratic model to capture curvature and interaction effects. CCD was chosen over the Box–Behnken design (BBD) to include extreme conditions (axial and factorial points), which were critical for identifying optimal leaching parameters, particularly given the significant interaction between acid concentration and temperature (AC). A three-factor CCD with $\alpha = 1.682$ comprised 20 runs: 8 factorial points, 6 axial points, and 6 center points. Center points were repeated six times to estimate experimental error, while factorial and axial points were run once, with selected conditions (e.g., optimal parameters) validated through three replicates. The response variable was molybdenum leaching recovery (%), defined as the percentage of molybdenum extracted relative to the initial content in the concentrate. Table 3 lists all CCD experiments, their conditions, and measured recovery values.

Table 3. CCD experimental design and molybdenum leaching recovery.

Run	A (g/dm³)	B (min)	C (°C)	Type	Recovery (%)
1	50.00	30.00	50.00	Factorial	20
2	75.00	60.00	60.00	Factorial	25
3	100.00	90.00	70.00	Factorial	30
4	12.00	120.00	80.00	Factorial	35
5	150.00	150.00	90.00	Factorial	40
6	175.00	180.00	100.00	Factorial	45
7	250.00	240.00	95.00	Factorial	50
8	225.00	240.00	55.00	Factorial	42
9	250.00	270.00	95.00	Axial (−α, 0, 0)	48
10	100.00	120.00	85.00	Axial (+α, 0, 0)	38
11	150.00	90.00	75.00	Axial (0, −α, 0)	44
12	200.00	180.00	65.00	Axial (0, +α, 0)	46
13	50.00	240.00	55.00	Axial (0, 0, −α,)	22
14	150.00	240.00	95.00	Axial (0, 0, +α,)	50
15	250.00	60.00	105.00	Center	55
16	175.00	200.00	100.00	Center	43
17	125.00	100.00	80.00	Center	33
18	180.00	140.00	60.00	Center	37
19	120.00	220.00	90.00	Center	47
20	250.00	240.00	95.00	Center	50

The optimal levels and value ranges determined based on temperature conditions, acid concentration ($HNO_3 + H_2SO_4$), and leaching time as independent variables are presented in Table 2. Table 4 presents the full range of parameter values tested in the experiments, including preliminary and control runs, while Table 2 specifies the levels used in the CCD for optimization.

Table 4. Leaching parameters and value ranges applied in the present experiments.

Parameter	Value
Concentration $HNO_3 + H_2SO_4$ (g/dm^3)	50, 100, 150, 200, 250 *
Time (min)	30, 60, 120, 180, 240 *
Temperature (°C)	25, 55, 75, 95 *

* These parameters were held constant.

During the experiments, control variables were applied to evaluate the influence of factors such as nitric and sulfuric acid concentrations, leaching time, and reaction temperature. Three series of experiments were conducted under identical experimental conditions, marked with an asterisk in Table 4. To ensure the reliability of the results, each experiment was repeated at least three times, and the average values of the obtained data were used for analysis.

3. Results and Discussion

Molybdenite (MoS_2) exhibits a white color, pronounced bireflectance, and strong anisotropy, occurring as free, platy, and anhedral grains with irregular, contorted boundaries, reaching up to 183.2 µm (Figure 1a). Pyrite (FeS_2) displays a yellow color and high reflectivity, present as isotropic, irregularly shaped grains up to 67.2 µm (Figure 1a). Chalcopyrite ($CuFeS_2$), a brass-yellow, anisotropic mineral, is observed as anhedral grains up to 44.8 µm, closely associated with molybdenite (Figure 1a,b).

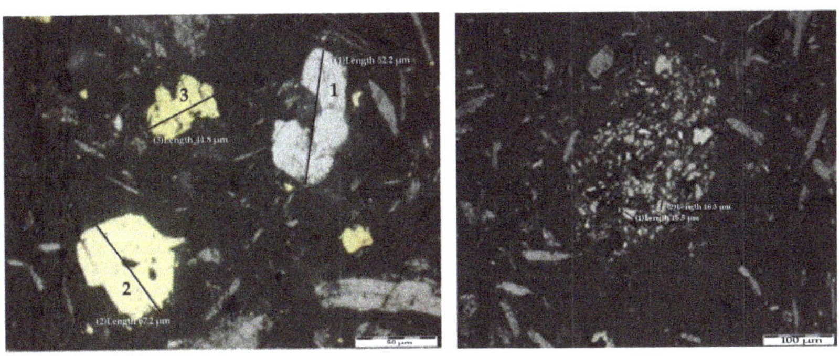

(a) (b)

Figure 1. (**a**) Molybdenite (1), pyrite (2), and chalcopyrite (3). (**b**) Cluster of chalcopyrite grains.

X-ray diffraction (XRD) analysis (Table 5, Figure 2) confirms molybdenite as the dominant phase (83.6%), with quartz (12.9%) and chalcopyrite (3.5%). The absence of pyrite in XRD, despite microscopic identification, suggests separation during beneficiation, consistent with porphyry deposit characteristics.

Table 5. Phase composition of the molybdenite concentrate.

Phase Number in the Diffractogram	Compound Name	Chemical Formula	wt.%
1	Molybdenite-2H	MoS_2	83.6
2	Quartz, α-form	SiO_2	12.9
3	Chalcopyrite	$CuFeS_2$	3.5

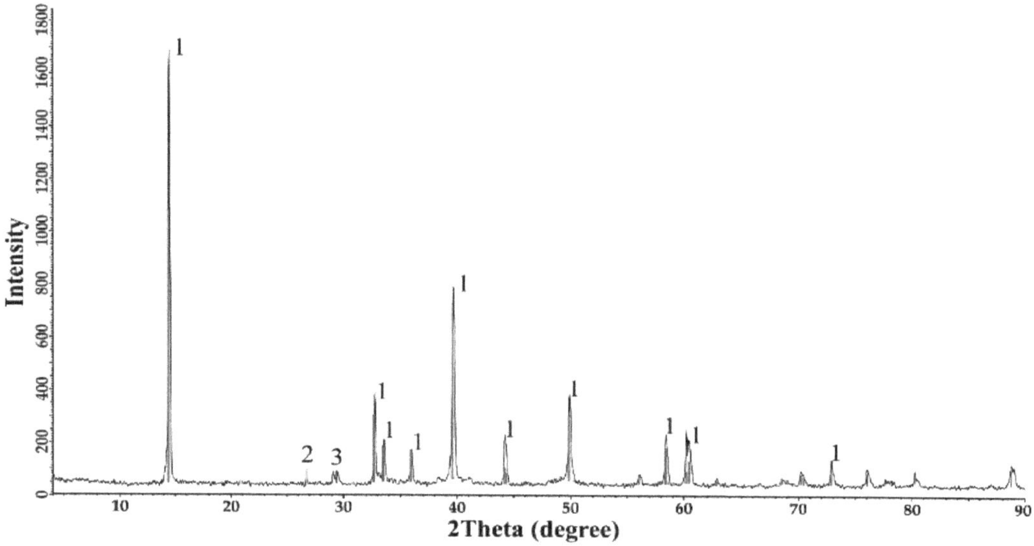

Figure 2. XRD analysis of molybdenite concentrate.

Elemental and phase composition mapping of the samples was performed using an electron probe microanalyzer JXA-8230 (JEOL, Tokyo, Japan). Electron probe microanalysis of the molybdenite concentrate was conducted in the following modes: 1. COMPO—imaging of minerals in back-scattered electrons; 2. WDS—wavelength-dispersive spectroscopy, providing clearer and more sensitive imaging. Electron probe microanalysis enabled the examination of several mineral points, specifically molybdenite with chalcopyrite inclusions (Figure 3a), pyrite (Figure 3b), and molybdenite (Figure 3c).

The investigation of acid leaching of molybdenite concentrate represents a critical aspect of modern metallurgy, aimed at extracting molybdenum, which is widely utilized in high-strength steels, the aerospace industry, and chemical engineering. Molybdenite (MoS_2), the primary mineral source for molybdenum extraction, exhibits a layered structure similar to graphite, with strong covalent bonds within the layers and weak van der Waals forces between them. This structure renders molybdenite relatively inert to acids, complicating its processing. However, studies indicate that mechanoactivation, as one of the most effective approaches, can significantly enhance this process by increasing the material's reactivity. Mechanical activation, a process involving high-energy grinding in systems such as planetary or vibratory mills, introduces structural defects and increases surface area, thereby significantly enhancing the chemical reactivity of mineral phases. Numerous studies have shown that this pretreatment accelerates leaching kinetics and improves metal recovery.

From a physicochemical perspective, defects such as vacancies or dislocations induced by mechanoactivation serve as active sites for the adsorption of acid molecules or other reagents. The increase in specific surface area due to particle size reduction provides more contact points between the mineral and the solution, directly influencing the leaching efficiency. Furthermore, amorphization caused by mechanoactivation results in a more disordered structure, often associated with a higher energy state, making the material more prone to reactions as the system seeks to lower its energy by forming more stable compounds.

Experimental data obtained using the IV-1 pulverizer (Tekhnolit, Kazakhstan) demonstrate the high efficiency of the process. Mechanoactivation significantly enhanced molybdenite reactivity, with sieving post-treatment showing a 91.6% fine fraction (−0.045 mm), a 29.3% increase compared to the initial concentrate (Table 6). XRD revealed increased molybdenite content (91.8%) and amorphization, indicating structural changes that improved leaching efficiency (Table 7, Figure 4).

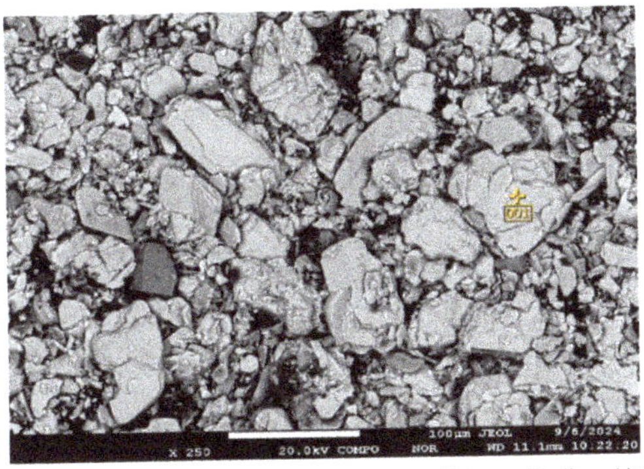

Chemical formula	ms%	mol%	Sigma	Net	K ratio	Line
S	41.77	67.28	0.05	7675120	0.4567080	K
Fe	1.52	1.40	0.12	99052	0.0154736	K
Cu	2.82	2.30	0.24	113481	0.0289432	K
Mo	53.90	29.02	0.14	6646754	0.4794907	L
Total	100.00	100.00				

(a)

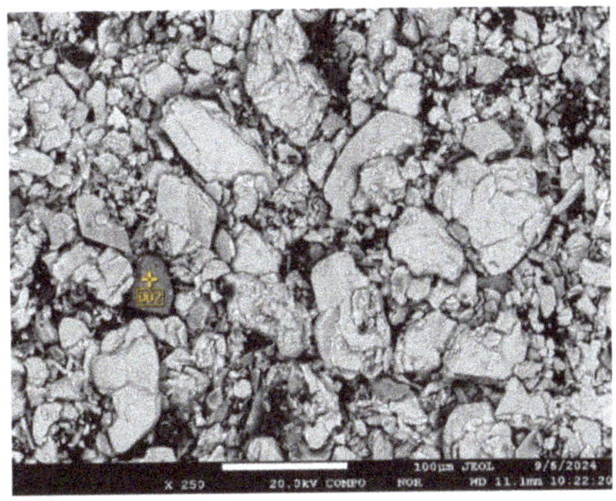

Chemical formula	ms%	mol%	Sigma	Net	K ratio	Line
S	53.52	66.73	0.07	8001429	0.4771450	K
Fe	46.48	33.27	0.26	2732460	0.4277700	K
Total	100.00	100.00				

(b)

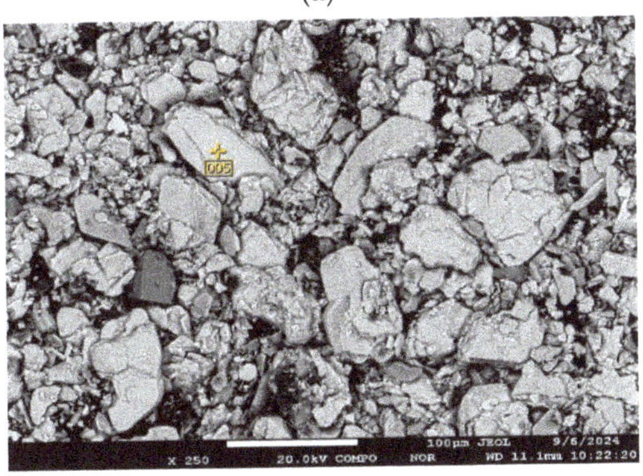

Chemical formula	ms%	mol%	Sigma	Net	K ratio	Line
S	42.58	68.93	0.05	7839582	0.4738505	K
Mo	57.42	31.07	0.14	7096299	0.5199930	L
Total	100.00	100.00				

(c)

Figure 3. (**a**) Point with the dominant minerals molybdenite (MoS_2) and chalcopyrite ($CuFeS_2$); (**b**) point of the pyrite (FeS_2) mineral; (**c**) point of the molybdenite (MoS_2) mineral.

Table 6. Particle size distribution of molybdenite concentrate before and after mechanoactivation.

Particle Size Fraction	Before Mechanoactivation (% wt.)	After Mechanoactivation (% wt.)
+0.045 mm	37.7	8.4
−0.045 mm	62.3	91.6

Table 7. Phase analysis of the ground concentrate.

Phase Number in the Diffractogram	Compound Name	Chemical Formula	wt.%
1	Molybdenite-2H	MoS_2	91.8
2	Quartz, α-form	SiO_2	5.8
3	Chalcopyrite	$CuFeS_2$	2.7

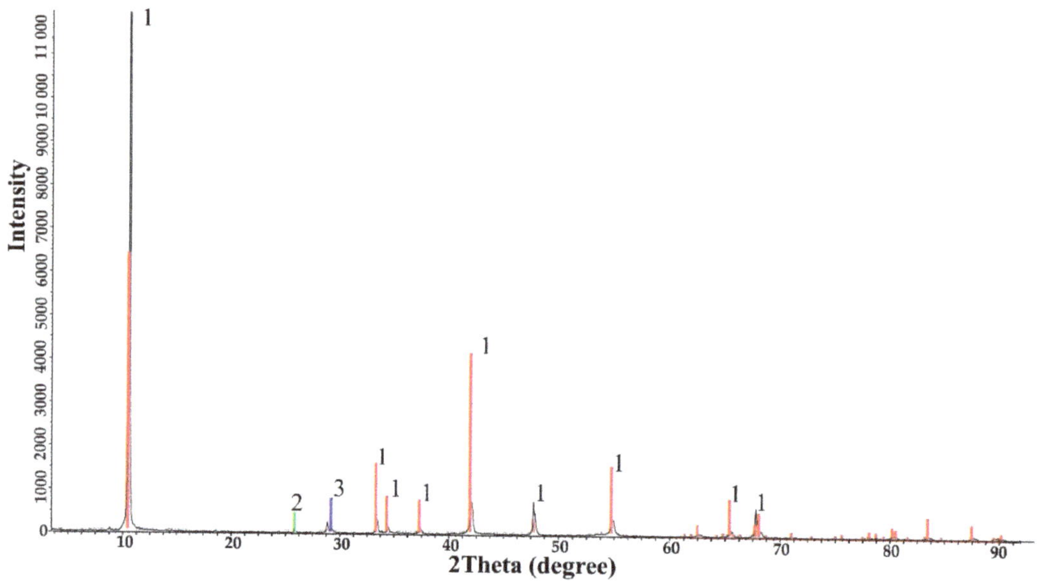

Figure 4. XRD analysis of milled molybdenite concentrate.

From Table 6, it is evident that the molybdenite phase increased from 83.6% to 91.8%, the quartz phase decreased from 12.9% to 5.8%, and the chalcopyrite phase also slightly decreased from 3.5% to 2.7%. Amorphization of the crystalline lattice of minerals in the molybdenite concentrate was observed.

To evaluate the structural transformations occurring in the molybdenite concentrate as a result of mechanoactivation, studies were conducted using electron probe microanalysis. Comparative images of the initial concentrate and post-mechanoactivation reveal significant changes in particle morphology (Figure 5). In the COMPO mode (back scattered electrons), the images (Figure 5a,b) exhibit a characteristic transformation of the material's microtexture accompanied by an increase in dispersity. In the WDS mode (wavelength-dispersive spectroscopy), the images (Figure 5c,d) show pronounced particle comminution, indicating an increase in specific surface area and a potential enhancement in the material's reactivity.

Figure 5. (**a**,**b**)—Images of the initial Mo concentrate, (**c**,**d**)—after mechanical activation.

Various hydrometallurgical leaching methods have been explored in the literature, primarily using sulfuric acid (H_2SO_4), hydrochloric acid (HCl), and nitric acid (HNO_3), each offering distinct advantages. Studies [43] have demonstrated that leaching with H_2SO_4 and H_2O_2 effectively extracts molybdenum but requires prolonged processing and results in higher dissolution of impurities. In the work [42], HCl leaching was improved, achieving efficient metal recovery with lower impurity levels and the potential for acid reuse, albeit at elevated temperatures. Research [44,45] has highlighted HNO_3 leaching as a potent oxidative system but noted the need for controlling gas emissions. Nitric acid (HNO_3) is the primary driver of molybdenum extraction due to its strong oxidative properties, which efficiently convert MoS_2 to soluble molybdate ($MoO_4{}^{2-}$) as per Equation (6). Sulfuric acid (H_2SO_4) plays a secondary role, enhancing ionic stability by forming molybdenyl–sulfate complexes (Equation (9)), but its contribution is less significant [28].

$$MoS_2 + 3H_2O_2 + H_2SO_4 \rightarrow MoO_3 + 2S + 2H_2O + H_2SO_4, \quad (2)$$

Further dissolution of MoO_3 (partially soluble) in acid may form molybdate ions:

$$MoO_3 + H_2O \rightarrow H_2MoO_4 \rightleftharpoons MoO_4{}^{2-} + 2H^+, \quad (3)$$

In the presence of oxygen or another oxidizer (like air or H_2O_2), MoS_2 can be oxidized in HCl, thus:

$$MoS_2 + 6HCl + 1.5O_2 \rightarrow MoCl_6 + 2S + 3H_2O, \quad (4)$$

$MoCl_6$ may hydrolyze in aqueous systems, forming molybdic acid or molybdate ions depending on pH:

$$MoCl_6 + 3H_2O \rightarrow MoO_3 + 6HCl, \quad (5)$$

HNO_3 acts as both an acid and an oxidizing agent, effectively oxidizing MoS_2:

$$MoS_2 + 6HNO_3 \rightarrow MoO_3 + 2S + 6NO_2 + 3H_2O, \quad (6)$$

Emissions like NO_2 must be controlled, as described above.

Our study, incorporating nitric acid, sulfuric acid, and their combination, optimizes efficiency, impurity control, and environmental impact. The oxidative strength of HNO_3, combined with the stabilizing effect of H_2SO_4, enhances molybdenum recovery while mitigating the limitations of single-acid systems.

This dual-acid system leverages the oxidizing power of HNO_3 and the ionic stability provided by H_2SO_4.

$$MoS_2 + 3HNO_3 + 2H_2SO_4 \rightarrow MoO_3 + 2S + 3NO_2 + 2H_2O + H_2SO_4, \quad (7)$$

In the presence of sulfate ions, molybdenum may form molybdenyl–sulfate complexes such as the following:

$$MoO_3 + H_2SO_4 \rightarrow MoO_2(SO_4) + H_2O, \quad (8)$$

$$Mo^{6+} + SO_4^{2-} \rightarrow [MoO_2(SO_4)_2]^{2-} \quad (9)$$

To determine the optimal conditions for the decomposition of molybdenite concentrate, experimental studies on acid leaching were conducted, involving variations in the composition and concentration of leaching agents.

3.1. *Statistical Analysis and Model Selection: Data Analysis*

Table 8 presents the results of the analysis of variance (ANOVA) for the response surface model of the molybdenum leaching process.

Table 8. ANOVA for the quadratic response surface model.

Source	Sum of Squares	df	Mean Square	F Value	p-Value Prob > F	Standard Error	95% CI (Lower–Upper)
Model	1789.61	9	198.85	26.73	<0.0001	–	–
A-A	363.71	1	363.71	48.89	<0.0001	±0.067	(0.33–0.61)
B-B	86.94	1	86.94	11.69	0.0066	±0.082	(−0.47–−0.09)
C-C	28.23	1	28.23	3.79	0.0800	±0.15	(−0.05–1.92)
AB	113.96	1	113.96	15.32	0.0029	±0.30	(−1.85–−0.47)
AC	174.64	1	174.64	23.47	0.0007	±0.66	(−6.88–−3.96)
BC	68.98	1	68.98	9.27	0.0124	±0.51	(2.01–4.77)
A^2	99.59	1	99.59	13.39	0.0044	±0.74	(6.13–9.85)
Residual	105.63	13	8.13			±	
Lack of Fit	105.63	12	8.80			±	
Pure Error	0.000	1	0.000			±	
Cor Total	1864.00	19				⊥	

The F-value of 26.73 indicates the statistical significance of the model ($p < 0.0001$). p-values less than 0.0500 signify the statistical significance of the corresponding model terms. Initially, the model included all linear, interaction, and quadratic terms (A, B, C, AB, AC, BC, A^2, B^2, and C^2). However, terms C ($p = 0.0800$), B^2 ($p = 0.0506$), and C^2 ($p = 0.5173$) were found to be insignificant and were removed to improve model parsimony and predictive accuracy. The reduced model, comprising A, B, AB, AC, BC, and A^2, yielded an adjusted R^2 of approximately 0.90 and a predicted R^2 of approximately 0.80–0.85, indicating improved agreement. The lack-of-fit F-value could not be calculated due to insufficient pure error data (mean square = 0.000, df = 1). Additional replicates are recommended to assess model adequacy. The "Adeq Precision" metric, which measures the signal-to-noise ratio, is desirable when exceeding 4. Here, the obtained ratio of 18.885 indicates a sufficiently strong signal.

Consequently, this model is statistically robust and can be reliably used for optimization and prediction within the studied range of factors.

The following regression equation was derived from the reduced model:

$$E(Mo) = -28.93 + 0.47A - 0.28B - 1.16AB - 5.42AC + 3.39BC + 7.99A^2 \quad (10)$$

Table 9 summarizes the optimized leaching parameters, predicted molybdenum leaching recovery, and experimental results, including replicates for conditions with and without oxygen sparging. The close agreement between predicted and experimental values (e.g., 55% predicted vs. 50% experimental without sparging and 70% predicted vs. 72.6% with sparging) validates the model's predictive accuracy. The higher recovery with oxygen sparging (72.6%) highlights its critical role in enhancing oxidative decomposition.

Table 9. Optimized leaching parameters and experimental replicates.

Run	(HNO_3 + H_2SO_4), (g/dm^3)	Time (h)	Temperature °C	Oxygen Sparging	Predicted Recovery (%)	Experimental Recovery (%)	Replicate 1 (%)	Replicate 2 (%)	Replicate 3 (%)
1	250 (50 + 200)	4	95	No	55	50	49.8	50.1	50.2
2	250 (50 + 200)	4	95	Yes (0.5 dm^3/min)	70	72.6	72.4	72.7	72.5

To assess the adequacy of the developed model, key diagnostic plots were constructed and analyzed, as shown in Figure 6. The plot (Figure 6a) illustrates the distribution of experimental points along the diagonal axis, confirming the absence of significant systematic errors. Most points align well with the theoretical line of normal distribution, indicating a Gaussian distribution of model errors and supporting the model's validity. The residuals are randomly distributed around the y = 0 axis, without apparent patterns or trends, suggesting the absence of autocorrelation and confirming the model's adequacy (Figure 6b). The lack of systematic deviation indicates that the model does not suffer from the omission of critical variables. The absence of a clear dependency or trend in the distribution of points further confirms that the errors are random (Figure 6c). The range of values from −3.00 to 3.00 demonstrates that all residuals fall within acceptable limits, further validating the model's stability. The experimental data points are aligned along the diagonal (Figure 6d), confirming the high accuracy of the model's predictions. The distribution of points shows minimal deviations, indicating a strong agreement between theoretical and experimental values. Thus, the analysis of diagnostic plots confirms the model's adequacy, its high predictive capability, and its accurate representation of the relationships among the studied parameters.

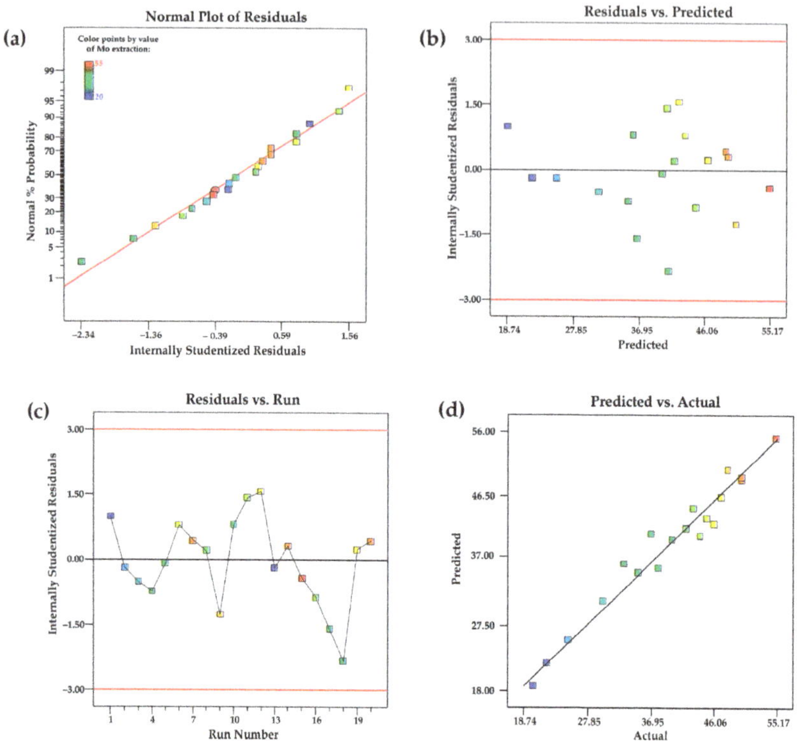

Figure 6. (**a**) Graph of the normal probability depending on internally identified residuals, (**b**) internally identified residuals depending on predicted responses, (**c**) internally identified residuals depending on the number of runs, and (**d**) predicted responses depending on actual values.

Interactions Between Factors

The coefficients for factors A and B in the reduced regression model (Equation (10)) are +0.47 and −0.28, respectively, reflecting their significant influence on molybdenum leaching recovery (Table 8). These values align with theoretical assumptions [46,47]. The standalone temperature term (C) was excluded from the reduced model due to its statistical insignificance ($p = 0.0800$), though it contributes through interaction terms (AC, BC). This reduction improves model parsimony while maintaining predictive accuracy (adjusted $R^2 \approx 0.90$).

Analysis of the coefficients indicates that acid concentration (A) exerts the most significant effect on molybdenum leaching recovery, followed by interaction terms (AC, AB, and BC). Temperature (C) has a positive coefficient (+0.935), suggesting a favorable effect, but its statistical insignificance ($p = 0.0800$) indicates a limited impact within the tested range (20–95 °C). This may be attributed to the narrow temperature range, which is below typical autoclave conditions (150–200 °C) for molybdenum leaching, and the compensating effects of mechanoactivation and oxygen sparging. Mechanoactivation enhances reactivity by increasing surface area and defects, while oxygen sparging accelerates oxidation, reducing reliance on thermal energy. Additionally, potential passivation at higher temperatures may limit the temperature's effect. The order of influence is thus as follows: A > AC > AB > BC > B > C.

Three-dimensional response surfaces, derived from the quadratic model, provide a comprehensive analysis of the interrelationship between the main process parameters and the degree of molybdenum leaching recovery.

To evaluate the impact of key parameters on molybdenum extraction (Mo extraction, %), three-dimensional response surfaces were constructed, as shown in Figure 7a–c. The plot (Figure 7a) illustrates the dependence of molybdenum extraction on temperature (C, °C) and concentration (A, $HNO_3 + H_2SO_4$, g/dm^3) at a fixed leaching time (B = 177.57 min). The response surface reveals that increasing both temperature and concentration significantly enhances molybdenum extraction, particularly at high values of A and C, indicating a strong synergistic effect (AC interaction). At lower temperatures, increasing concentration has a limited effect, likely due to kinetic constraints in the process.

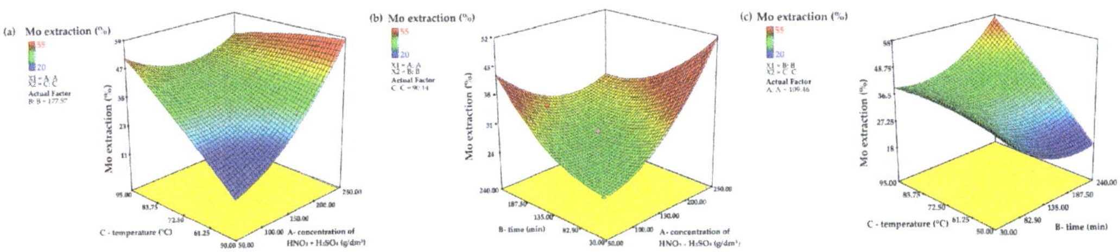

Figure 7. Three-dimensional response surfaces (the remaining parameters are stored at the center level) showing the combined effect of C and A (**a**); B and A (**b**); and C and B (**c**) (A is acid concentration, B is time, C is temperature).

Figure 7b depicts the interaction between time (B) and concentration (A), demonstrating the influence of leaching time (B, min) and concentration (A, $HNO_3 + H_2SO_4$, g/dm^3) on molybdenum extraction at a fixed temperature (C = 90.14 °C). Increasing both time and acid concentration leads to a rise in Mo extraction; however, the effect stabilizes in the saturation zone. The optimal range lies in the region of high acid concentrations and extended time, but further increases in time result in a slowdown of the process, likely due to the formation of passivating layers. Acid concentration exerts a dominant influence, while the effect of increasing time is less pronounced. Figure 7c shows the influence of temperature (C, °C) and time (B, min) on molybdenum extraction at a fixed concentration (A = 109.46 g/dm^3). The plot indicates that temperature enhances molybdenum extraction, especially at longer leaching times, reflecting the role of the BC interaction. At low temperatures, prolonged reaction time does not substantially increase extraction, highlighting temperature-related kinetic limitations. Maximum extraction is achieved at high temperatures and extended times, confirming the critical role of temperature in ensuring a high degree of molybdenum dissolution.

Analysis of the three-dimensional response surfaces reveals that the interaction between temperature and acid concentration (AC) is critical for maximizing molybdenum extraction. While temperature's direct effect is limited, its combination with high acid concentration significantly boosts recovery. Acid concentration (A) remains the most influential single factor, followed by the AC interaction, with leaching time (B) having a lesser direct impact but contributing through interactions.

Thus, optimizing the acid concentration and leveraging the synergistic effect of temperature and concentration is crucial for enhancing process efficiency, while the regulation of reaction time must account for potential passivation.

Based on the analysis of F-values and interactions, the order of factor influence is as follows: A > AC > AB > BC > B > C.

Each interaction factor exhibits its own extremal points, enabling the prediction of optimal process conditions, as illustrated in Figure 7. Design Expert 7.0 software was utilized for modeling and optimizing the molybdenum leaching process. According to the calculations, the optimal parameters included an acid solution concentration (A: HNO_3

50 g/dm^3 + H_2SO_4 200 g/dm^3) of 250 g/dm^3, a leaching time of 4 h, and a temperature of 95 °C. Under these conditions, the predicted molybdenum extraction was 55%.

Experimental validation conducted under the optimized conditions confirmed the model's high accuracy: the actual molybdenum extraction reached 50%, closely matching the predicted value. This agreement demonstrates the reliability and predictive capability of the response surface-based model. Additionally, incorporating oxygen sparging under the same conditions further enhanced molybdenum extraction to 72.6%, underscoring the importance of oxidative conditions in the leaching process.

The study results confirm that mechanoactivation enhances the reactivity of molybdenite, which, when combined with the selection of an optimal acid reagent, intensifies the decomposition process and improves the selectivity of valuable component extraction.

3.2. Effect of Oxygen Sparging on Molybdenum Leaching

The investigation of various acid leaching regimes for molybdenite concentrate revealed that process intensification through oxygen sparging significantly enhances the extraction of target components, such as molybdenum, into the solution [48,49]. Oxygen sparging was performed using a fine-pore gas diffuser connected to a high-purity oxygen cylinder, delivering a flow rate of 0.5 dm^3/min at atmospheric pressure (1 atm). Sparging was continuous throughout the 4 h leaching experiments conducted at 95 °C, ensuring sufficient dissolved oxygen to enhance the oxidative decomposition of MoS_2 into soluble molybdate (MoO_4^{2-}) as per Equation (11). The setup maintained consistent oxygen saturation in the leaching solution, contributing to the observed 22.6% increase in molybdenum leaching recovery. This is of critical importance in metallurgy, as molybdenum is widely utilized in alloys, steels, and the aerospace industry due to its high strength and corrosion resistance. Oxygen acts as an oxidant in the acid leaching process, accelerating the oxidation reaction of molybdenum sulfide. The primary reaction involves the conversion of MoS_2 into soluble forms, such as molybdate (MoO_4^{2-}), with the formation of sulfate (SO_4^{2-}), which can be represented as follows:

$$MoS_2 + 6O_2 + 2H_2O \rightarrow MoO_4^{2-} + 2SO_4^{2-} + 4H^+ \quad (11)$$

Oxygen introduced via sparging or under pressure increases the concentration of dissolved oxygen, thereby enhancing oxidation and, consequently, improving the process kinetics. Studies indicate [50] that without oxygen, molybdenum extraction can be extremely low (less than 2% under acidic and neutral conditions), but with the addition of oxygen, extraction can reach 85–86.4%, significantly improving efficiency.

Experimental studies on the preliminary leaching of molybdenite concentrate with oxygen sparging, combined with mathematical modeling techniques, enabled the determination of optimal technological parameters for the leaching process, achieving a molybdenum leaching recovery of 72.6%. Figure 8 presents results illustrating molybdenum leaching recovery and the impact of oxygen sparging on process efficiency. Due to the intensification of oxidative reactions, molybdenum leaching recovery increased by 22.6% compared to the baseline process without sparging, where the recovery was 50%.

The experimental data presented in the plot "Mo leaching recovery over time with and without O_2" demonstrate the significant influence of oxygen on the molybdenum leaching process. These results underscore the catalytic role of oxygen in oxidative reactions, accelerating molybdenum dissolution, which aligns with the principles of physical chemistry related to the kinetics and thermodynamics of heterogeneous processes.

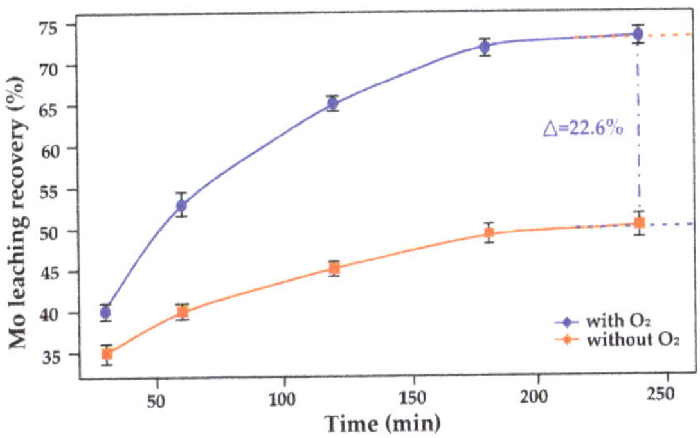

Figure 8. Molybdenum leaching efficiency over time with and without O_2.

4. Conclusions

This study achieved a molybdenum leaching recovery of 72.6% under optimal conditions (acid concentration: 250 g/dm^3, HNO_3 50 g/dm^3 + H_2SO_4 200 g/dm^3; leaching time: 4 h; temperature: 95 °C; S/L ratio: 1:6) with oxygen sparging, demonstrating significant improvements over single-acid systems. The interaction between acid concentration and temperature (AC) was critical, with nitric acid driving oxidative decomposition. These findings offer a sustainable framework for molybdenum extraction, with implications for industrial metallurgy.

Statistical data processing using response surface methodology (RSM) and central composite design (CCD) enabled the construction of a robust quadratic model that accurately captured the influence of leaching parameters. The model facilitated the identification of key factor interactions and optimal conditions, offering a reliable framework for process optimization and predictive control in molybdenum hydrometallurgy.

The dissolved molybdenum, present as molybdate (MoO_4^{2-}) or molybdenyl–sulfate complexes, can be recovered from the electrolyte via solvent extraction using a tertiary amine (e.g., Alamine 336), followed by stripping with ammonium hydroxide to precipitate ammonium molybdate ($(NH_4)_2MoO_4$). Calcination at 500–600 °C yields high-purity molybdenum trioxide (MoO_3) for industrial applications. The electrolyte, containing residual nitric and sulfuric acids and impurities (e.g., copper, iron), can be partially recycled to the leaching stage to reduce reagent costs. Residual acids should be neutralized with calcium hydroxide to form gypsum, and impurities removed via precipitation as hydroxides at pH 6–8 to ensure environmental compliance.

Future research should focus on scaling up the process to industrial conditions, evaluating economic feasibility, reagent recycling, and energy efficiency. Additionally, optimizing downstream processes for molybdenum recovery and electrolyte management will enhance overall process economics and sustainability. Additionally, the exploration of alternative oxidants and solvent systems, as well as advanced simulations, could further enhance process sustainability and support the transition toward greener metallurgical technologies.

Author Contributions: Conceptualization: B.K. and A.U.; Methodology: A.U., A.M. and N.L.; Software: K.K., A.M. and N.L.; Validation: B.K. and A.U.; Formal analysis: A.U.; Investigation: A.U., N.L. and B.K.; Resources: K.K., A.M. and N.L.; Data curation: B.K. and A.U.; Writing—original draft preparation: B.K.; Writing—review and editing: B.K. and A.U.; Visualization: K.K. and A.M.; Supervision: A.U.; Project administration: B.K.; Funding acquisition: B.K. All authors have read and agreed to the published version of the manuscript.

Funding: This research is funded by the Science Committee of the Ministry of Science and Higher Education of the Republic of Kazakhstan, program targeted funding BR24992757.

Data Availability Statement: The original contributions presented in this study are included in the article. Further inquiries can be directed to the corresponding author.

Conflicts of Interest: The authors declare no conflict of interest.

References

1. El-Sharkawy, M.; Alotaibi, M.O.; Li, J.; Du, D.; Mahmoud, E. Heavy metal pollution in coastal environments: Ecological implications and management strategies: A review. *Sustainability* **2025**, *17*, 701. [CrossRef]
2. Liu, R. The pollution and the ecological influence of chemical and industrial waste. *Sci. Technol. Eng. Chem. Environ. Prot.* **2024**, *1*, 1–4. [CrossRef]
3. U.S. Geological Survey. Mineral Commodity Summaries 2025. 2025. Available online: https://pubs.usgs.gov/publication/mcs2025 (accessed on 16 March 2025).
4. Panichkin, A.; Uskenbayeva, A.; Kenzhegulov, A.; Mamaeva, A.; Imbarova, A.; Kshibekova, B.; Alibekov, Z.; Nurhadiyanto, D.; Yunita, I. Assessment of the effect of small additions of some rare earth elements on the structure and mechanical properties of castings from hypereutectic chromium white irons. *AIMS Mater. Sci.* **2023**, *10*, 517–540. [CrossRef]
5. Molybdenum processing | Extraction, Applications & Uses | Britannica. Published 2023-06-16. Available online: https://www.britannica.com/science/molybdenum (accessed on 7 May 2025).
6. Jiang, K.; Wang, Y.; Zou, X.; Zhang, L.; Liu, S. Extraction of Molybdenum from Molybdenite Concentrates with Hydrometallurgical Processing. *JOM* **2012**, *64*, 1285–1289. [CrossRef]
7. Liu, B.; Zhang, B.; Han, G.; Wang, M.; Huang, Y.; Su, S.; Xue, Y.; Wang, Y. Clean separation and purification for strategic metals of molybdenum and rhenium from minerals and waste alloy scraps–A review. *Resour. Conserv. Recycl.* **2022**, *181*, 106232. [CrossRef]
8. Hesami, R.; Ahmadi, A.; Hosseini, M.; Torabi, M. Effect of mechanical activation on the hypochlorite leaching of Sarcheshmeh molybdenite concentrate. *Sep. Sci. Technol.* **2022**, *57*, 1966–1977. [CrossRef]
9. Zhang, L.; Peng, W.; Wang, W.; Cao, Y.; Qi, M.; Huang, Y. A green method for selective separation of molybdenite and pyrite via electrochemical oxidation pretreatment-flotation and its mechanism. *Colloids Surf. A Physicochem. Eng. Asp.* **2024**, *687*, 133508. [CrossRef]
10. Yessengaziyev, A.M.; Toishybek, A.M.; Mukangaliyeva, A.O.; Abdyldayev, N.N.; Yersaiynova, A. A Study of the leaching process for dust chamber sublimates followed by the extraction of niobium and zirconium into solution. *Kompleks. Ispolz. Miner. Syra Complex Use Miner. Resour.* **2024**, *331*, 117–126. [CrossRef]
11. Lin, X.; Jiao, F.; Wei, Q.; Wu, Y. Application and depression mechanism of sodium humate on flotation separation of molybdenite and pyrite. *Sep. Sci. Technol.* **2024**, *59*, 449–463. [CrossRef]
12. Harvey, J.; Courchesne, W.; Vo, M.; Oishi, K.; Robelin, C.; Mahue, U.; Leclerc, P.; Al-Haiek, A. Greener reactants, renewable energies and environmental impact mitigation strategies in pyrometallurgical processes: A review. *MRS Energy Sustain.* **2022**, *9*, 212–247. [CrossRef]
13. Gao, H.; Jin, Z.; Hu, K.; Zhang, Y.; Zhang, C.; Jin, X. A Fast and Environment-friendly Method for Preparing Nanopowders Mo through Electro- desulfidation of MoS_2 in LiCl melt. *Sep. Purif. Technol.* **2023**, *327*, 124823. [CrossRef]
14. Giannopoulou, I.; Panias, D.; Paspaliaris, I. Electrochemical modeling and study of copper deposition from concentrated ammoniacal sulfate solutions. *Hydrometallurgy* **2009**, *99*, 58–66. [CrossRef]
15. Zhang, H.; Li, S.; Lv, Z.; He, J.; Fan, Y.; Song, J. The role of zinc sulfide in the electrochemical extraction of molybdenum. *Sep. Purif. Technol.* **2023**, *311*, 123290. [CrossRef]
16. Tumen-Ulzii, N.; Batnasan, A.; Gunchin, B. Selective dissolution of copper and iron from molybdenite concentrate using acidic sodium nitrate solution. *Miner. Eng.* **2022**, *185*, 107715. [CrossRef]
17. Chanturiya, V.; Matveeva, T.; Getman, V.; Karkeshkina, A.; Gromova, N. Substantiation of New Reagent Compositions for the Effective Extraction of Rhenium in the Processing of Complex Molybdenum Ores. *Minerals* **2023**, *13*, 372. [CrossRef]
18. Subasinghe, H.; Ratnayake, A. Mechanical activation and physicochemical factors controlling pyrometallurgical, hydrometallurgical, and electrometallurgical processing of titanium ore: A review. *J. South. Afr. Inst. Min. Metall.* **2023**, *123*, 399–414. [CrossRef]
19. Panichkin, A.; Kenzhegulov, A.; Mamaeva, A.; Uskenbayeva, A.; Kshibekova, B.; Imbarova, A.; Alibekov, Z. Effect of Carbon and Cooling Rate on the Structure of Hypereutectic High Chromium Cast Iron in the Cast State and after Heat Treatment. *J. Compos. Sci.* **2023**, *7*, 403. [CrossRef]
20. Baizakov, R.; Ivanova, E. Development and research of means and methods for improving the efficiency of heat stations in the Republic of Kazakhstan. *Bull. Innov. Univ. Eurasia* **2021**, *3*, 64–71. [CrossRef]
21. Batyrbekov, E.; Vityuk, V.; Zarva, D.; Sharipov, M. Conceptual View of the Implementation of the Nuclear Energy Program in the Republic of Kazakhstan. *Energies* **2024**, *17*, 5788. [CrossRef]

22. Woo, S.; Chirayath, S.; Fuhrmann, M. Nuclear fuel reprocessing: Can pyro-processing reduce nuclear proliferation risk? *Energy Policy* **2020**, *144*, 111601. [CrossRef]
23. Soltangazinov, A.; Smagulova, Z.; Amirova, M.; Kashuk, L.; Karimbergenova, M.; Kadyrova, A.; Zhaltyrova, O. Energy Efficiency as a Factor of Sustainable Development in Kazakhstan. *Int. J. Energy Econ. Policy* **2020**, *10*, 325–330. [CrossRef]
24. Yessengaziyev, A.; Mukangaliyeva, A.; Toishybek, A.; Karshyga, Z.; Abdyldayev, N.; Yersaiynova, A.; Bakhytuly, N. Studies the crucial role of selection of extractant to extract niobium from sulfuryl fluoride solution and optimization of extraction conditions. *Acta Metall. Slovaca* **2024**, *30*, 120–126. [CrossRef]
25. Abdulvaliyev, R.; Ultarakova, A.; Mukangaliyeva, A.; Lokhova, N.; Kassymzhanov, K. Comparative Analysis of Acid Leaching for the Efficient Recovery of Lanthanum and Cerium from Phosphate. *Separations* **2024**, *11*, 288. [CrossRef]
26. Sheybani, K.; Paydar, M.; Shariat, M. Investigation on the Kinetics and Mechanism of Aluminothermic Reduction of Molybdenum Trioxide: Non-isothermal Kinetics. *Trans. Indian Inst. Met.* **2020**, *73*, 2875–2888. [CrossRef]
27. Wang, Y.; Wang, C.; Liu, H.; Meng, Y. Extraction of Molybdenum and Uranium from Low-Grade Molybdenum Bearing Ore Containing Uranium through Mechanical Activation Following Acid Leaching. *Miner. Process. Extr. Metall. Rev.* **2022**, *45*, 304–313. [CrossRef]
28. Vizsolyi, A.; Peters, E. Nitric acid leaching of molybdenite concentrates. *Hydrometallurgy* **1980**, *6*, 103–119. [CrossRef]
29. Khoshnevisan, A.; Yoozbashizadeh, H.; Mozammel, M.; Sadmezhaad, S.K. Kinetics of pressure oxidative leaching of molybdenite concentrate by nitric acid. *Hydrometallurgy* **2012**, *111*, 58–64. [CrossRef]
30. Behmadi, R.; Mirzaei, M.; Afshar, R.; Najafi, H. Influence of chalcopyrite removal and mechanical exfoliation on the performance of molybdenite catalysts supported on activated bauxite for alcohol synthesis by Fischer-Tropsch process. *Fuel* **2023**, *357*, 129772. [CrossRef]
31. Paphane, B.; Nkoane, B.; Oyetunji, O. Kinetic studies on the leaching reactions in the autoclave circuit of the Tati Hydrometallurgical Demonstration Plant. *J. South Afr. Inst. Min. Metall.* **2013**, *113*, 485–489.
32. Rasulova, S.; Guro, V.; Safarov, E.; Adinaev, X. Metals Recovery from Molybdenite Concentrate by Electrooxidation and Leachingiop Conference Series. *Mater. Sci. Eng.* **2020**, *848*, 012076. [CrossRef]
33. Kenzhaliyev, B.; Koizhanova, A.; Fischer, D.; Magomedov, D.; Yerdenova, M.; Smailov, K.; Abdyldayev, N. Study of efficiency of organic activator application to process difficult-to-beneficiate polymetallic ridder ores. *Transit. Met. Chem.* **2025**, 1–13. [CrossRef]
34. Cao, Z.; Zhong, H.; Jiang, T.; Wang, S.; Liu, G.; Xia, L. A novel hydrometallurgy of molybdenite concentrate and its kinetics. *J. Chem. Technol. Biotechnol.* **2012**, *87*, 938–942. [CrossRef]
35. Parsons, G.; Brimacombe, J.; Peters, E. Computer simulation of a molybdenite leaching process using dilute nitric acid. *Hydrometallurgy* **1987**, *17*, 133–154. [CrossRef]
36. Ospanov, K.; Smailov, K.; Nuruly, Y. Patterns of Non-Traditional Thermodynamic Functions ΔrG0/n and ΔfG0 (Averaged) Changes for Cobalt Minerals. *Chem. Bull. Kazakh Natl. Univ.* **2020**, *96*, 22–30. [CrossRef]
37. Lyapin, S.; Guro, V.; Parpiev, N.; Rasulova, S. Photometric determination of rhenium in mixed hydrochloric-nitric acidic solutions formed upon processing of molibdenite concentrate. *Ind. Lab. Diagn. Mater.* **2020**, *86*, 23–29. [CrossRef]
38. Medvedev, A.; Aleksandrov, P. Investigations on processing low-grade molybdenum concentrate by the nitric-acid method. *Russ. J. Non-Ferr. Met.* **2009**, *50*, 353–356. [CrossRef]
39. Liu, R.; Zhao, Z.; Li, Y. Acid leaching–extraction–circulation process based on Mo(VI) coordination with H_3PO_4 to efficiently extract molybdenum from different components of molybdenum calcine. *Sep. Purif. Technol.* **2023**, *322*, 124269. [CrossRef]
40. KAZ Minerals. Aktogay Mining and Processing Plant. Available online: https://www.kazminerals.com/our-business/aktogay/ (accessed on 23 April 2025).
41. Kenzhaliyev, B.; Surkova, T.; Berkinbayeva, A.; Baltabekova, Z.; Smailov, K.; Abikak, Y.; Saulebekkyzy, S.; Tolegenova, N.; Omirbek, T.; Dosymbaeva, Z. Innovative Methods for Intensifying the Processing of Zinc Clinker: Synergy of Microwave Treatment and Ultrasonic Leaching. *Metals* **2025**, *15*, 246. [CrossRef]
42. Liu, W.; Xu, H.; Yang, X.; Shi, X. Extraction of molybdenum from low-grade Ni–Mo ore in sodium hypochlorite solution under mechanical activation. *Miner. Eng.* **2011**, *24*, 1580–1585. [CrossRef]
43. Banda, R.; Nguyen, T.H.; Sohn, S.H.; Lee, M.S. Recovery of valuable metals and regeneration of acid from the leaching solution of spent HDS catalysts by solvent extraction. *Hydrometallurgy* **2013**, *133*, 161–167. [CrossRef]
44. Barik, S.P.; Kyung-Ho Park, P.K.; Parhi, J.T. Park, Direct leaching of molybdenum and cobalt from spent hydrodesulphurization catalyst with sulphuric acid. *Hydrometallurgy* **2012**, *111*, 46–51. [CrossRef]
45. Karimova, L.; Kairalapov, Y.; Tussupbekova, T.; Oleinikova, T.; Makasheva, G. Hydrometallurgical processing of molybdenum middlings from Shatyrkul-Zhaysan cluster ore. *J. Min. Metall. Sect. B Metall.* **2024**, *60*, 71–83. [CrossRef]
46. Peng, H.; Wang, F.; Li, G.; Guo, J.; Li, B. Highly Efficient Recovery of Vanadium and Chromium: Optimized by Response Surface Methodology. *ACS Omega* **2019**, *4*, 904–910. [CrossRef] [PubMed]
47. Tian, J.J.; Wu, D.D.; Li, S.Y.; Ma, W.H.; Wang, R.Z. Effect of process variables on leaching behavior and kinetics of silver element from waste photovoltaic modules. *Sep. Purif. Technol.* **2024**, *335*, 126062. [CrossRef]

48. Shoghian-Alanaghi, A.; Zamharir, A.J.; Aghajani, H.; Tabrizi, A.T. Improving the Leaching Rate of Molybdenite Concentrate Using Oxidants by Adding Ethylene Glycol and Oxygen: Kinetic Study. *Min. Metall. Explor.* **2022**, *39*, 1753–1761. [CrossRef]
49. Yang, C.; Zhou, Y.; Liu, D.; Jiang, W.; Liu, F.; Liu, Z. *Preparation of Molybdenum Powder from Molybdenite Concentrate Through Vacuum Decomposition-Acid Leaching Combination Process*; Springer: Berlin/Heidelberg, Germany, 2017; pp. 235–246. [CrossRef]
50. Li, Y.; Li, Z.; Wang, B.; Dong, Z.; Song, S. A fundamental study of leaching kinetics and mechanisms of molybdenite assisted by mechanical activation. *Miner. Eng.* **2019**, *131*, 376–384. [CrossRef]

processes MDPI

Article

Optimizing Energy Storage Participation in Primary Frequency Regulation: A Novel Analytical Approach for Virtual Inertia and Damping Control in Low-Carbon Power Systems

Wentian Lu [1], Enkai Tan [1], Lefeng Cheng [1,*], Kuozhen Zhang [2,*] and Wenjie Liu [1]

1 School of Mechanical and Electrical Engineering, Guangzhou University, Guangzhou 510006, China; luwentian@gzhu.edu.cn (W.L.); 2112207055@e.gzhu.edu.cn (E.T.); lwj1993@gzhu.edu.cn (W.L.)
2 Law School, Shantou University, Shantou 515063, China
* Correspondence: chenglefeng@gzhu.edu.cn (L.C.); kzhzhang@stu.edu.cn (K.Z.)

Abstract: As renewable energy penetration increases, maintaining grid frequency stability becomes more challenging due to reduced system inertia. This paper proposes an analytical control strategy that enables distributed energy resources (DERs) to provide inertial and primary frequency support. A reduced second-order model is developed based on aggregation theory to simplify the multi-machine system and facilitate time-domain frequency analysis. Building on this model, we design virtual inertia and damping coefficients for the frequency response, ensuring that it meets acceptable limits for both overshoot and steady-state deviation. To address energy storage constraints, an adaptive strategy is introduced to adjust control parameters dynamically based on the state of charge (SOC). Simulation results validate the accuracy of the aggregation model, showing that it closely approximates the full multi-machine system with minimal error. The proposed method significantly enhances frequency stability under varying load conditions while maintaining efficient SOC utilization. This study provides a practical framework for integrating DERs into grid frequency regulation by combining analytical control design with SOC-aware adaptation. The approach offers a computationally efficient alternative to detailed models, supporting more responsive and stable low-inertia power systems.

Keywords: distributed energy resources (DERs); frequency modulation; virtual inertia; primary frequency regulation; adaptive control strategy; energy storage optimization

1. Introduction

Renewable energy has made remarkable progress during the last decade. From 2013 to 2023, the total installed wind power expanded nearly fivefold [1]. As the advances of the commercialization process of offshore wind power, this upward momentum is projected to persist over the coming decades. Nevertheless, as renewable sources like wind and photovoltaic energy become more prevalent [2], many units utilizing maximum power point tracking (MPPT) techniques lack intrinsic frequency response functions, resulting in a significant reduction in system inertia [3]. To address these challenges, energy storage systems can be controlled to emulate the inertial response of synchronous generators by providing virtual inertia, thereby enhancing the frequency stability of power systems [4]. This approach has been widely recognized and adopted in modern low-inertia power systems. In China, an estimated 138 GW of electrochemical energy storage may be required by 2030 to mitigate the challenges posed by the increasing penetration of renewable energy sources [5]. Therefore, many power system operators are trying to find ways to enhance the

https://doi.org/10.3390/pr13041146

auxiliary role of new energy sources, such as wind, photovoltaics and storage, in frequency control, while safeguarding power quality [6–8].

Numerous studies have investigated control strategies that enable distributed energy resources (DERs), such as wind turbines, photovoltaic systems, and energy storage, to contribute to primary frequency regulation. These studies encompass various methods including virtual inertia control, MPPT deviation, and coordinated virtual synchronous generator (VSG) strategies. In Ref. [9], wind turbines are typically configured to operate under the MPPT algorithm in standard conditions, but frequency regulation can be improved by temporarily deviating from the MPPT operation. Specifically, by discharging rotor kinetic energy and employing virtual inertia control, wind turbines can actively contribute to frequency regulation. VSG control is a more comprehensive control strategy, and it has been proposed to further enhance frequency regulation. It integrates virtual inertia, virtual damping, and frequency droop control to simulate the behavior of conventional synchronous generators. This enables DERs to not only respond quickly to frequency disturbances (via virtual inertia) but also maintain long-term grid stability (via virtual damping and droop control). By mimicking the dynamic behavior of synchronous generators, VSG control allows DERs to provide effective frequency support in systems with high renewable energy penetration. In Ref. [10], a flexible control approach targeting virtual parameters within a virtual synchronous generator (VSG) system incorporating energy storage is proposed to improve wind power's responsiveness to primary frequency regulation. In Ref. [11], the integrated inertia control parameters are optimized, showing that increasing virtual inertia control can worsen the system's frequency response when frequency modulation resources are inadequate. In Ref. [12], an advanced wind turbine control strategy is presented, utilizing a real-time reduced-order model of the power grid. This strategy takes into account transient grid dynamics and enables accurate coordination among multiple wind farms for synchronized frequency regulation.

Current research on energy storage control strategies primarily focuses on whether energy storage systems participate in frequency regulation independently or in coordination with wind farms and photovoltaic power plants [13]. Integrated inertia control strategies [14–17] include the following: (i) direct setting of the proportional and differential coefficients and (ii) dynamically adjusting the integrated inertia control parameters using adaptive methods. However, the necessity and optimality of adopting this strategy for energy storage have yet to be fully explored. Regarding the first approach, Ref. [18] presents a refined control scheme tailored for battery-based energy storage systems (BESSs), aimed at mitigating wind power fluctuations and their impact on grid frequency. Ref. [19] proposes a coordinated VSG control method for photovoltaic (PV) systems and BESSs, which not only optimize PV output but also enhance grid frequency stability. Ref. [20] proposes an integrated planning approach that accounts for frequency-related limitations, optimizing the configuration of generation as well as storage systems to ensure stable grid frequency performance under conditions of high renewable energy penetration. Using the latter approach, Ref. [21] proposes a flexible VSG (FVSG) control strategy with adaptive inertia. Unlike conventional fixed-parameter designs, the proposed method dynamically adjusts the inertia coefficient based on a nonlinear function of the system's rate of change of frequency (RoCoF). This approach allows the system to provide faster and smoother inertial response during frequency disturbances, as validated through Matlab/Simulink simulations. However, the specific method for selecting coefficients is not discussed. Ref. [22] proposes a virtual inertia regulation method incorporating a fuzzy-based secondary controller to enhance microgrids' voltage/frequency dynamic response. However, no theoretical analysis is provided on the fuzzy decision table, which is somewhat complex. Refs. [23,24] propose a BESS planning approach constrained by operational conditions, aimed at ensuring adequate primary frequency support. Nevertheless, these models overlook critical stability metrics, including frequency nadir, rate-of-change-of-frequency (ROCOF), and quasi-steady-state frequency deviation, thereby failing to fully address system frequency stability assurance.

Based on the principle of aggregation and compensation, this study introduces an innovative analytical control approach for the coordinated integration of wind and photovoltaic energy storage systems into inertial and primary frequency modulation, fully leveraging the fast, flexible, and adaptable nature of power electronic components. Firstly, a simplified aggregated second-order model is established, in which the system frequency and combined governor dynamics are treated as state variables, while load variations serve as external inputs. This formulation preserves the core characteristics of traditional synchronous generators. Subsequently, by utilizing the analytical tractability of the reduced-order model, the transfer function for the integrated frequency regulation process is developed to match the desired dynamic frequency profile. The participation of wind photovoltaic storage-assisted primary frequency modulation optimization is optimally quantified by incorporating the known parameters of the conventional generator unit. The contributions of this paper to the research field are as follows:

- Establishing the second-order model of load frequency control based on aggregation theory to replace the original complex multi-unit model and enabling analysis of the time-domain frequency trajectory.
- Developing a fast analytical method based on the compensation principle to calculate DERs' virtual inertia and damping coefficients and guaranteeing a desired frequency overshoot and steady-state offset.
- Proposing a flexible regulation scheme for energy storage systems involved in frequency control, and dynamically adjusting synthetic inertia and damping coefficients according to state of charge (SOC) levels.

The structure of this paper is arranged as follows: Section 2 introduces the dynamic modeling framework for individual components, including generators, wind power, and energy storage systems, and establishes a simplified second-order representation for the entire system. This model facilitates the determination of overall inertia and damping parameters, as elaborated in Section 3. Section 4 demonstrates the effectiveness of the proposed approach through simulation-based case studies. Finally, Section 5 concludes this paper with key findings and suggestions for future research.

2. System Dynamical Models

This section presents the relevant mathematical formulations used to model the behavior of generators and DERs. In this study, we used MATLAB R2023b software for the simulation research, with the software version number 23.2.0.2365128 (64-bit), released on 23 August 2023. MATLAB is developed and distributed by MathWorks, located in Natick, MA, USA.

2.1. Design Ideas

As shown in Figure 1, the power network comprises two traditional synchronous generators alongside two distributed energy resources (DERs). In general, three-phase grid-connected inverters are equipped with power controllers that receive predefined commands for active and reactive power and incorporate phase-locked loops (PLLs) to track grid frequency variations [25]. By tracking the condition of grid frequency, DER calculates and regulates the active power in response to the variation in frequency offset relative to the synchronous frequency ω_s after the current load ΔP_{load} to match the dynamic process of grid demand.

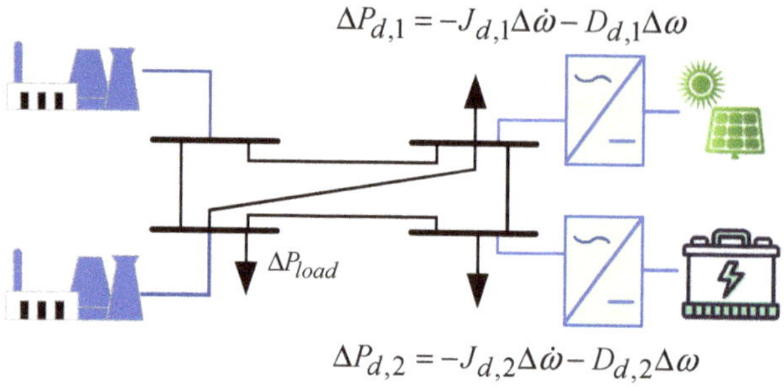

Figure 1. Schematic diagram of power system components.

In this study, a simplified second-order model is developed, where the system frequency and integrated governor dynamics are treated as state variables, with load disturbances serving as the input. This model's analytical simplicity is leveraged to derive closed-form expressions for the time-domain frequency trajectories. These trajectories are parameterized in terms of the equivalent damping and inertia parameters, which are characterized as the combined contributions of the damping and inertia terms from both the generator and the photovoltaic-storage-based DERs. Finally, the total damping and inertia contributions from the DER are decomposed to isolate the specific contribution of the DER alone.

2.2. Synchronous-Generator Dynamics

Frequency is a critical parameter in synchronous generator (SG)-based power systems, ensuring the synchronization of power equipment with the grid [26]. As such, requirements related to frequency performance are fundamental to grid regulations and must be carefully considered when designing frequency support strategies for DERs. Traditionally, the regulation of grid frequency is achieved via the rotor dynamics of synchronous generators, with mechanical inertia and damping properties influencing the system's dynamic frequency behavior, following the swing equation.

As illustrated in Figure 2, the inertia parameter H and damping factor D are the key factors determining the dynamics of active power and frequency. Specifically, the inertia constant primarily influences the ROCOF, while the damping gain has a more significant effect on the deviation in steady-state frequency. In large-scale power systems, ROCOF is primarily governed by the system's total inertia, while the steady-state frequency deviation is more closely associated with the system's total damping. Therefore, to design effective virtual inertia regulation (VIR) and frequency attenuation control (FAC) strategies for DERs, it is essential to first study the dynamics of synchronous generators. Typically, the dynamics of power system stabilizers (PSSs) are neglected when analyzing phenomena related to primary frequency response, and the dynamics of the automatic voltage regulator (AVR) are assumed to be stable. Under these assumptions, the set G is defined as the collection of synchronous generator units participating in the DER aggregation model. Specifically, $G = \{1, 2, \ldots, |G|\}$, where $|G|$ represents the total number of synchronous generators included in the aggregated dispatch. Consequently, the behavior of each individual synchronous generator is described as follows:

$$\dot{\theta}_g = \omega_g - \omega_s \tag{1a}$$

$$J_g \dot{\omega}_g = P_g^m - D_g(\omega_g - \omega_s) - P_g^e \tag{1b}$$

$$\tau_g \dot{P}_g^m = -P_g^m - R_g(\omega_g - \omega_s) \tag{1c}$$

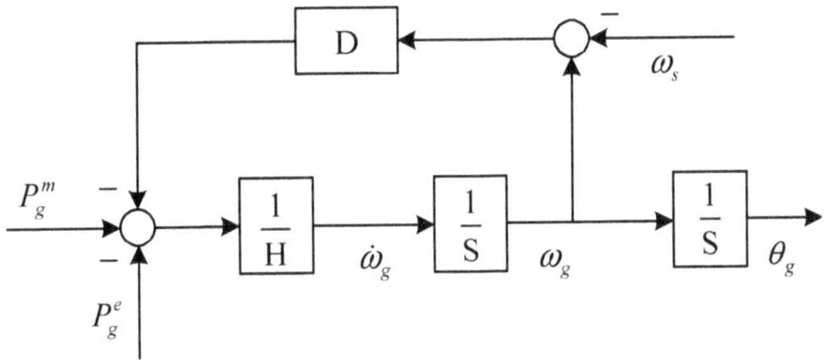

Figure 2. Schematic diagram of the oscillation equation.

Equations (1a) and (1b) together describe the rotor dynamics of the synchronous generator. Specifically, (1a) represents the evolution of the rotor electrical angular position, and (1b) is the swing equation that models the generator's frequency response under the influence of mechanical input power, load damping, and electrical output. (1c) describes the governor dynamics, where the mechanical power output is adjusted in response to the frequency deviation between the generator and the nominal grid. These three equations form a coupled dynamic model for conventional synchronous generators under primary frequency control. θ_g, ω_g, and ω_s are the rotor electrical angular position, machine operating frequency, and synchronous frequency, respectively; J_g denotes the moment of inertia, while D_g represents the damping factor associated with load variations; R_g corresponds to the reciprocal of the speed-droop coefficient used in frequency-power regulation, and τ_g is the turbine time constant; P_g^m is turbine mechanical power, and P_g^e denotes the total electrical power at bus g given by $P_g^e = P_{g,\text{load}} + \sum_{l \in N_g} P_{g,l}$, where $P_{g,\text{load}}$ denotes the power demand at bus g, and $P_{g,l}$ indicates the active power transmitted from bus g to l.

2.3. Frequency-Responsive DER Model

Due to the flexible and adaptable nature of the power output of energy storage, distributed energy resources, in conjunction with energy storage, can output an arbitrarily shaped power curve when interfaced with the grid via a power electronic unit [27]. Consider the following dynamic model for DER $d \in D$:

$$\dot{\theta}_d = \omega_d - \omega_s \tag{2}$$

$$J_d \dot{\omega}_d = -D_d(\omega_d - \omega_s) - P_d^e \tag{3}$$

The dynamic frequency behavior of the DER located at bus d is described as the droop coefficient D_d, and the inertial response is determined by the synthetic-inertia constant J_d; P_d^e denotes the total electrical power at bus d given by $P_d^e = P_{d,\text{load}} + \sum_{l \in N_d} P_{d,l}$, where $P_{d,\text{load}}$ is the load at bus d, and $P_{d,l}$ represents the active power transferred from bus d to l.

2.4. Reduced Second-Order Model

In this section, we aggregate the inertia coefficients, damping coefficients, and time constants to derive a simplified second-order model that effectively captures the dynamics of system frequency.

(1) Aggregation of inertia and damping coefficients:

Assume the system initially operates at the steady-state equilibrium point with $\omega_g = \omega_d = \omega_s, \forall g \in G, d \in D$. Defining $\Delta\omega = \omega_g - \omega_s = \omega_d - \omega_s$, (1b) and (3) can be reformed as

$$J_g \Delta\dot{\omega} = P_g^m - D_g \Delta\omega - P_g^e \tag{4}$$

$$J_d \Delta\dot{\omega} = -D_d \Delta\omega - P_d^e \tag{5}$$

The frequency deviation $\Delta\omega$ is assumed to be uniform across all nodes, which is a standard simplification in synchronous systems. While small local variations may exist in practice, they are typically negligible for primary frequency response analysis. Aggregating Equation (4) across $g \in G$ and Equation (5) over all $d \in D$, we can obtain the following:

$$J_{\text{net}} \Delta\dot{\omega} = -D_{\text{net}} \Delta\omega + \sum_{g \in G} P_g^m - \sum_{g \in G} P_g^e - \sum_{d \in D} P_d^e \tag{6}$$

where J_{net} denotes the equivalent system inertia, and D_{net} represents the overall damping coefficient

$$J_{\text{net}} = \sum_{g \in G} J_g + \sum_{d \in D} J_d \tag{7}$$

$$D_{\text{net}} = \sum_{g \in G} D_g + \sum_{d \in D} D_d \tag{8}$$

Analyze the total electrical power, we have the following:

$$\begin{aligned} \sum_{g \in G} P_g^e + \sum_{d \in D} P_d^e = &\sum_{g \in G} P_{g,\text{load}} + \sum_{d \in D} P_{d,\text{load}} \\ &+ \sum_{g \in G} \sum_{l \in N_g} P_{g,l} + \sum_{d \in D} \sum_{l \in N_d} P_{d,l} \end{aligned} \tag{9}$$

Notice that the right-hand term in (9) is equal to the total electrical load of the whole network, i.e.,

$$P_{\text{load}} = \sum_{g \in G} P_{g,\text{load}} + \sum_{d \in D} P_{d,\text{load}} + \sum_{g \in G} \sum_{l \in N_g} P_{g,l} + \sum_{d \in D} \sum_{l \in N_d} P_{d,l} \tag{10}$$

Then, the rotor equation dynamic expression (6) can be integrated as follows:

$$J_{\text{net}} \Delta\dot{\omega} = -D_{\text{net}} \Delta\omega + \sum_{g \in G} P_g^m - P_{\text{load}} \tag{11}$$

The order of the rotor motion equation in (11) is 1, which achieves the reduction from $|G|$ in (4) for multi-synchronous generators.

(2) Derivation of the aggregated model:

Suppose that thermal power stations comprise multiple synchronous generators with varying characteristics, all of which are involved in grid frequency regulation. The aggregated frequency response model for the FPP system is illustrated in Figure 3.

As illustrated in the dashed box of Figure 3, the transfer function corresponding to a single SG is capable of approximating the behavior of multiple SGs. Observing the dynamic characteristics (1c) of the generator, τ_{red} is not obtained by a simple summation of the aggregated precursor group τ_g. Ref. [28] proposes an efficient and accurate calculation method to integrate τ_g of multiple SGs as follows:

For notational convenience, we define

$$\boldsymbol{P}_G^m = [P_1^m, \ldots, P_{|G|}^m], \boldsymbol{R}_G = [R_1, \ldots, R_{|G|}], \boldsymbol{\tau} = [\tau_1, \ldots, \tau_{|G|}] \tag{12}$$

Now, we can list copies of (1c) $\forall g \in G$ as

$$diag(\boldsymbol{\tau}) \dot{\boldsymbol{P}}_G^m = -\boldsymbol{P}_G^m - \boldsymbol{R}_G \Delta\omega \tag{13}$$

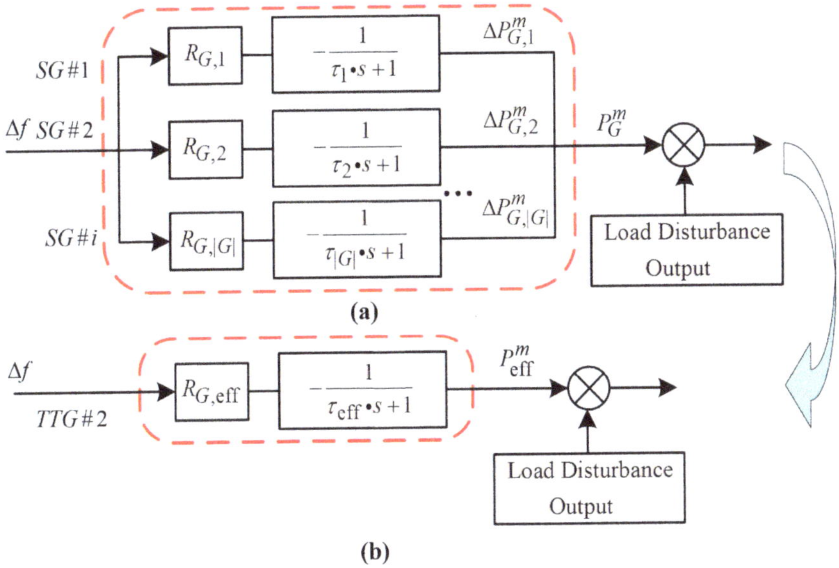

Figure 3. Aggregation of the synchronous generator frequency response model. (**a**) A multi-machine primitive model; (**b**) An aggregated equivalent model.

Combining (11) and (13) yields the state-space model:

$$\dot{x} = Ax + Bu \tag{14}$$

where the state vector $x \in \mathbb{R}^{|G|+1}$, the input vector $u \in \mathbb{R}^1$ and system matrices, $A, B \in \mathbb{R}^{(|G|+1)\times(|G|+1)}$ are given by

$$\begin{aligned} x &= \left[\Delta\omega, P^m_G\right], \qquad u = P_{\text{load}} \\ A &= \begin{bmatrix} -D_{\text{net}}J^{-1}_{\text{net}} & J^{-1}_{\text{net}}1^T_{|G|} \\ -diag(\tau)^{-1}R_G & -diag(\tau)^{-1} \end{bmatrix} \\ B &= \left[-J^{-1}_{\text{net}}, 0_{|G|}\right] \end{aligned} \tag{15}$$

The original model described in (14) has an order of $|G|+1$. Next, to enable an analysis of the time-domain frequency trajectory, we need to aggregate the original governor model (13) to reduce its order from $|G|$ to 1. Define the aggregated governor time coefficient as τ_{red}, and introduce a reduced second-order representation to describe the system's frequency dynamics:

$$\dot{x}_{\text{red}} = A_{\text{red}}x_{\text{red}} + B_{\text{red}}u_{\text{red}} \tag{16}$$

where the state vector $x_{\text{red}} \in \mathbb{R}^2$, the input vector $u_{\text{red}} \in \mathbb{R}$, and system matrices, $A_{\text{red}}, B_{\text{red}} \in \mathbb{R}^{2\times 2}$ are given by

$$\begin{aligned} x_{\text{red}} &= \left[\Delta\omega_{\text{red}}, P^m_{\text{red}}\right], \qquad u_{\text{red}} = P_{\text{load}} \\ A_{\text{red}} &= \begin{bmatrix} -D_{\text{net}}J^{-1}_{\text{net}} & J^{-1}_{\text{net}} \\ -\tau^{-1}_{\text{red}}R_{\text{red}} & -\tau^{-1}_{\text{red}} \end{bmatrix} \\ B_{\text{red}} &= \left[-J^{-1}_{\text{net}}, 0\right] \\ R_{\text{red}} &= \sum_{g\in G} R_g \end{aligned} \tag{17}$$

Since the time constants of the turbines differ from one another, a deviation arises between $\Delta\omega_{\text{red}}(t)$ and $\Delta\omega(t)$, and this error has an upper bound. To quantify the error generated in the simplification of the model dimensions, we define the matrix $\boldsymbol{\Gamma}$ as

$$\boldsymbol{\Gamma} = diag\left\{1, \tau_{\text{red}}^{-1} diag(\boldsymbol{\tau})\right\} \tag{18}$$

According to Equation (12), $\boldsymbol{\tau}$ denotes the collection of time constants corresponding to all synchronous generators involved in the aggregated model, characterizing the differences in the dynamic response speeds of individual generators within the aggregation framework. Then, an auxiliary dynamic model can be constructed with the state vector $\overline{\boldsymbol{x}} \in \mathbb{R}^{|G|+1}$ and system matrix $\overline{\boldsymbol{A}}, \overline{\boldsymbol{B}} \in \mathbb{R}^{(|G|+1)\times(|G|+1)}$.

$$\dot{\overline{\boldsymbol{x}}} = \overline{\boldsymbol{A}}\overline{\boldsymbol{x}} + \overline{\boldsymbol{B}}u \tag{19}$$

$$\begin{cases} \overline{\boldsymbol{x}} = \left[\Delta\overline{\omega}, (\overline{\boldsymbol{P}}_G^m)^{\mathbf{T}}\right], \quad u = P_{\text{load}} \\ \overline{\boldsymbol{A}} = \boldsymbol{\Gamma}\boldsymbol{A} = \begin{bmatrix} -D_{\text{net}}J_{\text{eff}}^{-1} & J_{\text{aff}}^{-1}\mathbf{1}_{|G|}^{\mathbf{T}} \\ -\tau_{\text{red}}^{-1}\boldsymbol{R}_G\boldsymbol{I}_{|G|} & -\tau_{\text{red}}^{-1}\boldsymbol{I}_{|G|} \end{bmatrix} \\ \overline{\boldsymbol{B}} = \boldsymbol{\Gamma}\boldsymbol{B} = \left[-J_{\text{eff}}^{-1}, \mathbf{0}_{|G|}^{\mathbf{T}}\right] \end{cases} \tag{20}$$

Combining Equations (14) and (19) and considering the dynamics of $\Delta\boldsymbol{x}(t) = \overline{\boldsymbol{x}}(t) - \boldsymbol{x}(t)$, the following can be deduced:

$$\begin{aligned} \Delta\dot{\boldsymbol{x}} &= \dot{\overline{\boldsymbol{x}}} - \dot{\boldsymbol{x}} \\ &= \boldsymbol{\Gamma}\boldsymbol{A}\overline{\boldsymbol{x}} - \boldsymbol{A}\boldsymbol{x} + \boldsymbol{\Gamma}\boldsymbol{B}u - \boldsymbol{B}u \\ &= \boldsymbol{\Gamma}\boldsymbol{A}\overline{\boldsymbol{x}} - \boldsymbol{A}\boldsymbol{x} + (\boldsymbol{\Gamma} - \boldsymbol{I}_{|G|+1})\boldsymbol{B}u \\ &= \boldsymbol{\Gamma}\boldsymbol{A}\overline{\boldsymbol{x}} - \boldsymbol{A}\boldsymbol{x} + (\boldsymbol{\Gamma} - \boldsymbol{I}_{|G|+1})(\dot{\boldsymbol{x}} - \boldsymbol{A}\boldsymbol{x}) \\ &= \boldsymbol{\Gamma}\boldsymbol{A}(\overline{\boldsymbol{x}} - \boldsymbol{x}) + (\boldsymbol{\Gamma} - \boldsymbol{I}_{|G|+1})\dot{\boldsymbol{x}} \\ &= \boldsymbol{\Gamma}\boldsymbol{A}\Delta\boldsymbol{x} + (\boldsymbol{\Gamma} - \boldsymbol{I}_{|G|+1})\dot{\boldsymbol{x}} \end{aligned} \tag{21}$$

By considering $\dot{\boldsymbol{x}}$ as an external input to the system defined in (21), its solution can be expressed as follows:

$$\Delta\boldsymbol{x}(t) = \int_{s=0}^{t} e^{\boldsymbol{\Gamma}\boldsymbol{A}(t-s)}(\boldsymbol{\Gamma} - \boldsymbol{I}_{|G|+1})\dot{\boldsymbol{x}}(s)ds \tag{22}$$

There exist $k, \lambda > 0$ such that we can bound $||e^{\boldsymbol{\Gamma}\boldsymbol{A}(t-\bar{t})}||_2 \leq ke^{-\lambda(t-\bar{t})}, \forall 0 \leq \bar{t} \leq t$. Based on this relationship, Equation (22) yields

$$\begin{aligned} ||\Delta\boldsymbol{x}(t)||_2 &\leq \textstyle\int_{s=0}^{t} ke^{-\lambda(t-s)} \cdot ||(\boldsymbol{\Gamma} - \boldsymbol{I}_{|G|+1})(\boldsymbol{A}\boldsymbol{x}(s) + \boldsymbol{B}u(s)||_2 ds \\ &\leq \frac{k}{\lambda}||(\boldsymbol{\Gamma} - \boldsymbol{I}_{|G|+1})\boldsymbol{A}||_2 \cdot \sup_{0\leq s\leq t}(||\boldsymbol{x}(s)||_2 + ||\boldsymbol{A}^{-1}\boldsymbol{B}u(s)||_2 \end{aligned} \tag{23}$$

We require that the dynamic characteristics of the system before and after aggregation are as consistent as possible, i.e., we require that $\Delta\omega(t)$ and $\Delta\overline{\omega}(t)$ are as equal as possible at any time t. For this purpose, the error function is established as follows:

$$f_\Delta(t) = |\Delta\overline{\omega}(t) - \Delta\omega(t)| \tag{24}$$

Recognizing that

$$|\Delta\overline{\omega}(t) - \Delta\omega(t)| = |\Delta\omega_{\text{red}}(t) - \Delta\omega(t)| \leq ||\Delta\boldsymbol{x}(t)||_2 \tag{25}$$

Combining Equations (23) and (25), in order to keep the error before and after aggregation as small as possible, τ_{red} should be chosen so that $\left\|(\boldsymbol{\Gamma} - \boldsymbol{I}_{|G|+1})\boldsymbol{A}\right\|_2$ is as small as possible, which can be obtained as

$$\tau_{\text{red}} = \underset{\overline{\tau}\geq 0}{\operatorname{argmin}}\left\|(\boldsymbol{\Gamma}(\overline{\tau}) - \boldsymbol{I}_{|G|+1})\boldsymbol{A}\right\|_2 \tag{26}$$

where $\boldsymbol{\Gamma}(\overline{\tau}) = \boldsymbol{X}diag\{\boldsymbol{X}^{-1}, diag(\boldsymbol{\tau})\}, \boldsymbol{X} = \overline{\boldsymbol{\tau}}^{-1}$. This is because the first row of the matrix $(\boldsymbol{\Gamma}(\overline{\tau}) - \boldsymbol{I}_{|G|+1})A$ has all zero entries. Therefore, it follows from (26) that

$$\tau_{\text{red}} = \underset{\boldsymbol{X}\geq 0}{\operatorname{argmin}}\left\|(\boldsymbol{X}diag(\boldsymbol{\tau}) - \boldsymbol{I}_{|G|})\widetilde{A}\right\|_2 \tag{27}$$

where $\widetilde{A} = \left[-diag(\boldsymbol{\tau})^{-1}R_{\text{red}} \; -diag(\boldsymbol{\tau})^{-1}\right]$. For any matrix $\boldsymbol{A}$, $\|\boldsymbol{A}\|_2 \leq \|\boldsymbol{A}\|_{\text{F}}$ [29]. We solve

$$\tau_{\text{red}} = \underset{\boldsymbol{X}\geq 0}{\operatorname{argmin}}\left\|(\boldsymbol{X}diag(\boldsymbol{\tau}) - \boldsymbol{I}_{|G|})\widetilde{A}\right\|_{\text{F}} \tag{28}$$

Next, using

$$\begin{aligned}
&\min_{\boldsymbol{X}\geq 0}\left\|(\boldsymbol{X}diag(\boldsymbol{\tau}) - \boldsymbol{I}_{|G|})\widetilde{A}\right\|_{\text{F}}^2 \\
&= \min_{\boldsymbol{X}\geq 0} Tr((\boldsymbol{X}diag(\boldsymbol{\tau})\widetilde{A} - \widetilde{A})(\boldsymbol{X}diag(\boldsymbol{\tau})\widetilde{A} - \widetilde{A})^{\text{T}}) \\
&= \min_{\boldsymbol{X}\geq 0} Tr(\boldsymbol{X}^2 diag(\boldsymbol{\tau})\widetilde{A}\widetilde{A}^{\text{T}} diag(\boldsymbol{\tau}) - 2\boldsymbol{X}diag(\boldsymbol{\tau})\widetilde{A}\widetilde{A}^{\text{T}} + \widetilde{A}\widetilde{A}^{\text{T}}) \\
&= \min_{\boldsymbol{X}\geq 0} \boldsymbol{X}^2 Tr(diag(\boldsymbol{\tau})\widetilde{A}\widetilde{A}^{\text{T}} diag(\boldsymbol{\tau})) - 2\boldsymbol{X}Tr(diag(\boldsymbol{\tau})\widetilde{A}\widetilde{A}^{\text{T}}) + Tr(\widetilde{A}\widetilde{A}^{\text{T}})
\end{aligned}$$

By invoking the first-order optimality criterion, the optimal $\boldsymbol{X}$ can be determined, and, consequently, the solution minimizing Equation (26) is given by

$$\tau_{\text{red}} = \frac{Tr(diag(\boldsymbol{\tau})\widetilde{A}\widetilde{A}^{\text{T}} diag(\boldsymbol{\tau}))}{Tr(diag(\boldsymbol{\tau})\widetilde{A}\widetilde{A}^{\text{T}})} \tag{29}$$

In this chapter, the dynamic models of a single conventional generating unit and a single Distributed Energy Resource (DER) are first developed. Based on these models, the corresponding state-space equations are formulated. A comprehensive derivation of the aggregation process is then conducted for key parameters that influence the system's dynamic behavior, including inertia coefficients, damping coefficients and unit time constants. The final result is the dynamic model of the system after aggregation. Through careful aggregation of key parameters, the complexity of the original detailed power system model is significantly reduced. Consequently, this aggregation strategy preserves key dynamic characteristics, ensuring clarity and computational efficiency in subsequent time-domain analyses.

3. Designing Inertia and Damping Coefficients

In this section, we first outline the development of a transfer function that maps net load disturbances to frequency deviations. Then, we introduce the configuration of damping parameters and inertia constants to achieve specified steady-state frequency performance and limit peak overshoot.

3.1. Transfer Function from Load Disturbance to Frequency Response

By applying the reduced second-order Model (17), one can derive the Laplace-domain transfer function that characterizes the system's response from load fluctuations to frequency variation.

$$\frac{\Delta\omega(s)}{P_{\text{load}}(s)} = -\frac{k(s+\zeta)}{s^2 + 2\xi\omega_n s + \omega_n^2} \tag{30}$$

and the expressions for the parameters k, ζ, ω_n and ξ are specified as follows

$$\begin{cases} k = \frac{1}{J_{\text{net}}}, & \zeta = \frac{1}{\tau_{\text{net}}} \\ \omega_n = \sqrt{\frac{D_{\text{eff}}}{\tau_{\text{net}} J_{\text{net}}}}, & \xi = \frac{1}{2}\frac{J_{\text{net}} + \tau_{\text{net}} D_{\text{net}}}{\sqrt{J_{\text{net}}\tau_{\text{net}} D_{\text{eff}}}} \end{cases} \tag{31}$$

where for notational compactness, we introduce the following definitions:

$$D_{\text{eff}} = D_{\text{net}} + R_{\text{net}},\ R_{\text{net}} = \sum_{g\in G} R_g \tag{32}$$

Under the assumption of an underdamped system, the inverse Laplace transform yields the following result:

$$\Delta\omega(t) = \Delta\omega_{\text{ss}}\left(1 - \frac{e^{-\xi\omega_n t}}{\sqrt{1-\xi^2}}(\sin(\omega_d t + \varphi)\,\frac{\omega_n}{\zeta}\sin(\omega_d t)\right) \tag{33}$$

where $\Delta\omega_{\text{ss}}$ denotes the steady-state frequency deviation after equilibrium under step perturbation, and we have

$$\Delta\omega_{\text{ss}} = -\frac{\Delta P_{\text{load}}}{D_{\text{eff}}} \tag{34a}$$

$$\omega_d = \omega_n\sqrt{1-\zeta^2} \tag{34b}$$

$$\varphi = \tan^{-1}\left(\zeta^{-1}\sqrt{1-\zeta^2}\right) \tag{34c}$$

3.2. Determine Damping and Inertia Coefficients

The above-formed inverse Laplace transform derived in (33) describes the time-domain evolution of frequency as determined by the system's damping and inertia parameters. As shown in Figure 4, the time-domain evolution exhibits two feature points: frequency nadir $\Delta\omega_{\text{nadir}}$ at the first time t_{nadir}, and equilibrium $\Delta\omega_{\text{ss}}$.

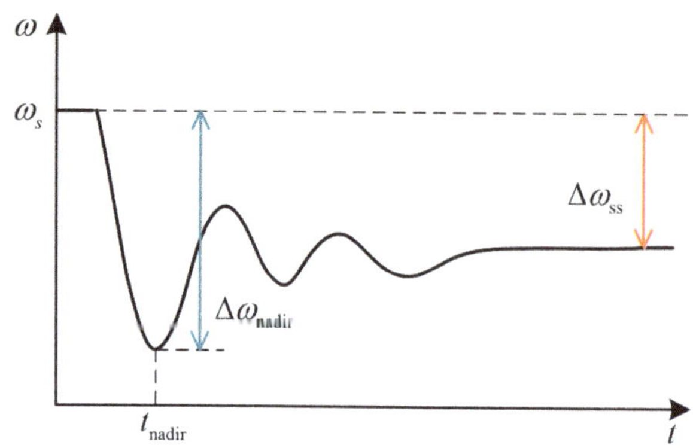

Figure 4. Frequency response time domain evolution.

The steady-state frequency response is defined as the proportion between the net active load variation and the steady-state frequency change, which is used to express the specified value of the frequency regulation, and the unit is usually MW/0.1 Hz, as listed in (35).

$$R_{P/\omega} = -\frac{\Delta P_{\mathrm{load}}}{\Delta \omega_{\mathrm{ss}}}[\mathrm{MW}/0.1\mathrm{Hz}] \tag{35}$$

Observe from (34a) and (35) that we can determine D_{eff} to ensure this frequency regulation specification is met. Given a specified D_{eff}, we can calculate the net damping coefficient D_{net} according to (32), and the damping coefficients of individual DERs should subsequently be modified to comply with Equation (8).

The peak overshoot criterion is defined as follows:

$$\omega_{\mathrm{peak}}^{\%} = \frac{\Delta \omega_{\mathrm{nadir}}}{\Delta \omega_{\mathrm{ss}}} \times 100 \tag{36}$$

Typically, it is difficult to correlate the rotational inertia and damping parameters with the peak overshoot due to the complexity and variety of the dynamic models of the generator and DER, whereas it can be solved through the time-domain relation (33), which is transformed from the second-order simplified model provided in this paper.

$$\begin{aligned} &e^{-\xi\omega_n t_{\mathrm{nadir}}}((\omega_n\xi \sin(\omega_d t_{\mathrm{nadir}} + \varphi) - \omega_d \cos(\omega_d t_{\mathrm{nadir}} + \varphi) \\ &-\zeta^{-1}\omega_n^2\xi \sin(\omega_d t_{\mathrm{nadir}}) + \zeta^{-1}\omega_n\omega_d \cos(\omega_d t_{\mathrm{nadir}})) = 0 \end{aligned} \tag{37}$$

The preceding expression can be reduced through trigonometric transformation to

$$\tan(\omega_d t_{\mathrm{nadir}}) = \frac{\omega_d}{\xi\omega_n - \zeta} \tag{38}$$

from this, we obtain t_{nadir} by substituting $\omega_d, \omega_n, \zeta, \xi$ from (31) and (34b) as

$$\begin{aligned} t_{\mathrm{nadir}} &= \omega_d^{-1} \tan^{-1}\left(\frac{\omega_d}{\xi\omega_n - \zeta}\right) \\ &= 2\rho^{-1}\tau_{\mathrm{net}}^{-1} J_{\mathrm{net}} \tan^{-1}\left(\frac{\rho}{J_{\mathrm{net}} + \tau_{\mathrm{net}}^{-1} D_{\mathrm{net}} - 2J_{\mathrm{net}}\tau_{\mathrm{net}}^{-2}}\right) \end{aligned} \tag{39}$$

where $\rho = \sqrt{4\tau_{\mathrm{net}}^{-1} J_{\mathrm{net}} D_{\mathrm{eff}} - (J_{\mathrm{net}} + \tau_{\mathrm{net}}^{-1} D_{\mathrm{net}})^2}$.

Then, we can obtain $\Delta\omega(t_{\mathrm{nadir}})$ by substituting t_{nadir} from (39) in (33) as

$$\Delta\omega_{\mathrm{nadir}} = \Delta\omega(t_{\mathrm{nadir}}) = \Delta\omega_{ss} e^{-\xi\omega_n t_{\mathrm{nadir}}} \sqrt{\omega n^2 + \zeta^2 - 2\zeta\xi\omega_n} \tag{40}$$

$$\begin{aligned} \omega_{\mathrm{peak}}^{\%} &= \frac{\Delta\omega_{\mathrm{nadir}}}{\Delta\omega_{\mathrm{ss}}} \times 100 \\ &= e^{\frac{2J_{\mathrm{net}}\tau_{\mathrm{net}}^{-2} - J_{\mathrm{net}} - \tau_{\mathrm{net}}^{-1} D_{\mathrm{net}}}{2J_{\mathrm{net}}\tau_{\mathrm{net}}^{-1}} t_{\mathrm{nadir}}} \sqrt{\frac{R_{\mathrm{net}}}{\tau_{\mathrm{net}} J_{\mathrm{net}}}} \times 100 \end{aligned} \tag{41}$$

Substituting t_{nadir} from (39) in (41), the peak overshoot can expressed as a function of the two aggregated parameters J_{net} and D_{net}.

When given D_{eff} to satisfy specification (32) and given specification $\omega_{\mathrm{peak}}^{\%}$, (41) will be a nonlinear expression involving J_{net}. Determining J_{net} from this equation, individual DER inertia coefficients are subsequently modified to meet (7).

The above design strategy was made for DER total damping $\sum_{d \in D} D_d$ and inertia coefficient $\sum_{d \in D} J_d$. In fact, this sum can be decomposed into individual values D_d, J_d according to the actual requirements of the grid. In this paper, we adopt the optimization-based perspective provided by Ref. [30], i.e., decompose the DER total effective inertia and damping in proportion to their power ratings:

$$J_d = \frac{J_{\text{net}} - \sum\limits_{g \in G} J_g}{\sum\limits_{d \in D} \overline{P}_d} \overline{P}_d \tag{42}$$

$$D_d = \frac{D_{\text{net}} - \sum\limits_{g \in G} D_g}{\sum\limits_{d \in D} \overline{P}_d} \overline{P}_d \tag{43}$$

where $\overline{P}_d$ is the base rating of the DER located at buses d.

4. SOC-Based Adaptive Optimization

In this section, we propose an adaptive optimization framework considering the energy storage SOC to dynamically optimize the synthetic inertia and droop control parameters of the storage device. This approach leverages the real-time SOC to enable the energy storage device to provide stable and sustained auxiliary support for grid frequency modulation over extended periods.

4.1. Response Mode Incorporating SOC

Energy storage devices are capable of significantly improving the system's equivalent inertia and damping via virtual inertia and droop control, thereby improving grid frequency response performance. However, in real-world scenarios, the capacity of energy storage systems is subject to inherent limitations. Using the maximum droop coefficient in both charge and discharge modes during the initial frequency control phase can easily cause the SOC of the energy storage device to exceed its operational limits.

This study introduces a strategy for dynamically adjusting virtual inertia and droop parameters based on SOC levels. When the SOC becomes excessively high (during charging) or drops too low (during discharging), the virtual parameters are accordingly tuned to limit the output power of the energy storage unit, thereby alleviating negative impacts on grid frequency resulting from SOC limit violations. In addition, the proposed method dynamically regulates the proportion of virtual inertia and droop control synergistically. During the inertia response phase, virtual inertia serves as the primary control mechanism, with droop control playing a supplementary role. Conversely, in the primary frequency modulation phase, droop control becomes the dominant mechanism, while virtual inertia provides secondary support. This adaptive control strategy ensures the efficient management of energy storage output, minimizing stress on the storage device and maintaining system stability.

$$\Delta P_E = c_1 J_{\text{net}} \frac{d\Delta\omega}{dt} + c_2 D_{\text{net}} \Delta\omega \tag{44}$$

where c_1 and c_2 are the scaling factors of virtual inertia mode and virtual droop mode, and J_{net}, D_{net} are the virtual inertia parameter and droop parameter of the energy storage device calculated based on the aggregation principle described in Section 3.2.

4.2. Design of Scaling Factors

(1). The inertial response stage: By taking into account both the frequency deviation and its rate of change, the proportional scaling factor for this stage can be formulated as follows:

$$\begin{cases} c_1 = e^{n\Delta f} \\ c_2 = 1 - e^{n\Delta f} \end{cases} \quad 0 \geq \Delta f \geq \ln(\tfrac{1}{2})/n \\ \Delta f = \Delta\omega / 2\pi \tag{45}$$

where n represents the scaling factor. It is used to adjust the alignment between the asymptotic curves of the scaling factors c_1, c_2 and the frequency change characteristics during the inertia response stage. If n is too small, the changes in c_1 and c_2 remain minimal, even when the rate of frequency change increases significantly, or the frequency deviation decreases considerably. In such cases, the benefits of inertial response and damping response cannot be fully leveraged, leading to an excessively large maximum frequency deviation $\Delta f_{\max}$. Conversely, if n is too large, c_1 and c_2 may fluctuate drastically in response to frequency variations, making it difficult to effectively suppress the rate of change of frequency deviation. Based on the above analysis, we selected $n = 100$ as a compromise for the optimal performance of frequency dynamic response.

(2). The primary frequency regulation stage: When $\frac{\mathrm{d}\Delta f}{\mathrm{d}t} = 0$, control switches to virtual droop as the dominant mechanism, with virtual inertia providing supplementary control. The scaling factor equation for this phase is given as follows:

$$\begin{cases} c_1 = \frac{1}{2}\left(\frac{\Delta f}{\Delta f_{\max}}\right)^n \\ c_2 = 1 - \left(\frac{\Delta f}{\Delta f_{\max}}\right)^n \end{cases} \tag{46}$$

where $\Delta f_{\max}$ is the maximum frequency deviation value in one frequency modulation.

4.3. Parameter Optimization Involving SOC

While the previous section addressed the dynamic adjustment of the scaling factors c_1 and c_2, this section focuses on dynamically adjusting the virtual inertia and droop parameters based on the SOC to effectively regulate the energy storage device's output.

The state of charge (SOC) of the battery represents the ratio of the remaining energy to the rated energy capacity, indicating the battery's ability to continue operating. In this study, the SOC is computed by integrating the output power of the energy storage system over time and subtracting the accumulated energy from the initial SOC value. This approach allows real-time tracking of the energy state and reflects the dynamic charging and discharging behavior of the system. All SOC values are expressed in per-unit (p.u.) relative to the rated capacity, and system-specific SOC bounds (e.g., 0.2–0.6) are defined according to the operational characteristics of the energy storage device.

As shown in Figure 5, D_{net} represents the maximum adjustable virtual droop coefficient of the energy storage system, which corresponds to the total virtual droop of the DER, calculated based on the method described in the previous section. Q_{SOC} denotes the state of charge (SOC) of the energy storage system, expressed as a per-unit (p.u.) value, defined as the ratio of current energy content to its rated capacity. Specifically, $Q_{\text{SOC\_min}}$ and $Q_{\text{SOC\_max}}$ represent the lower and upper SOC thresholds for adaptive control activation, which are adjustable according to device specifications. For illustration purposes, the minimum and maximum SOC values are set to 0.2 and 0.6, respectively, but these can be tuned based on operational constraints.

The virtual inertia parameters and virtual droop parameter for energy storage, considering SOC, are defined as follows:

$$J_{\text{net}}(Q_{\text{SOC}}) = \begin{cases} \alpha K_c & \mathrm{d}\Delta f/\mathrm{d}t > 0 \\ \alpha K_d & \mathrm{d}\Delta f/\mathrm{d}t < 0 \end{cases} \tag{47}$$

$$D_{\text{net}}(Q_{\text{SOC}}) = \begin{cases} K_c & \Delta f \geq -0.03\text{Hz} \\ K_d & \Delta f < -0.03\text{Hz} \end{cases} \tag{48}$$

$$\alpha = J_{\text{net}}/D_{\text{net}} \tag{49}$$

where α represents the scaling factor that defines the linkage between the synthetic inertia parameter and the corresponding droop setting. K_c and K_d are the charging and discharging coefficients in the droop regulation process of the energy storage system, respectively. To prevent issues arising from SOC limits, as shown in Figure 5, a piecewise linear function is employed to define the charging and discharging curves. This approach not only ensures smooth output but also avoids the control difficulties associated with more complex functions, making it more suitable for practical engineering applications [31]. The values of K_c and K_d are as follows:

$$K_c = \begin{cases} D_{\text{net}} & Q_{\text{SOC}} \in [0, 0.45] \\ \frac{0.6 - Q_{\text{SOC}}}{0.15} D_{\text{net}} & Q_{\text{SOC}} \in [0.45, 0.6] \\ 0 & Q_{\text{SOC}} \in [0.6, 1.0] \end{cases} \tag{50}$$

$$K_d = \begin{cases} 0 & Q_{\text{SOC}} \in [0, 0.2] \\ \frac{Q_{\text{SOC}} - 0.2}{0.15} D_{\text{net}} & Q_{\text{SOC}} \in [0.2, 0.35] \\ D_{\text{net}} & Q_{\text{SOC}} \in [0.35, 1.0] \end{cases} \tag{51}$$

The energy storage-assisted frequency modulation output under adaptive control, which accounts for SOC, is expressed as follows:

$$\Delta P_{\text{E}}(Q_{\text{SOC}}) = c_1 J_{\text{net}}(Q_{\text{SOC}}) \frac{\text{d}\Delta f}{\text{d}t} + c_2 D_{\text{net}}(Q_{\text{SOC}}) \Delta f \tag{52}$$

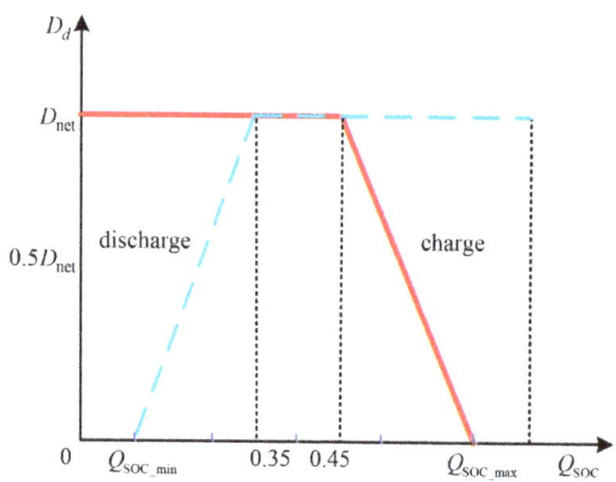

Figure 5. Relationship between unit regulation power of energy storage and SOC. The blue line represents the discharge power curve, indicating the reduction in power as the state of charge (SOC) decreases. The red line represents the charge power curve, showing the increase in power as SOC rises. The dashed lines correspond to the adjustable virtual droop coefficient (D_{net}) for different SOC thresholds ($Q_{\text{SOC\_min}}$ and $Q_{\text{SOC\_max}}$), with the power regulation behavior illustrated between these limits.

5. Simulation Results and Analysis

In this section, the effectiveness of the proposed aggregation model is evaluated by comparison with the original multi-machine configuration. Then, the influence of DER units' inertia and damping characteristics on the system's primary frequency regulation is investigated. Lastly, a model of the integrated energy system incorporating wind and storage is constructed to examine how distributed energy sources contribute to primary frequency control.

5.1. Aggregation Model Validation

We conduct simulations on the 10-unit, 39-bus New England test system, where generators are located on buses $G = \{1, 2, \cdots, 10\}$ [32], as shown in Figure 6. The example is implemented in the MATLAB/SIMULINK environment and is run on a computer equipped with an Intel Core i5-8300H CPU and 16 GB of RAM. The total simulation time spans 70 s. A step change in load is introduced at 2 s. The synchronous generator (SG) parameters are provided in Table 1. The dynamic characteristics of the single synchronous generator set in this system are modeled in Section 2. We selected the synchronous generator model parameters for a set of cases exhibiting the largest deviations within the normal stability range to validate the accuracy of the aggregation model.

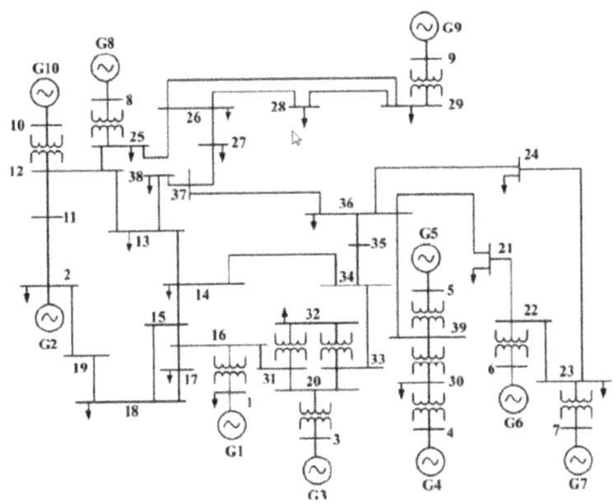

Figure 6. Single-line diagram of IEEE 39-bus.

Table 1. Parameters of 10 synchronous generator units in IEEE39 system.

Type	Number	$\tau(s)$	R_g	$J_g(s)$	D_g
SG	G1	4	0.217	0.1302	0.0434
	G2	4	0.217	0.1302	0.0434
	G3	5	0.1984	0.1203	0.0343
	G4	6	0.1798	0.1203	0.0343
	G10	4	0.217	0.1302	0.0434
	G5	7	0.1612	0.1203	0.0343
	G6	8	0.1426	0.1203	0.0343
	G7	9	0.1240	0.1302	0.0434
	G8	10	0.1054	0.1302	0.0434
	G9	10	0.0868	0.1302	0.0434

According to Equations (7), (8) and (27), we calculate the equivalent parameters of the aggregation model $\{\tau_{net}, R_{net}, J_{net}, D_{net}\} = \{5.8683, 1.6492, 1.2624, 0.3976\}$. Next, we build simulation models both before and after aggregation in Simulink, using the same load fluctuation. Figure 7 presents a comparison between the frequency and output power responses, along with the corresponding response errors, considering both the introduced aggregation framework and the conventional multi-machine model. The frequency and power response trajectories of the aggregation model are very close to those of the multi-machine primitive model, with the maximum deviations observed in frequency and power outputs remaining under 0.01. These results imply that the aggregation model can effectively substitute the multi-machine primitive model with high accuracy.

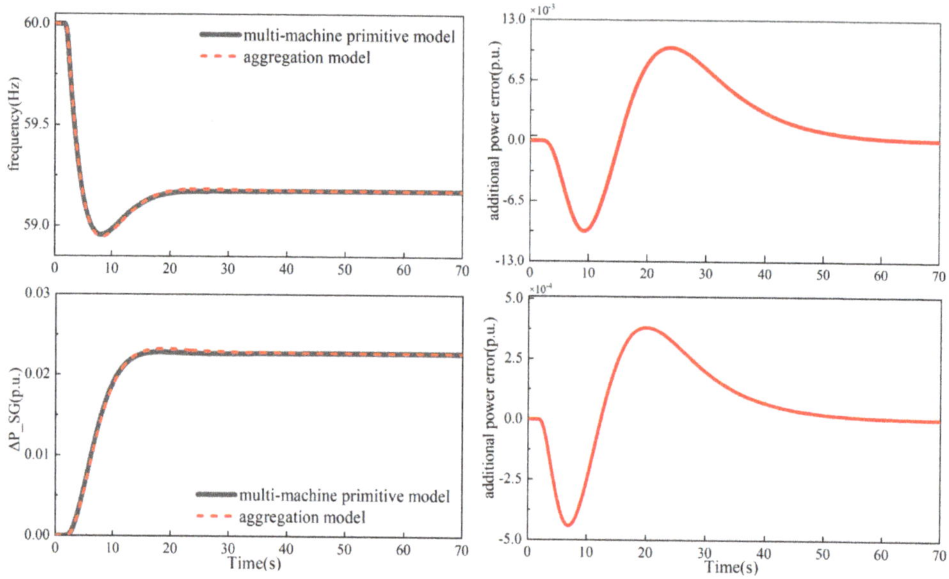

Figure 7. Comparison of the aggregation model with the multi-machine primitive model.

5.2. Impact of DER in Inertial and Primary Frequency Response

We add five frequency response DERs to the IEEE 39-bus system on the buses $D = \{13, 17, 19, 20, 22\}$, to evaluate the effectiveness of DERs involved in inertial and primary frequency regulation. The test platform is PSAT [33]. A load step perturbation of $\Delta P_{load} = 0.01$ is set at time $t = 0$ at bus 17. The test results are illustrated in Figure 8. The red curves represent the scenario where the DERs do not participate in inertial and primary frequency regulation. In comparison, we employed an analytical approach to calculate the DERs' inertial and dynamic coefficients outlined in Section 3 ($\sum_{d \in D} D_d = 64.7$ and $\sum_{d \in D} J_d = 59.38$) and depict the trajectories of the frequency response in black line. It can be found that the frequency response with DER support shows a reduced frequency nadir and decreased steady-state frequency deviation, which verifies the effectiveness and practicality of the DERs in inertial and primary frequency regulation.

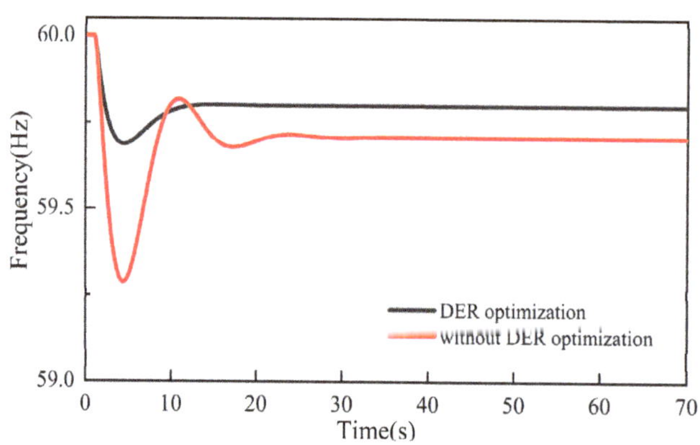

Figure 8. Comparison of frequency response before and after DER involvement in optimization.

In order to analyze the effect of the DER participation level (measured by the proportion of total system inertia attributed to DERs) on the system's frequency behavior, three cases with the permeability $\eta = \{5, 50, 90\}$ are simulated. As shown in Figure 9, increasing the permeability level can significantly suppress frequency overshoot and enhance system stability; however, when DER permeability arrives high, e.g., 90%, increasing the damping cannot further reduce the amount of frequency response overshoot.

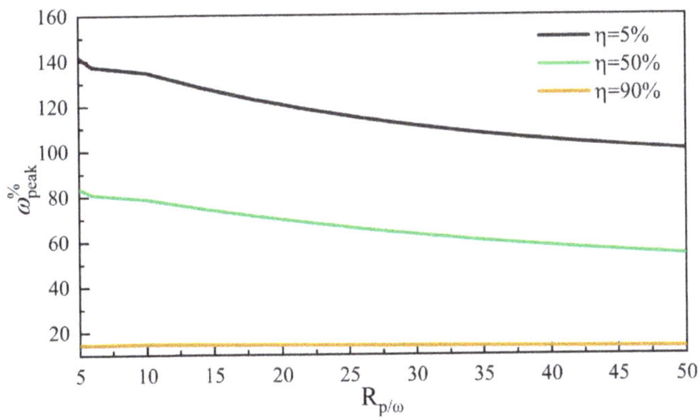

Figure 9. Comparison of frequency response overshoot for different renewable energy penetration rates.

5.3. *Adaptive Auxiliary Frequency Modulation with Energy Storage Considering SOC*

To assess the performance of the proposed adaptive control scheme under SOC considerations for auxiliary frequency regulation, we modify the model outlined in Section 5.2 by replacing all five DERs with energy storage units. For analytical purposes, the initial SOC of the energy storage system is initialized at 60%, with a minimum SOC threshold set to 20%. When the SOC reaches this lower threshold, the operating voltage will drop to the minimum discharge voltage and no further discharge operations will be conducted. This section compares the frequency response, output power and SOC changes under the three control methods: (1) the original control without DER integration, (2) the optimized control with DER-based auxiliary frequency support and (3) the virtual inertia and droop adaptive control method considering the SOC. In the middle case, the calculated inertia and droop parameters D, J without considering the SOC remain constant. The proportionality coefficients, c_1 and c_2, are both set to 0.5.

Figure 10 depicts the curves of frequency deviation corresponding to the three examined control schemes. The black curve represents the no-storage participation frequency modulation strategy, the red curve corresponds to the storage-assisted frequency modulation strategy without considering the SOC, and the blue curve depicts the adaptive storage-assisted optimization strategy that accounts for the SOC. After the energy storage system engages in supporting auxiliary frequency control, all performance indicators of the frequency response demonstrate improvement. However, at 40 s, the system frequency experiences a further drop of 0.1094 Hz due to the SOC of the energy storage exceeding its limit and halting discharge. The control strategy introduced in this work significantly improves the dynamic performance of the frequency response while maintaining system stability. Moreover, it avoids sudden frequency changes caused by the storage SOC exceeding its limits.

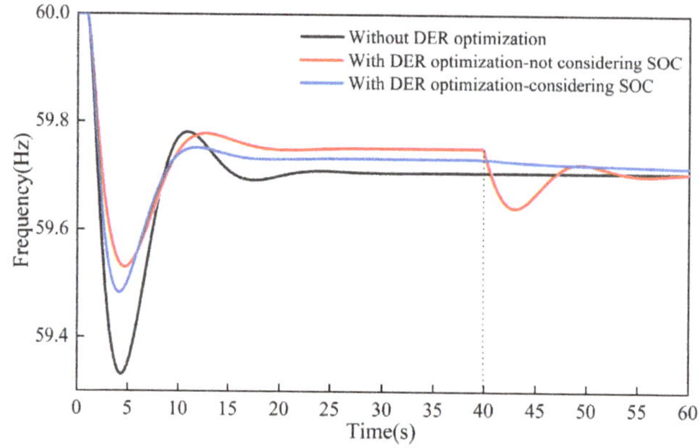

Figure 10. A comparison of system frequency deviations under various control strategies.

As illustrated in Figure 11, the SOC-based adaptive control strategy lowers the rated power demand of the energy storage system by 14.3% in comparison to the method that does not account for the SOC while also yielding a smoother discharge profile. Due to the limited capacity of the energy storage device, the SOC reaches its lower limit at 40 s, causing the device to stop discharging and thereby ceasing its auxiliary regulation role. This results in an instantaneous load deficit and triggers a frequency drop. The SOC-based adaptive control method proposed in this paper takes into account the dynamic changes in the SOC of the energy storage device. By adjusting its virtual inertia and droop parameter in real time according to Equations (47)–(52), the method adaptively regulates the power output. This ensures that, while providing auxiliary support for grid frequency modulation, it also prevents over-charging and over-discharging, thereby mitigating the risk of sudden frequency fluctuations.

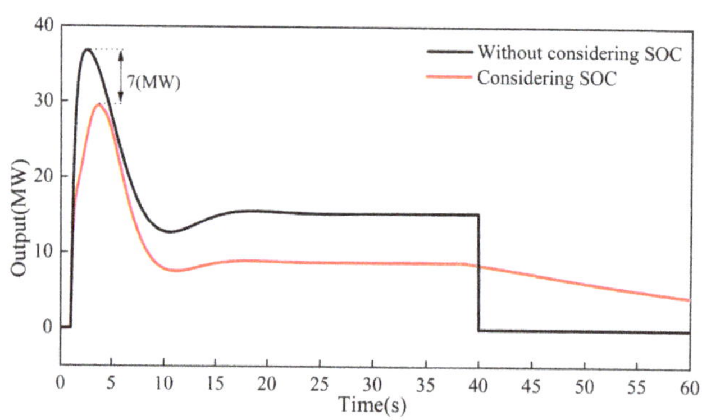

Figure 11. The curves of output power of energy storage devices under load surging.

For the purpose of analysis, the initial SOC is set to 60% in this study. As shown in Figure 12, under a sudden load disturbance in the grid, the energy storage device discharges continuously and promptly to fulfill its role in auxiliary frequency modulation. In contrast, under the control method, which does not account for the SOC state, the SOC reaches its lower limit of 20% at 40 s, limiting further discharge. The adaptive control method proposed in this paper demonstrates superior SOC maintenance performance, improving SOC utilization by 13.5% compared to the method without considering the SOC.

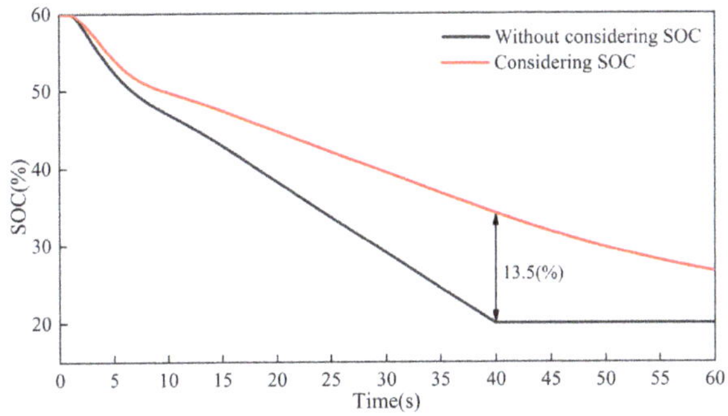

Figure 12. The value of the SOC of energy storage devices under load surging.

6. Conclusions

This paper proposes an analytical method targeting energy storage systems involved in inertial and primary frequency regulation. Initially, a second-order equivalent model is developed using aggregation theory, which reduces the multi-machine system into a single generator representation. Based on the concept of compensation, a fast-response strategy for DERs to participate in primary frequency regulation is proposed, which derives the virtual inertia and damping coefficients that DERs must provide to achieve acceptable levels of frequency overshoot and steady-state deviation. Finally, the virtual inertia and damping coefficients within the energy storage module are adaptively tuned using a control framework that incorporates SOC considerations. The case study demonstrates that the proposed analytical approach for DER participation in primary frequency regulation exhibits high accuracy. Additionally, the inclusion of SOC considerations enhances the efficiency of energy storage resource utilization, improves the performance of DERs in primary frequency regulation, and significantly boosts the overall frequency stability of the power system.

Currently, the analytical method is applied on the assumption of a common frequency and does not take into account multiple equilibrium regions and settings. Future work should expand the optimization settings to consider power flow settings, reserves of inertia, and the primary frequency response, as well as distribution network-level power flow, enabling more comprehensive participation of DERs in the frequency regulation loop and enhancing the overall control performance.

Author Contributions: Conceptualization, W.L. (Wentian Lu), E.T. and L.C.; methodology, W.L. (Wentian Lu), E.T. and L.C.; formal analysis, W.L. (Wentian Lu), E.T. and L.C.; investigation, W.L. (Wentian Lu), E.T., L.C., K.Z. and W.L. (Wenjie Liu); resources, W.L. (Wentian Lu), E.T., L.C. and K.Z.; data curation, W.L. (Wentian Lu), E.T., L.C. and K.Z.; writing—original draft preparation, W.L. (Wentian Lu), E.T., L.C., K.Z. and W.L. (Wenjie Liu); writing—review and editing, W.L. (Wentian Lu), E.T., L.C. and K.Z.; visualization, W.L. (Wentian Lu), E.T., L.C. and K.Z.; supervision, W.L. (Wentian Lu), L.C. and K.Z.; project administration, W.L. (Wentian Lu), L.C. and K.Z.; funding acquisition, W.L. (Wentian Lu), L.C. and K.Z. All authors have read and agreed to the published version of the manuscript.

Funding: This research was funded in part by the National Natural Science Foundation of China, grant number 52107093 (funder: W.L.), in part by the Guangdong Basic and Applied Basic Research Foundation, grant number 2022A1515240038 (funder: W.L.), in part by the Guangzhou Education Bureau University Research Project - Graduate Research Project, grant number 2024312278 (funder: L.C.), and in part by the STU Scientific Research Initiation Grant (SRIG), grant number STF23021 (funder: K.Z.).

Data Availability Statement: The original contributions presented in this study are included in this article. Further inquiries can be directed to the corresponding authors.

Conflicts of Interest: The authors declare no conflicts of interest.

References

1. Sadorsky, P. Wind energy for sustainable development: Driving factors and future outlook. *J. Clean. Prod.* **2021**, *289*, 125779. [CrossRef]
2. Cheng, L.; Yu, T. A new generation of AI: A review and perspective on machine learning technologies applied to smart energy and electric power systems. *Int. J. Energy Res.* **2019**, *43*, 1928–1973. [CrossRef]
3. Shabestari, P.M.; Mehrizi-Sani, A. Frequency response improvement of PMSG wind turbines using a washout filter. *Energies* **2020**, *13*, 4797. [CrossRef]
4. Li, M.; Huang, W.; Tai, N.; Yang, L.; Duan, D.; Ma, Z. A dual-adaptivity inertia control strategy for virtual synchronous generator. *IEEE Trans. Power Syst.* **2019**, *35*, 594–604. [CrossRef]
5. Zhang, Y.; Weng, Y.; Li, H.; Zhou, Z.; Li, X. Research on China's Electricity Market and Photovoltaic and Electrochemical Energy Storage Industry. In Proceedings of the 2023 8th Asia Conference on Power and Electrical Engineering (ACPEE), Tianjin, China, 14–16 April 2023; pp. 949–953.
6. Ranjan, M.; Shankar, R. A literature survey on load frequency control considering renewable energy integration in power system: Recent trends and future prospects. *J. Energy Storage* **2022**, *45*, 103717. [CrossRef]
7. Cheng, L.; Huang, P.; Zou, T.; Zhang, M.; Peng, P.; Lu, W. Evolutionary game-theoretical approaches for long-term strategic bidding among diverse stakeholders in large-scale and local power markets: Basic concept, modelling review and future vision. *Int. J. Electr. Power Energy Syst.* **2025**, *166*, 110589. [CrossRef]
8. Cheng, L.; Peng, P.; Lu, W.; Sun, J.; Wu, F.; Shi, M.; Yuan, X.; Chen, Y. The evolutionary game equilibrium theory on power market bidding involving renewable energy companies. *Int. J. Electr. Power Energy Syst.* **2025**, *167*, 110588. [CrossRef]
9. Hamid, B.; Hussain, I.; Iqbal, S.J.; Singh, B.; Das, S.; Kumar, N. Optimal MPPT and BES control for grid-tied DFIG-based wind energy conversion system. *IEEE Trans. Ind. Appl.* **2022**, *58*, 7966–7977. [CrossRef]
10. Xu, B.; Gao, X.; Jing, S.; Yang, Z.; Guo, R. A Coordinated Control Strategy for the Wind Turbine Generator-Energy Storage Device Integrated System. In Proceedings of the 2023 International Conference on Power System Technology (PowerCon), Jinan, China, 21–22 September 2023; pp. 1–6.
11. Sun, M.; Min, Y.; Chen, L.; Hou, K.; Xia, D.; Mao, H. Optimal auxiliary Frequency control of wind turbine generator and coordination with synchronous generator. *CSEE J. Power Energy Syst.* **2021**, *7*, 78–85.
12. Thakallapelli, A.; Nair, A.R.; Biswas, B.D.; Kamalasadan, S. Frequency regulation and control of grid-connected wind farms based on online reduced-order modeling and adaptive control. *IEEE Trans. Ind. Appl.* **2020**, *56*, 1980–1989. [CrossRef]
13. Li, X.; Huang, Y.; Huang, J.; Tan, S.; Wang, M.; Xu, T.; Cheng, X. Modeling and control strategy of battery energy storage system for primary frequency regulation. In Proceedings of the 2014 International Conference on Power System Technology, Chengdu, China, 20–22 October 2014; pp. 543–549.
14. Engels, J.; Claessens, B.; Deconinck, G. Combined stochastic optimization of frequency control and self-consumption with a battery. *IEEE Trans. Smart Grid* **2017**, *10*, 1971–1981. [CrossRef]
15. Rajan, R.; Fernandez, F.M.; Yang, Y. Primary frequency control techniques for large-scale PV-integrated power systems: A review. *Renew. Sustain. Energy Rev.* **2021**, *144*, 110998. [CrossRef]
16. Rajan, R.; Fernandez, F.M. Power control strategy of photovoltaic plants for frequency regulation in a hybrid power system. *Int. J. Electr. Power Energy Syst.* **2019**, *110*, 171–183. [CrossRef]
17. Li, J.; Wen, B.; Wang, H. Adaptive virtual inertia control strategy of VSG for micro-grid based on Iimproved bang-bang control strategy. *IEEE Access* **2019**, *7*, 39509–39514. [CrossRef]
18. De Siqueira, L.M.S.; Peng, W. Control strategy to smooth wind power output using battery energy storage system: A review. *J. Energy Storage* **2021**, *35*, 102252. [CrossRef]
19. Liu, Y.; Wang, Y.; Wang, M.; Xu, Z.; Peng, Y.; Li, M. Coordinated VSG control of photovoltaic/battery system for maximum power output and grid supporting. *IEEE J. Emerg. Sel. Top. Circuits Syst.* **2022**, *12*, 301–309. [CrossRef]
20. Zhang, C.; Liu, L.; Cheng, H.; Liu, D.; Zhang, J.; Li, G. Frequency-constrained co-planning of generation and energy storage with high-penetration renewable energy. *J. Mod. Power Syst. Clean Energy* **2021**, *9*, 760–775. [CrossRef]
21. Meng, J.; Wang, Y.; Fu, C.; Wang, H. Adaptive virtual inertia control of distributed generator for dynamic frequency support in microgrid. In Proceedings of the 2016 IEEE Energy Conversion Congress and Exposition (ECCE), Milwaukee, WI, USA, 18–22 September 2016; pp. 1–5.
22. Andalib-Bin-Karim, C.; Liang, X.; Zhang, H. Fuzzy-secondary-controller-based virtual synchronous generator control scheme for interfacing inverters of renewable distributed generation in microgrids. *IEEE Trans. Ind. Appl.* **2018**, *54*, 1047–1061. [CrossRef]

23. Zhang, Y.J.; Zhao, C.; Tang, W.; Low, S.H. Profit-maximizing planning and control of battery energy storage systems for primary frequency control. *IEEE Trans. Smart Grid* **2018**, *9*, 712–723. [CrossRef]
24. Carrion, M.; Dvorkin, Y.; Pandzic, H. Primary frequency response in capacity expansion with energy storage. *IEEE Trans. Power Syst.* **2018**, *33*, 1824–1835. [CrossRef]
25. Purba, V.; Dhople, S.V.; Jafarpour, S.; Bullo, F.; Johnson, B.B. Reduced-order structure-preserving model for parallel-connected three-phase grid-tied inverters. In Proceedings of the 2017 IEEE 18th Workshop on Control and Modeling for Power Electronics (COMPEL), Stanford, CA, USA, 9–12 July 2017; pp. 1–7.
26. Cheng, L.; Yu, T. Smart dispatching for energy internet with complex cyber-physical-social systems: A parallel dispatch perspective. *Int. J. Energy Res.* **2019**, *43*, 3080–3133. [CrossRef]
27. Blaabjerg, F.; Yang, Y.; Kim, K.A.; Rodriguez, J. Power electronics technology for large-scale renewable energy generation. *Proc. IEEE* **2023**, *111*, 335–355. [CrossRef]
28. Guggilam, S.S.; Zhao, C.; Dall, E.; Chen, Y.C.; Dhople, S.V. Engineering inertial and primary-frequency response for distributed energy resources. In Proceedings of the 2017 IEEE 56th Annual Conference on Decision and Control (CDC), Melbourne, Australia, 12–15 December 2017; pp. 5112–5118.
29. Horn, R.A.; Johnson, C.R. *Matrix Analysis*; Cambridge University Press: Cambridge, UK, 2012.
30. Guggilam, S.S.; Zhao, C.; Dall'Anese, E.; Chen, Y.C.; Dhople, S.V. Optimizing DER participation in inertial and primary-frequency response. *IEEE Trans. Power Syst.* **2018**, *33*, 5194–5205. [CrossRef]
31. Zhao, C.; Sun, D.; Nian, H.; Wang, M.; Yang, L. Analysis on the contribution of virtual synchronous generator controlled energy storage system to primary frequency modulation. In Proceedings of the 2021 4th International Conference on Energy, Electrical and Power Engineering (CEEPE), Chongqing, China, 23–25 April 2021; pp. 869–874.
32. Pai, A. *Energy Function Analysis for Power System Stability*; Springer Science & Business Media: Berlin/Heidelberg, Germany, 1989.
33. Milano, F.; Vanfretti, L.; Morataya, J.C. An open source power system virtual laboratory: The PSAT Case and Experience. *IEEE Trans. Educ.* **2008**, *51*, 17–23. [CrossRef]

processes

MDPI

Article

Evolutionary Game Theory-Based Analysis of Power Producers' Carbon Emission Reduction Strategies and Multi-Group Bidding Dynamics in the Low-Carbon Electricity Market

Jianlin Tang [1,2,*], Bin Qian [1,2], Yi Luo [1,2,*], Xiaoming Lin [1,2], Mi Zhou [1,2], Fan Zhang [1,2] and Haolin Wang [1,2]

1 Electric Power Research Institute of China Southern Power Grid, Guangzhou 510663, China; qianbin@csg.cn (B.Q.); linxm4@csg.cn (X.L.); zhoumi2@csg.cn (M.Z.); zhangfan4@csg.cn (F.Z.); wanghl1@csg.cn (H.W.)
2 Guangdong Provincial Key Laboratory of Intelligent Measurement and Advanced Metering of Power Grid, Guangzhou 510663, China
* Correspondence: jianlintang_csg@163.com (J.T.); luoyi1@csg.cn (Y.L.)

Abstract: China's power generation system has undergone reforms, leading to a competitive electricity market where independent producers participate through competitive bidding. With the rise of low-carbon policies, producers must optimize bidding strategies while reducing carbon emissions, creating complex interactions with local governments. Evolutionary game theory (EGT) is well-suited to analyze these dynamics. This study begins by summarizing the fundamental concepts of electricity trading markets, including transaction models, bidding mechanisms, and carbon reduction strategies. Existing research on the application of evolutionary game theory in power markets is reviewed, with a focus on theoretical constructs such as evolutionary stable strategies and replicator dynamics. Based on this foundation, the study conducts a detailed mathematical analysis of symmetric and asymmetric two-group evolutionary game models in general market scenarios. Building upon these models, a three-group evolutionary game framework is developed to analyze interactions within power producer groups and between producers and regulators under low-carbon mechanisms. A core innovation of this study is the incorporation of a case study based on China's electricity market, which examines the evolutionary dynamics between local governments and power producers regarding carbon reduction strategies. This includes analyzing how regulatory incentives, market-clearing prices, and demand-side factors influence producers' bidding and emission reduction behaviors. The study also provides a detailed analysis of the bidding strategies for small, medium, and large power producers, revealing the significant impact of carbon pricing and market-clearing prices on strategic decision-making. Specifically, the study finds that small producers tend to adopt more conservative bidding strategies, aligning closely with market-clearing prices, while large producers take advantage of economies of scale, adjusting their strategies at higher capacities. The study explores the conditions under which carbon emission reduction strategies achieve stable equilibrium, as well as the implications of these equilibria for both market efficiency and environmental sustainability. The study reveals that integrating carbon reduction strategies into power market dynamics significantly impacts bidding behaviors and long-term market stability, especially under the influence of governmental penalties and incentives. The findings provide actionable insights for both power producers and policymakers, contributing to the advancement of low-carbon market theories and supporting the global transition to sustainable energy systems.

 https://doi.org/10.3390/pr13040952

Keywords: bidding strategies; evolutionary game theory; carbon emission reduction; electricity market; low-carbon mechanism; power producers

1. Introduction

The reform of China's power generation market system is still in its developmental stages. In an effort to invigorate the electricity industry, the original planned generation system has gradually transitioned into a market-based bidding mechanism, characterized by the establishment of centralized electricity trading platforms and numerous independent power generation enterprises [1]. Under this new market-oriented mechanism, power generation companies have become more competitive than under the previous planned economy system. The maximum electricity output awarded through competitive bidding and the minimum electricity price offered in these bids are critical factors that determine the profitability of power producers. Consequently, power producers must rely on various strategies, such as predicting competitors' bidding behaviors and simulating future market demand fluctuations, to analyze market trends and devise bidding strategies that align with their financial interests [2].

For power producers, the optimal strategy typically involves setting bidding prices marginally below the future market-clearing price. This approach maximizes profitability by ensuring bids are competitive yet sufficiently rewarding. On the other hand, from the perspective of the government, the ideal market bidding scenario is one where power producers bid according to their marginal costs. Such bidding behavior enables optimal resource allocation and improves the overall operational efficiency of the electricity market. Whether it is power producers aiming to maximize profits or regulators seeking to design policies that promote sustainable market development, understanding and analyzing the evolutionary dynamics of multi-group bidding strategies in power markets is of paramount importance.

One effective method for predicting bidding behaviors and market dynamics involves developing practical models of power market interactions based on evolutionary game theory. Although the application of evolutionary game theory to power market competition is still in the exploratory phase, significant progress has been made. For example, Cheng et al. (2020) [3] focused on the general N-population multi-strategy evolutionary games, and used them to investigate the generation-side long-term bidding issues in electricity markets. Based on this, Ref. [4] adopted evolutionary game theory to analyze the bidding equilibrium in electricity markets under the marginal cost pricing (MCP) and pay-as-bid (PAB) mechanisms. They further compared the bidding behaviors of power producers under different mechanisms. Building on this, some researchers have conducted detailed investigations into multi-group strategic interactions in power markets operating under the MCP mechanism, including the long-term evolutionary stability of bidding equilibria.

In the renewable energy sector, the application of evolutionary game theory (EGT) in renewable electricity markets has become a pivotal approach to understanding stakeholder interactions and promoting sustainable energy transitions. Jamali et al. (2022) [5] used EGT to analyze the technological transformation of industries toward renewable electricity procurement, focusing on both technological-based (TB) and non-technological-based (NTB) strategies. They demonstrated how subsidies and regulatory incentives significantly influence industries' adoption of renewable energy and identified equilibrium conditions that favor renewable electricity. In a subsequent study, Jamali et al. (2023) [6] extended this analysis to the long-term behavior of industries purchasing renewable and non-renewable energy in Iran. This research highlighted the importance of dynamic regulatory mechanisms, such as subsidies and carbon taxes, in driving tipping points where industries shift to renewable energy. Both studies emphasize that aligning industry strategies with regulatory frameworks is critical to achieving low-carbon objectives. Building on this foundation, Sun et al. (2024) [7] explored the coordination between renewable energy suppliers and grid operators using EGT. Their study constructed a game model to analyze strategies for renewable energy integration and peak shaving, revealing that cooperation between these stakeholders is vital for improving grid stability and efficiency. Key factors, such as grid capacity, market-clearing prices, and renewable penetration rates, significantly affect strategy outcomes, underscoring the need for dynamic pricing models and infrastructure investments. Huang et al. (2024) [8] introduced a tripartite evolutionary game model to examine interactions among renewable energy suppliers, coal-fired power plants, and market users. Their findings demonstrated that market equilibrium could be achieved through optimized trading strategies, leading to increased renewable energy consumption. They further emphasized the role of green electricity demand and pricing behaviors in shaping market dynamics. These studies collectively highlight EGT's capability to model the complex interactions among stakeholders in renewable energy markets. By identifying evolutionary stable strategy (ESS) and equilibrium conditions, they provide actionable insights for policymakers, market participants, and regulators. Key takeaways include the importance of subsidies, penalties, and carbon pricing in accelerating renewable energy adoption, the necessity of cooperation among stakeholders to achieve stable market operations, and the critical role of investment in renewable technologies and grid infrastructure. Future research should integrate EGT with other modeling approaches, such as system dynamics, to capture the multi-dimensional complexities of renewable energy systems. Together, these findings contribute to the global understanding of renewable energy markets and offer practical strategies for fostering sustainable energy transitions. These studies highlight the utility of evolutionary game theory in uncovering critical developmental patterns within diverse power market scenarios. The ongoing exploration of evolutionary game theory in power markets continues to focus on the strategic behaviors of generation groups, with particular attention to identifying bidding strategies that maximize profitability. As this research domain matures, it is anticipated that more researchers will explore the dynamics of group bidding evolution in power markets.

For power producers, evolutionary game theory offers a novel approach to forecasting bidding behaviors and market trends. Meanwhile, for market regulators, evolutionary game studies in power bidding can provide valuable recommendations for improving relevant laws and regulations governing electricity markets. The integration of this approach not only enhances the theoretical understanding of electricity market dynamics but also offers practical tools for both market participants and policymakers, contributing to the efficient and sustainable operation of modern power systems. For power producers, EGT has emerged as a crucial tool for modeling dynamic interactions in competitive electricity markets. Unlike traditional models, EGT accommodates the continuous adaptation of strategies by market participants, capturing the iterative decision-making processes required in markets influenced by factors such as renewable energy integration, market-clearing prices, and regulatory policies. This adaptability makes EGT particularly effective in addressing uncertainties, including those arising from demand fluctuations and the variability of renewable energy sources.

Recent studies have validated the utility of EGT in electricity markets. Under this background, the integration of renewable energy into electricity markets has become a key area of research, particularly as countries aim to balance economic growth with carbon emission reduction targets. The following studies explore innovative frameworks and methodologies for addressing this challenge through game-theoretic approaches.

Zhang et al. (2023) [9] developed a hybrid game model to study the evolutionary dynamics of renewable energy bidding strategies in China's electricity-carbon integrated market. The paper incorporates a multi-agent framework that combines cooperative and non-cooperative game elements to analyze interactions between renewable energy producers, traditional power plants, and regulatory authorities. The study reveals that carbon pricing mechanisms and green certificate incentives significantly influence bidding strategies and market equilibrium. Importantly, the hybrid model provides insights into how renewable energy producers can optimize their bids under varying carbon pricing policies. However, the model assumes that all market participants possess complete information, which may limit its applicability in real-world scenarios where uncertainties and asymmetric information often exist. A key contribution of the paper lies in its ability to combine multiple market elements (electricity, carbon, and green certificates) into a single analytical framework, offering regulators a comprehensive tool for policy evaluation.

Perera (2018) [10] explored a two-population evolutionary game to examine carbon emission reduction strategies in electricity markets. The study contrasts the strategic behaviors of power producers using fossil fuels and those adopting cleaner renewable energy sources. The results indicate that stricter carbon penalties and subsidies for renewables drive the market toward a sustainable equilibrium. However, the study emphasizes that the evolutionary process is highly sensitive to the initial conditions of the market, such as the relative costs of renewable versus fossil fuel energy. A notable limitation is the study's lack of focus on policy mechanisms that could mitigate these sensitivities, leaving room for future research on adaptive regulatory frameworks. Despite these shortcomings, the research provides a foundational understanding of how carbon pricing and subsidies influence producer behaviors over time.

Wang et al. (2023) [11] proposed an evolutionary game model to optimize the joint operation of green certificates, carbon emission rights, and electricity markets in a system comprising thermal, wind, and photovoltaic power producers. The study highlights the interplay between carbon prices, renewable energy subsidies, and green certificate trading in shaping market dynamics. Results from numerical simulations suggest that higher carbon prices and stricter emission controls create favorable conditions for renewable energy consumption, leading to increased market penetration of wind and solar power. A strength of the paper is its incorporation of multi-dimensional market mechanisms, which reflects the complexity of real-world systems. However, the study assumes static carbon prices, which may not fully capture the dynamic nature of carbon markets. Additionally, while the model provides valuable insights into market optimization, its reliance on theoretical assumptions could benefit from validation through real-world case studies.

A common limitation across these studies is the lack of empirical validation. While their theoretical contributions are substantial, future research should focus on applying these models to real-world scenarios to enhance their reliability and policy relevance. Additionally, incorporating uncertainties, such as fluctuating carbon prices and renewable energy intermittency, would improve the robustness of these models. Despite these limitations, the studies collectively provide valuable tools for optimizing renewable energy integration into electricity markets and offer actionable insights for policymakers aiming to balance economic efficiency with environmental sustainability.

In competitive electricity markets, power producers and generation groups commonly adopt bidding prediction measures prior to submitting their bids. These measures are designed to help producers formulate optimal bidding strategies that maximize their profitability in dynamic market environments. Various researchers and power producers have proposed mathematical models to address the challenges of designing optimal bidding strategies from different perspectives. For example, one approach involves forecasting the unified market-clearing price and then submitting bids slightly lower but still close to this forecasted price. Researchers have employed different models to predict market-clearing prices. For instance, Peng et al. (2005) [12] constructed an optimization model for power generation strategies based on generation costs and market price variations. Zhao et al. [13] proposed a single-bid period profit model using genetic algorithms, while Conejos et al. [14] developed a profit model based on the probability density distribution of electricity trading.

A second approach focuses on predicting the bidding strategies of competing producers, allowing individual producers to adjust their own strategies accordingly. This approach has led to numerous mathematical models for power producer bidding, each based on distinct principles. For example, Wang et al. [15] utilized fuzzy algorithms to establish a bidding model for power producers, while Ma et al. [16] proposed a bidding strategy model tailored to specific market conditions. Similarly, Liu et al. [17] developed a model that incorporates the probability distribution of electricity market loads to predict bidding outcomes.

Despite the diversity of these methodologies, they share a critical limitation: both approaches rely heavily on extensive historical data about market changes and competitors' bidding behaviors. For instance, forecasting market-clearing prices necessitates a comprehensive understanding of market conditions, including transmission constraints and the historical bidding decisions of all producers. Similarly, predicting competitors' bidding strategies requires access to sensitive data such as cost functions and profit curves, which are typically confidential and not shared among competitors. Moreover, these methods often fail to account for the interdependence of multiple participants in the market. For instance, predicting market-clearing prices assumes a perfectly competitive market where prices remain constant and unaffected by individual bidding strategies—an assumption that rarely holds in real-world electricity markets characterized by complex and interdependent bidding behaviors.

Additionally, some power producers and researchers have opted to establish cost models based solely on their own historical data, using a cost-plus-margin approach to formulate bids. For example, Gountis et al. [18] developed a mathematical model based on self-reported generation costs and historical data. Zhang et al. [19] extended this approach by forecasting thermal power companies' bidding strategies using cost functions and historical profit data. However, these self-focused models overlook the competitive dynamics of the market, making them inadequate for today's complex and rapidly evolving electricity markets. Such models fail to consider the influence of competitors' bidding strategies, thereby limiting producers' ability to maximize profits in a highly competitive environment.

A more promising approach involves integrating EGT with the dynamics of modern electricity markets. By treating bidding groups as distinct game participants, EGT allows researchers to model interactions between multiple groups within a competitive bidding framework. Unlike traditional models, EGT accounts for the interdependence and iterative interactions of participants, making it well suited for analyzing multi-group evolutionary dynamics in electricity markets. The previously mentioned methods are primarily designed for studying competitive and non-cooperative bidding scenarios. In contrast, EGT enables the examination of bidding situations where participants influence and balance one another. For instance, Zhang et al. [20] applied EGT to develop an optimization strategy model for deregulated electricity markets, while Wen et al. [21] constructed an EGT model to analyze asymmetric information and bidding strategies among power producers. Similarly, Guo et al. [22] investigated power company bidding decisions using EGT, and Wang et al. [23] incorporated networked group behaviors into an evolutionary game model.

The key strength of EGT lies in its ability to model interconnected game participants, where some groups exert influence over the entire market. This framework is particularly advantageous for analyzing multi-group asymmetric cooperative games under incomplete information. For example, Wang et al. [24] introduced a networked EGT model under partial information constraints, where each group has limited knowledge of the market but can refine their strategies through repeated interactions. In such models, game participants iteratively adapt based on historical data and evolving strategies, leading to more realistic representations of market dynamics.

While EGT addresses many limitations of traditional bidding models, its practical application faces several challenges. First, the computational complexity of EGT-based models increases significantly with the number of participants and the intricacy of market interactions. Second, while EGT accommodates partial information, the quality of predictions still depends on the availability and accuracy of historical data. Finally, validating EGT-based models in real-world electricity markets remains an ongoing area of research. Future studies should focus on integrating real-time data, incorporating stochastic elements to capture market uncertainties, and testing these models in live market environments.

Therefore, EGT offers a dynamic framework for understanding the decision-making processes of power producers in competitive markets [25,26]. Unlike traditional optimization or forecasting models, EGT does not depend on fixed, deterministic inputs or extensive historical data. Instead, it focuses on how strategies evolve over time based on interactions between participants and the payoffs they receive. For instance, producers in emerging electricity markets can iteratively adjust their bidding strategies based on observed market outcomes, which allows them to gradually identify stable and profitable bidding behaviors.

Additionally, EGT's flexibility in incorporating external factors, such as regulatory changes, renewable energy integration, and demand fluctuations, makes it uniquely suited for markets that are still developing [27–31]. This flexibility addresses a key limitation of conventional bidding strategies, which often fail to account for the dynamic and uncertain nature of early-stage electricity markets. Studies have demonstrated that EGT provides robust insights into equilibrium conditions and can guide both market participants and regulators in designing strategies that promote market efficiency and sustainability [32–34].

The integration of EGT with electricity market dynamics represents a significant advancement over traditional bidding models [35]. By addressing the limitations of historical data reliance and incorporating multi-group interactions, EGT provides a robust framework for analyzing and optimizing bidding strategies [36–38]. However, further research is required to overcome practical challenges and fully realize the potential of EGT in modern electricity markets.

The bidding strategies discussed rely on extensive historical data, making them ideal for established markets with stable structures. However, EGT is particularly valuable for emerging electricity markets with limited data and evolving mechanisms. EGT provides a flexible framework to model and analyze strategic interactions in dynamic, uncertain environments.

In this paper, we propose that, for the power producers, employing appropriate evolutionary game models is one of the most effective tools for formulating optimal bidding strategies. In bidding models constructed based on evolutionary game theory, different groups possess varying levels of information, engage in sequential bidding, and influence one another. These characteristics align the models with the current development patterns of the domestic power bidding market. By establishing a game model through the initial market profit matrix, we can calculate the system's final stable equilibrium points under various influencing factors. This enables us to determine the stable market states that emerge when power groups adopt different bidding strategies. Power producers can use these insights to identify strategies that maximize their returns while ensuring stability. This study has conducted dynamic simulations of two-group and three-group two-strategy bidding models within the power market and elucidated the practical significance of the equilibrium points. Based on these findings, power producers can incorporate additional influencing factors to develop more complex market game models.

For regulatory authorities, the continuous transformation of the power market necessitates the refinement of bidding systems in China's power sector. Utilizing multi-group evolutionary game models allows for the analysis of how various factors—such as regulations, clearing prices, and the intensity of rewards and penalties—affect the equilibrium points in power market bidding. This understanding helps in discerning the current developmental trends of the power market and forecasting future market dynamics. Furthermore, by implementing measures such as fines and subsidies, malicious competitive behaviors like market monopolization and manipulation can be mitigated, thereby fostering healthy competition within the power market. This paper investigates the impact of governmental oversight, market-clearing prices, and varying market demands on the long-term stability of the market. These insights provide regulatory bodies with valuable recommendations for enhancing bidding systems.

Additionally, this study integrates a critical research focus: evolutionary game analysis between local governments and power companies under low-carbon mechanisms. This aspect examines how local governments and power producers interact and evolve their strategies in response to low-carbon policies. Understanding this dynamic is essential for both policymakers and power companies to collaboratively achieve low-carbon objectives. By analyzing the strategic adaptations of local governments and power companies, this research contributes to the formulation of more effective low-carbon policies and the optimization of corporate strategies to align with sustainability goals.

The paper is organized as follows: Section 1 reviews forecasting and bidding behaviors in domestic and international power markets, alongside a discussion of evolutionary game theory applications. Section 2 focuses on the core components of evolutionary game theory and develops a three-group evolutionary game model. Section 3 uses real data from the East China Power Market to analyze bidding strategies under varying load demands and market-clearing prices. Sections 4–6 discuss the role of local governments in low-carbon mechanisms, present further discussions and prospects, and summarize the conclusions.

2. Evolutionary Game Theory and Its Application in Power Generation Bidding

2.1. EGT and Its Main Conceptions

EGT combines the analysis of static game theory with dynamic processes, providing a theoretical framework to study the evolution of strategies over time. In contrast to traditional game theory, where equilibrium concepts such as Nash equilibrium assume rational players with complete knowledge, EGT emphasizes the process of strategic adaptation through repeated interactions among players within a population. The core elements of EGT models include the population size, the set of available strategies, and the payoff functions that define the fitness or success of each strategy. These factors govern how strategies evolve based on interactions among players.

In power generation bidding, EGT is particularly useful for modeling the behavior of multiple power producers (agents) as they participate in the market. It allows us to simulate the competitive dynamics of strategy selection, where agents aim to optimize their bids to maximize their profits. The evolutionary process, which takes into account how strategies spread or disappear over time, can lead to either stable equilibria or ongoing instability depending on the market conditions. Understanding the reasons for instability—such as misaligned incentives or external perturbations—becomes critical for system design.

2.1.1. Replicator Dynamics

Replicator dynamics (RD) form the cornerstone of evolutionary game theory. They describe how the proportion of a strategy within a population changes over time, based on its relative payoff compared to other strategies. The replicator equation is given as follows:

$$\frac{dx(p_i)}{dt} = p_i[u(p_i, p) - u(p, p)] \tag{1}$$

where p_i represents the proportion of the population adopting strategy i, $u(p_i, p)$ is the payoff for strategy i, and $u(p, p)$ is the average payoff across all strategies. This equation indicates that strategies yielding above-average payoffs will increase in prevalence, while less successful strategies will diminish.

In power generation bidding scenarios, RD are used to model how bidding strategies evolve over time among different producers. For instance, if a certain bidding strategy consistently yields higher profits, more participants are likely to adopt it, leading to a shift in the overall strategy distribution. Conversely, less profitable strategies will fade out as producers shift toward the more successful ones. While this framework is powerful for modeling adaptive behavior in competitive settings, it may oversimplify the dynamics of real-world markets, which are often influenced by stochastic factors like demand fluctuations or regulatory changes. Future research could consider incorporating stochastic elements into the RD or adopting agent-based modeling to capture these complexities more effectively.

This study assumes a competitive market with heterogeneous power producers, each operating under different cost structures and capacities. These differences lead to varied strategic interactions. The evolutionary dynamics ensure that strategies with higher benefits gradually dominate the market, while the system remains resistant to sudden, abrupt changes in strategy profiles. Instead, the evolution of strategies is typically incremental and adaptive.

2.1.2. Evolutionarily Stable Strategy (ESS)

An ESS is a key concept in evolutionary game theory. It represents a strategy that, once established within a population, cannot be invaded by alternative strategies. Mathematically, a strategy s is ESS if, for any alternative strategy $s' \neq s$, the following condition holds:

$$f(s, s_{\text{ms}}) > f(s', s_{\text{ms}}) \tag{2}$$

where $f(s, s_{\text{ms}})$ denotes the payoff for strategy s when mixed with the population adopting strategy $s_{\text{ms}} = es + (1-e)s'$, for all $e \in (0, 1)$.

In the context of power generation group bidding, an ESS corresponds to a bidding strategy that remains optimal even when a small proportion of participants deviate from it. If the system dynamics governed by the RD equation converge to an ESS, the market achieves evolutionary stability. At this point, all participants settle on a stable bidding strategy, maximizing collective and individual benefits.

The identification of ESS in power markets is crucial for designing policies that ensure market stability. For example, regulatory mechanisms can encourage convergence toward socially optimal strategies, such as bidding strategies that balance profitability with grid reliability and sustainability. However, achieving ESS in practical markets may be challenging due to external shocks, incomplete information, and heterogeneity among participants.

2.1.3. Lyapunov-Based Stability Analysis

Lyapunov's method offers a systematic approach to assess the stability of equilibria in dynamic systems. For evolutionary game systems governed by RD equations, Lyapunov stability analysis involves computing the Jacobian matrix at equilibrium points and evaluating its eigenvalues. The stability conditions are summarized as follows:

- If all eigenvalues of the Jacobian matrix at an equilibrium point have negative real parts, the equilibrium is locally stable and corresponds to an ESS.
- If any eigenvalue has a positive real part, the equilibrium is unstable, leading to oscillations or divergence in strategy distributions.
- If eigenvalues have zero real parts, further analysis is required to determine stability.

In power generation bidding markets, Lyapunov stability analysis can help predict whether market rules or pricing schemes will lead to stable equilibria or cause volatility. By identifying stable equilibria, regulators can design mechanisms that minimize price fluctuations and strategic uncertainty. While this approach is robust, it relies on linear approximations near equilibrium points, which may not fully capture the nonlinear dynamics present in complex real-world markets. To overcome this limitation, advanced methods such as bifurcation analysis or machine learning-based stability prediction could be applied.

Stability of Equilibrium Points: To analyze the stability of equilibrium points, we derive the Jacobian matrix from the RD equations. The equilibrium points are stable if the eigenvalues of the Jacobian matrix are negative, indicating that the population will converge to these points over time. It is described as $J = \frac{\partial \dot{x}_i}{\partial x_j}$, where J is the Jacobian matrix of the system and describes how the change in one strategy affects the dynamics of others. The stability condition is derived by analyzing the eigenvalues of this matrix at equilibrium points.

Overall, Section 2.1. highlights the theoretical foundations and practical applications of evolutionary game theory in modeling power generation group bidding strategies. By integrating concepts such as RD, ESS, and Lyapunov stability analysis, researchers can systematically analyze the evolution of strategies in competitive markets. The insights gained from these models are invaluable for designing market mechanisms that promote stability, efficiency, and fairness. These concepts are crucial for understanding the strategic behavior of power producers in a competitive electricity market. In our research, we apply these principles to model the bidding behavior of small, medium, and large power producers. Specifically, we build a three-group evolutionary game model to analyze the dynamics of bidding strategies under various market conditions, such as fluctuating market-clearing prices and load demands. The resulting equilibrium strategies provide insights into how power producers adapt their bids in response to market forces and regulatory policies.

2.2. Evolutionary Game Models Applied in Power Generation Bidding

The application of EGT to power generation bidding has garnered significant attention in recent years, primarily due to its ability to bridge the gap between theoretical equilibrium analysis and the practical complexities of market dynamics. Unlike traditional static game theory models that assume perfect rationality and instantaneous equilibrium, EGT captures the adaptive and iterative nature of strategy selection among competing power producers. This is particularly valuable in deregulated electricity markets, where numerous generators dynamically adjust their bidding strategies in response to market signals, policy interventions, and competition.

2.2.1. Current Developments in EGT for Power Generation Bidding

The development of EGT in power generation bidding has focused on modeling the evolutionary dynamics of competing strategies within the context of deregulated energy markets. Key advancements include the following:

(1) Dynamic modeling frameworks: Evolutionary models extend static Nash equilibrium concepts by incorporating time-dependent strategy adjustments, often represented through systems of differential equations such as RD. These models capture the gradual evolution of strategies as agents learn and adapt based on observed payoffs.
(2) Multi-agent systems and heterogeneity: Electricity markets consist of diverse participants, including fossil fuel plants, renewable energy sources, and storage systems. Recent research incorporates heterogeneous agent characteristics, such as generation costs, capacity constraints, and emissions penalties, into evolutionary models. This ensures that the models accurately reflect real-world complexities.
(3) Algorithmic advancements: Computational algorithms, such as agent-based simulations and reinforcement learning, have been integrated into EGT frameworks. These methods enable researchers to explore high-dimensional strategy spaces, simulate complex market dynamics, and identify stable bidding strategies under varying market conditions.
(4) Policy-oriented applications: Evolutionary models have been employed to study the impact of policy measures, such as carbon pricing, renewable energy subsidies, and capacity auctions, on bidding strategies and market outcomes. These applications provide valuable insights into the design of market mechanisms that promote stability, efficiency, and sustainability.

2.2.2. A Key Model: RD in Power Markets

One of the foundational models in EGT is the RD model, which describes the evolution of strategy proportions within a population over time. This model is particularly well-suited for analyzing power generation bidding, where generators iteratively adjust their strategies based on relative payoffs. The RD equation is given as follows:

$$\frac{\mathrm{d}x_i}{\mathrm{d}t} = x_i[u_i(x) - \overline{u}(x)] \tag{3}$$

where x_i is the proportion of the population adopting strategy i and $u_i(x)$ is the payoff associated with strategy iii in the current population distribution x. $\overline{u}(x) = \sum_j x_j u_j(x)$ is the average payoff of the population, weighted by strategy proportions. This equation highlights two key principles: Strategies with payoffs above the population average ($u_i(x) > \overline{u}(x)$) increase in prevalence over time. Conversely, strategies yielding below-average payoffs ($u_i(x) < \overline{u}(x)$) gradually decline.

2.2.3. Application to Power Generation Bidding

In the context of a deregulated electricity market, let us define x_i as the proportion of generators adopting bidding strategy i, and $u_i(x)$ as the profit associated with that strategy. The RD equation can be extended to incorporate specific market features, such as price caps, demand elasticity, and capacity constraints. A generalized form of the RD in power markets is the following:

$$\frac{\mathrm{d}x_i}{\mathrm{d}t} = x_i[\pi_i(p,q,c) - \overline{\pi}(p,q,c)] \tag{4}$$

where $\pi_i(p, q, c)$ is the profit function for generators using strategy i, dependent on market price (p), quantity sold (q), and generation cost (c). $\overline{\pi}(p,q,c)$ is the average profit across all strategies. In this formula, some key mathematical components are elaborated as follows.

(1) Profit function ($\pi_i(p, q, c)$): The profit function for a generator is typically expressed as $\pi_i = p \cdot q_i - c_i(q_i)$, where p: Market-clearing price; q_i: Quantity of electricity generated by strategy i; and $c_i(q_i)$: Cost of generating q_i, which may be linear ($c_i = c \cdot q_i$) or nonlinear (e.g., including startup or ramping costs).

(2) Market-clearing condition: The market-clearing price p is determined by balancing supply and demand:

$$\sum_i q_i = D(p) \tag{5}$$

where (p) represents the electricity demand as a function of price p. This condition ensures that the price dynamically adjusts to match supply with demand in each market iteration.

(3) Evolutionary stability criterion: An equilibrium strategy x^* is considered evolutionarily stable if it satisfies the following:

$$\forall i \neq j, u_i(x^*) > u_j(x^*) \tag{6}$$

This implies that no alternative strategy can achieve a higher payoff in the presence of x^*.

2.2.4. Explanation of the Mathematical Framework and Future Directions

The above introduced mathematical framework includes four main features:

(1) Dynamic adjustment of strategies: The RD equation models how generators adjust their bidding strategies based on profit differentials. If a specific strategy i consistently yields higher profits, it will attract more participants, leading to an increase in x_i. Over time, this mechanism drives the population toward an equilibrium distribution of strategies.

(2) Role of payoff functions: The payoff functions (x) capture the economic incentives for generators to adopt specific strategies. In power markets, these payoffs depend not only on direct profits but also on external factors, such as regulatory penalties for emissions or rewards for renewable generation.

(3) Market equilibrium and stability: The market-clearing condition ensures that prices dynamically adjust to reflect the interplay between supply and demand. By coupling this condition with the RD equation, the model captures both short-term pricing dynamics and long-term strategic evolution.

(4) Stability analysis: Stability is assessed by examining whether the RD system converges to an equilibrium distribution x^*. Evolutionary stability ensures that once an equilibrium is reached, no participant has an incentive to unilaterally deviate from the prevailing strategy.

Based on the above, while RD provide a robust framework for modeling power generation bidding, several challenges remain:

Stochastic factors: Real-world electricity markets are subject to uncertainty, such as fluctuating renewable output and demand variability. Integrating stochastic dynamics into EGT models would enhance their realism.

Incorporating grid constraints: Future models should account for transmission constraints, grid reliability, and reserve requirements, which significantly impact bidding behavior.

Algorithmic innovations: Machine learning techniques, such as reinforcement learning, could complement evolutionary models by enabling agents to learn optimal strategies in complex, dynamic environments.

In conclusion, the application of evolutionary game theory to power generation bidding offers a powerful toolset for understanding and predicting market behavior. By leveraging models like RD and integrating advanced computational methods, researchers can design mechanisms that ensure market stability, efficiency, and sustainability.

2.3. *Evolutionary Game Model Simulation for Two-Strategy Groups in Power Markets*

2.3.1. Typical Two-Group Two-Strategy Symmetric Evolutionary Game Model

(1) Basic assumptions: We consider two homogeneous groups A and B participating in the electricity market. The strategies available to group A are $S_{A1} = x$ and $S_{A2} = 1 - x$, where $x \in [0, 1]$, while the strategies for group B are $S_{B1} = y$ and $S_{B2} = 1 - y$, with $y \in [0, 1]$. Both groups operate under the assumption of incomplete information regarding the payoff matrix in the competitive market. This two-group evolutionary game is represented by the payoff matrix shown below:

$$\begin{array}{ccc} & S_{B1}(y) & S_{B2}(1-y) \\ S_{A1}(x) & (a,a) & (c,d) \\ S_{A2}(1-x) & (d,c) & (g,g) \end{array} \tag{7}$$

Here, a, c, d, g represent the payoff parameters for the combination of strategies adopted by the two groups. These parameters are flexible and depend on the specific market conditions and empirical data. Using Equation (7), the system's RD equations can be derived as follows:

$$\begin{cases} \frac{dx}{dt} = x(1-x)[(a-c-d+g)y+c-g] \\ \frac{dy}{dt} = y(1-y)[(a-c-d+g)x+c-g] \end{cases} \tag{8}$$

The system's equilibrium points are determined by setting $\frac{dx}{dt} = 0$ and $\frac{dy}{dt} = 0$. Solving this yields the following:

$$\begin{cases} x(1-x)[(a-c-d+g)y+c-g] = 0 \\ y(1-y)[(a-c-d+g)x+c-g] = 0 \end{cases} \tag{9}$$

The equilibrium points are as follows:

$$\varphi_{ESS} = \{(0,0),(0,1),(1,0),(1,1)\} \tag{10}$$

(2) Evolutionary stability analysis: To determine the stability of these equilibrium points, the Jacobian matrix for the evolutionary game model is constructed as follows:

$$J_1 = \begin{bmatrix} (1-2x)[(a-c-d+g)y+c-g] & x(1-x)(a-c-d+g) \\ y(1-y)(a-c-d+g) & (1-2y)[(a-c-d+g)x+c-g] \end{bmatrix} \tag{11}$$

The determinant and trace of the Jacobian matrix are computed as follows:

$$\begin{cases} \det(J_1) = (1-2x)(1-2y)[(a-c-d+g)x+c-g][(a-c-d+g)y+c-g] - \\ \qquad [x(1-x)(a-c-d+g)][y(1-y)(a-c-d+g)] \\ \mathrm{tr}(J_1) = (1-2x)[(a-c-d+g)y+c-g] + (1-2y)[(a-c-d+g)x+c-g] \end{cases} \tag{12}$$

In our study, we utilized MATLAB software, version 2019b, for the simulation research. The software was provided by MathWorks, located in Natick, Massachusetts, USA. Based on this, the payoff parameters (a, c, d, g) are set as (3, 6, 4, 2), and the dynamics of x and y over time are simulated with a time step of 1/60 s for 50 iterations using MATLAB 2019b (MathWorks, Natick, MA, USA). The resulting trajectories in the $x - y$, $x - t$, and $y - t$ spaces are depicted in Figure 1. The observations are summarized below:

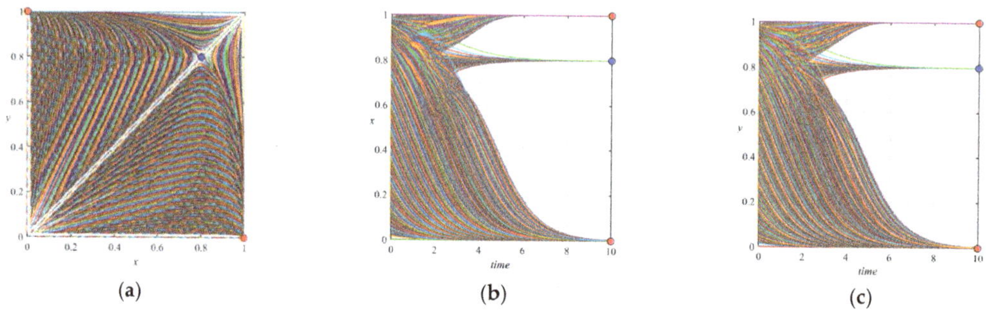

Figure 1. The phase trajectory simulation results for a typical symmetric two-group two-strategy evolutionary game when considering $a = 3$, $b = 6$, $c = 4$, and $d = 2$. (a) The phase trajectory of (x, y). (b) The phase trajectory of (x, t). (c) The phase trajectory of (y, t).

(1) **Stable equilibria:**

The system stabilizes at equilibrium points (0, 1) and (1, 0), where one group fully adopts a single strategy. These equilibria are evolutionarily stable and indicate that, under specific payoff conditions, the market will converge to a state where groups settle on distinct, stable strategies.

(2) **Intermediate states:** A saddle-point equilibrium exists at $\left(\frac{a-d}{a-c-d+g}, \frac{a-d}{a-c-d+g}\right)$, representing a transient state of partial strategy adoption. This equilibrium is unstable, meaning that the market will not remain in this intermediate state under normal dynamics.

(3) **Payoff impacts:** The analysis reveals that the stability of equilibria depends heavily on the relative magnitudes of the payoff parameters. Consider the following, for instance:
   - When $c > g$, equilibria (0, 0) and (1, 1) are unstable.
   - When $a > d$, the system stabilizes at (0, 1) or (1, 0).

(4) **Impact of incentives:** If market regulators introduce penalties or subsidies, the parameters a, c, d, g can be adjusted to guide the system toward desired equilibria. For example, the increasing g could make renewable energy strategies more competitive, encouraging their widespread adoption. This two-group two-strategy evolutionary model provides a valuable framework for analyzing the dynamics of bidding strategies in power generation markets. By incorporating payoff parameters that reflect real-world incentives, the model can simulate the impact of regulatory interventions, such as carbon pricing or renewable subsidies, on market stability.

2.3.2. Typical Two-Group Two Strategy Asymmetric Evolutionary Game Model

(1) Basic assumptions: Consider a heterogeneous market involving two groups, A and B, each with their own distinct strategies and payoffs. The strategy set for group A is defined as $S_{A1} = x$ and $S_{A2} = 1 - x$, where $x \in [0, 1]$. Similarly, the strategies for group B are $S_{B1} = y$ and $S_{B2} = 1 - y$, with $y \in [0, 1]$. Both groups operate under incomplete

information about the competitive environment, and their payoffs are governed by a non-symmetric payoff matrix:

$$\begin{array}{ccc} & S_{B1}(y) & S_{B2}(1-y) \\ S_{A1}(x) & (a,b) & (c,d) \\ S_{A2}(1-x) & (e,f) & (g,h) \end{array} \quad (13)$$

Here: a, b, c, d, e, f, g, h are the payoff parameters associated with the strategic interactions between groups A and B. These values depend on empirical data and market-specific characteristics, such as pricing mechanisms and regulatory constraints. Using Equation (13), the RD equations for the system are derived as follows:

$$\begin{cases} \dfrac{dx}{dt} = x(1-x)[(a-c-e+d)y+c-g] \\ \dfrac{dy}{dt} = y(1-y)[(b-d-f+h)x+f-h] \end{cases} \quad (14)$$

To find the system's equilibrium points, we set $\dfrac{dx}{dt} = 0$ and $\dfrac{dy}{dt} = 0$, resulting in the following:

$$\begin{cases} x(1-x)[(a-c-e+d)y+c-g] = 0 \\ y(1-y)[(b-d-f+h)x+f-h] = 0 \end{cases} \quad (15)$$

The equilibrium points are denoted as $\varphi_{ESS} = \{(0,0),(0,1),(1,0),(1,1)\}$.

(2) Evolutionary stability analysis: To analyze the stability of these equilibrium points, we construct the Jacobian matrix for the evolutionary game model as:

$$J_2 = \begin{bmatrix} (1-2x)[(a-c-e+d)y+c-g] & x(1-x)(a-c-e+d) \\ y(1-y)(b-d-f+h) & (1-2y)[(b-d-f+h)x+f-h] \end{bmatrix} \quad (16)$$

The determinant and trace of the Jacobian matrix J_2 are computed as follows:

$$\begin{cases} \det(J_2) = (1-2x)(1-2y)[(a-c-e+d)y+c-g][(b-d-f+h)x+f-h]- \\ \qquad xy(1-x)(1-y)(a-c-e+d)(b-d-f+h) \\ \mathrm{tr}(J_2) = (1-2x)[(a-c-e+d)y+c-g]+(1-2y)[(b-d-f+h)x+f-h] \end{cases} \quad (17)$$

The stability of the system is determined by evaluating $\det(J_2)$ and $\mathrm{tr}(J_2)$. The equilibrium points are classified as stable if $\det(J_2) > 0$ and $\mathrm{tr}(J_2) < 0$.

The payoff parameters (a,b,c,d,e,f,g,h) are set as (6, 2, 4, 5, 7, 6, 3, 2), and dynamic simulations of x and y are conducted using MATLAB with a time step of 1/60 s for 50 iterations. The trajectories in the $x-y$, $x-t$, and $y-t$ spaces are shown in Figure 2. The simulation results provide the following observations:

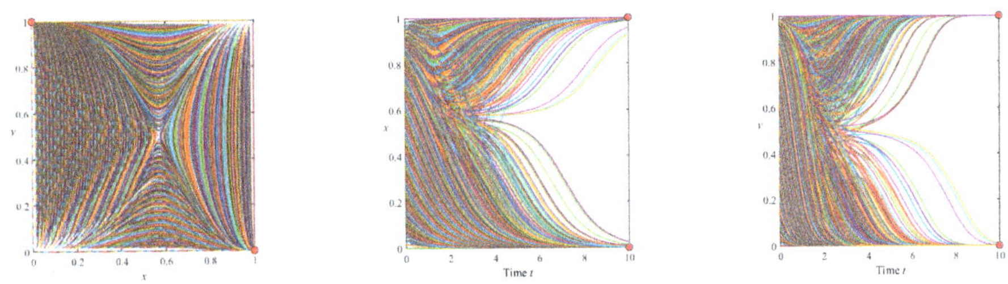

(a) (b) (c)

Figure 2. The phase trajectory simulation results for a typical asymmetric two-group two-strategy evolutionary game when considering $a = 6$, $b = 2$, $c = 4$, $d = 5$, $e = 7$, $f = 6$, $g = 3$, $h = 2$. (**a**) The phase trajectory of (x, y). (**b**) The phase trajectory of (x, t). (**c**) The phase trajectory of (y, t).

(1) **Stable equilibria:** The system converges to two stable equilibria: (0, 1) and (1, 0). These represent scenarios where one group fully adopts a dominant strategy while the other group selects an alternative strategy. This aligns with practical market dynamics, where different groups settle on distinct bidding behaviors over time.
(2) **Absence of saddle points:** The simulation reveals that no saddle points exist in the phase trajectories. This is due to the condition $a - c - e + d = 0$, which eliminates the possibility of intermediate equilibrium states.
(3) **Regulatory impacts:** If S_{A1}, S_{A2} represent low-cost and high-cost bidding strategies, regulators aiming to achieve balanced competition can introduce penalty mechanisms. For instance, ensuring $c \geq g, f \geq h, a > e, b > d$ restricts the system to the equilibrium (1, 1), encouraging both groups to adopt low-cost strategies.
(4) **Impact of parameter variations:** Stability is influenced by the signs of $(a - e)$, $(b - d)$, $(c - g)$, $(f - h)$. By systematically analyzing 16 different cases, it is evident that stability outcomes vary based on the relative magnitudes of these parameters. This highlights the importance of fine-tuning regulatory policies to guide market dynamics effectively.
(5) **Critical evaluation and expansion:** The non-symmetric evolutionary game model presented here extends the classical symmetric model by incorporating heterogeneity between groups. This framework is particularly relevant to power generation markets, where different participants—such as renewable energy producers, fossil fuel plants, and storage providers—operate under distinct cost structures and strategic priorities.

To this end, we give a summary and analysis of the four scenarios for unique evolutionary stable equilibria.

Scenario 1: Unique Equilibrium at (0, 0). The parameter settings are as follows: $a = 1$, $b = 1$, $c = 2$, $d = 1$, $e = 2$, $f = 2$, $g = 3$, $h = 2$. Purpose and Motivation: In this scenario, the system is designed such that (0, 0) becomes the only evolutionary stable equilibrium (ESE). This equilibrium implies that both groups, A and B, predominantly select their respective S_{A2} and S_{B2} strategies. The primary purpose of this setup is to investigate conditions under which both groups avoid adopting aggressive or high-cost strategies. Conditions to Ensure Stability: The payoff parameters c and g dominate, ensuring that deviations from $x = 0$ or $y = 0$ incur greater penalties compared to maintaining $x = 0$ and $y = 0$. The interactions between groups (represented by $a - c - e + d$ and $b - d - f + h$) are balanced to disincentivize any deviation. Analysis: By ensuring that $c > g$ and $f > h$, the payoff matrix penalizes any attempt to increase x or y. As a result, the system's trajectories converge to (0, 0). This setting reflects a scenario where cooperation or conservative strategies dominate, leading to stability at low strategic engagement levels.

Scenario 2: Unique Equilibrium at (0, 1), as illustrated in Figure 3. Parameter Settings: $a = 1$, $b = 3$, $c = 2$, $d = 1$, $e = 2$, $f = 2$, $g = 3$, $h = 1$. Purpose and Motivation: In this scenario, the system converges to (0, 1), where group A predominantly selects S_{A2}, and group B predominantly selects S_{B1}. This configuration simulates conditions where one group prefers a conservative strategy ($x = 0$), while the other adopts a more aggressive strategy ($y = 1$). Conditions to Ensure Stability: $b > d > f$ ensures that $y = 1$ is the dominant strategy for group B, while $a < c$ and $g > e$ ensure that $x = 0$ is the preferred strategy for group A. The interaction terms between the two groups reinforce the asymmetry, with b heavily incentivizing $y = 1$ while c discourages $x > 0$. Analysis. This scenario reflects an asymmetric market dynamic where group B gains a competitive advantage by adopting an aggressive strategy (S_{B1}), while group A settles into a supporting or complementary role by avoiding S_{A1}. Such a setup can model situations where one group leverages a cost advantage or market power to dominate.

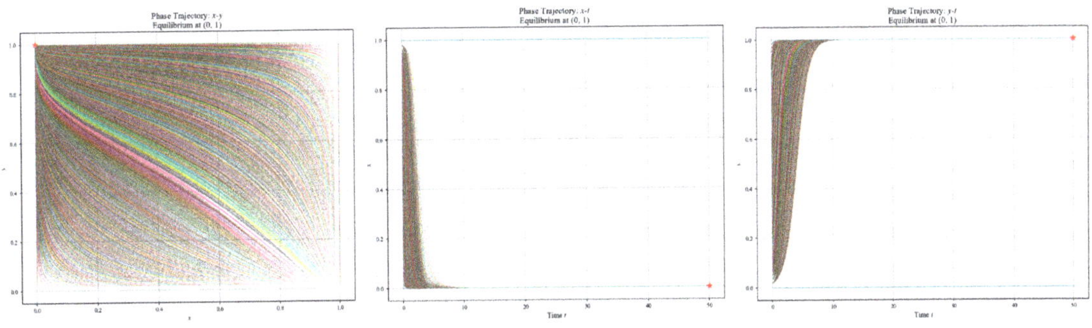

(a) (b) (c)

Figure 3. The phase trajectory simulation results for a typical asymmetric two-group two-strategy evolutionary game achieving a unique ESE at (0, 1) when considering $a = 1, b = 3, c = 2, d = 1, e = 2, f = 2, g = 3, h = 1$. (**a**) The phase trajectory of (x, y). (**b**) The phase trajectory of (x, t). (**c**) The phase trajectory of (y, t).

Scenario 3: Unique Equilibrium at (1, 0). Parameter Settings: $a = 3, b = 1, c = 2, d = 1, e = 1, f = 2, g = 3, h = 2$. Purpose and Motivation: Here, the system converges to (1, 0), where group A predominantly selects S_{A1}, and group B predominantly selects S_{B2}. This scenario represents a reversed asymmetry compared to Scenario 2, with group A now adopting an aggressive strategy ($x = 1$) while group B adopts a conservative strategy ($y = 0$). Conditions to Ensure Stability: $a > e > c$ ensures that $x = 1$ is the dominant strategy for group A, while $b < f$ and $h > d$ disincentivize any deviation from $y = 0$. The interaction terms balance the stability, with $a - c - e + d$ and $b - d - f + h$ adjusted to reinforce the dominance of the respective strategies. Analysis: This configuration models a situation where group A drives market dynamics by adopting an aggressive strategy, while group B acts more passively or defensively. Such a scenario could represent a market with a dominant player and a supporting or regulated competitor.

Scenario 4: Unique Equilibrium at (1, 1), as demonstrated in Figure 4. Parameter Settings: $a = 3, b = 3, c = 2, d = 1, e = 1, f = 2, g = 3, h = 1$. Purpose and Motivation: This scenario forces the system to converge to (1, 1), where both groups A and B predominantly select their respective aggressive strategies (S_{A1} and S_{B1}). The purpose is to study conditions under which mutual aggression or competition leads to stable outcomes. Conditions to Ensure Stability: $a > c > e$ and $b > d > f$ ensure that $x = 1$ and $y = 1$ are the dominant strategies for groups A and B, respectively. The interaction terms $a - c - e + d$ and $b - d - f + h$ are balanced to disincentivize deviations from (1, 1). Analysis: This setting models a highly competitive market where both groups adopt aggressive strategies to maximize their payoffs. The stability at (1, 1) suggests that aggressive competition is sustainable under these specific conditions, which may reflect markets with high stakes or significant rewards for dominating strategies.

Overall, the parameters a, b, c, d, e, f, g, h play a critical role in determining the system's dynamics and equilibria. By carefully adjusting these parameters, we can control the stability and location of the evolutionary stable equilibria (ESS). Increasing Payoff for Specific Strategies: To ensure stability at a specific equilibrium, increase the payoff for the desired strategies (e.g., A and B for (1, 1)) while reducing the payoffs for competing strategies. Balancing Interaction Terms: The terms $a - c - e + d$ and $b - d - f + h$ determine the interaction dynamics between the groups. Adjusting these terms ensures that the system strongly favors one equilibrium over others. Eliminating Saddle Points: To avoid unstable intermediate equilibria, ensure that the payoff differences $a - c - e + d$ and $b - d - f + h$ do not cancel out, which could create transient states. By fine-tuning the parameters, the system's behavior is modeled through evolutionary dynamics, allowing for detailed analysis of strategic interactions, market outcomes, and carbon reduction strategies under different regulatory environments.

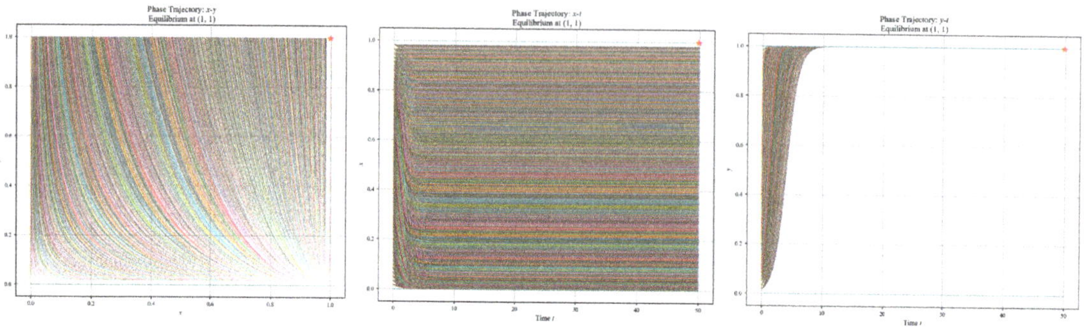

(a) (b) (c)

Figure 4. The phase trajectory simulation results for a typical asymmetric two-group two-strategy evolutionary game achieving a unique ESE at (1, 1) when considering $a = 3$, $b = 3$, $c = 2$, $d = 1$, $e = 1$, $f = 2$, $g = 3$, $h = 1$. (**a**) The phase trajectory of (x, y). (**b**) The phase trajectory of (x, t). (**c**) The phase trajectory of (y, t).

2.3.3. Typical Three-Group Two-Strategy Asymmetric Evolutionary Game Model

Basic Assumptions: This section considers a market region comprising three categories of power generation groups with distinct capacities: small capacity (SP), medium capacity (MP), and large capacity (LP). Here, the producers are classified based on their installed capacity (i.e., the total capacity of their power generation units) and market share within the relevant electricity market. In the real-world electricity market, these factors are significant because they affect producers' ability to influence market-clearing prices and their bidding behavior. Larger producers, with more capacity, have a greater ability to affect market outcomes, while smaller producers often have less influence on price formation.

- Large producers (LP): Typically, these are the dominant players in the market with a significant share of total market capacity (e.g., large utility companies or multinational power generation firms). These companies tend to have greater financial resources and can employ more sophisticated bidding strategies, such as strategic bidding to influence market prices. In our model, we define large producers as those with a capacity that constitutes over 30% of the total market capacity.
- Medium producers (MP): These producers have a moderate market share, generally between 10-30% of the total market capacity. They may have some ability to influence prices but are not as dominant as large producers. Their bidding strategies often focus on optimizing profits while responding to the strategies of larger producers and market conditions.
- Small producers (SP): These are usually independent producers or renewable energy producers (e.g., wind farms or solar power plants) with smaller market shares (typically less than 10%). These producers are price takers, and their bidding strategies are often more passive compared to larger producers.

To better reflect the practical dynamics of the electricity market, it is assumed that these three groups operate with incomplete information regarding each other's strategies, and each group has only two bidding strategies, S_H (high-price bidding) and S_L (low-price bidding). The evolutionary game model prioritizes the allocation of capacity, with larger-capacity groups bidding first, followed by smaller-capacity groups. Assuming that each group SP, MP, and LP consists of N_A, N_B, and N_C power producers, respectively, each producer's generation capacity is determined by its individual strategy S_L or S_H. The cost of production, $C(P_j)$, is defined by $C(P_j) = a_j + b_jP_j + c_jP_j^2, j \in \{1, 2, \cdots, N_i\}$, where P_j is the actual power generation output of producer j, and a_j, b_j, c_j are cost coefficients specific to the producer within each group.

The expected market-clearing price is denoted as P_{MCP}, and the revenue for each producer, $f(P_j)$, is expressed as follows:

$$f(P_j) = P_{MCP}P_j - C(P_j) \tag{18}$$

Under the market-clearing price (MCP) mechanism, the optimal strategy for maximizing profit $f(P_j)$ is determined by the following optimization problem:

$$\begin{cases} \max f(P_j) \\ s.t. \sum P_j = Q_{Ma}, B_j(S_j, P_j) = B_{MCP} \end{cases} \tag{19}$$

where S_j is strategy set for the power producer, Q_{Ma} is market demand during trading, and $B_j(S_j, P_j)$ is the cost–benefit equation for bidding in the MCP market. The payoffs for the SP, MP, and LP groups are summarized in Table 1.

Table 1. The payoffs for the SP, MP, and LP groups.

Group SP and Group MP				Group LP	
				S_H	S_L
SP	S_H	MP	S_H	(a_1, a_2, a_3)	(b_1, b_2, b_3)
			S_L	(c_1, c_2, c_3)	(d_1, d_2, d_3)
	S_L		S_H	(e_1, e_2, e_3)	(f_1, f_2, f_3)
			S_L	(g_1, g_2, g_3)	(h_1, h_2, h_3)

Based on Table 1, each payoff reflects the profit parameters for power producers bidding within each group, derived from actual operational data. The expected payoffs for executing strategies S_H and S_L for each group are given by E_{SP1}, E_{SP2}, E_{MP1}, E_{MP2}, E_{LP1}, and E_{LP2}, while the average payoffs for these groups are denoted as E_{Spav}, E_{Mpav}, and E_{Lpav}. These are expressed by the following equations:

$$\begin{cases} E_{SP1} = y[za_1 + (1-z)b_1] + (1-y)[zc_1 + (1-z)d_1] \\ E_{SP2} = y[ze_1 + (1-z)f_1] + (1-y)[zg_1 + (1-z)h_1] \\ E_{MP1} = z[xa_2 + (1-x)e_2] + (1-z)[xb_2 + (1-x)f_2] \\ E_{MP2} = z[xc_2 + (1-x)g_2] + (1-z)[xd_2 + (1-x)h_2] \\ E_{LP1} = x[ya_3 + (1-y)c_3] + (1-x)[ye_3 + (1-y)g_3] \\ E_{LP2} = x[yb_3 + (1-y)d_3] + (1-x)[yf_3 + (1-y)h_3] \end{cases} \tag{20}$$

$$\begin{cases} E_{SPav} = xE_{SP1} + (1-x)E_{SP2} \\ E_{MPav} = yE_{MP1} + (1-y)E_{MP2} \\ E_{LPav} = zE_{LP1} + (1-z)E_{LP2} \end{cases} \tag{21}$$

These equations capture the interaction dynamics between groups, where the variables x, y, and z denote the probabilities of selecting low-price bidding strategies for SP, MP, and LP groups, respectively. Critical expansion and discussion are shown as follows.

Economic rationale: The three-group $2 \times 2 \times 2$ evolutionary game model reflects the real-world dynamics of electricity markets, where generators of varying capacities adopt different strategies based on cost structures and market-clearing prices. Larger-capacity producers often dominate the bidding process, influencing the strategic choices of smaller-capacity producers. By incorporating two strategy options (S_H and S_L) for each group, this model captures the competitive interplay among heterogeneous participants.

Policy implications: The payoffs in Table 1 suggest that strategic adjustments to the parameters a, b, c, d, e, f, g, h can be used to influence market outcomes. For instance, regulators can incentivize low-cost bidding by providing subsidies (reducing c_j) or imposing penalties on high-cost strategies (increasing a_j). The MCP-based optimization ensures that the market operates efficiently by aligning power generation output with demand while maximizing individual profits.

Model limitations: While the model accounts for the heterogeneity in generation capacities, it assumes uniform behavior within each group (SP, MP, LP). This may oversimplify the complexities of real-world markets, where individual producers often have unique constraints and objectives.

Future directions include incorporating stochastic elements into the MCP mechanism to account for demand variability and renewable energy intermittency, as well as extending the model to include additional strategy options (S_H, S_L, S_M) or dynamic capacity adjustments, reflecting real-time grid conditions. The three-group $2 \times 2 \times 2$ evolutionary game model provides a robust theoretical framework for understanding the strategic bidding behaviors of heterogeneous power producers. By linking individual profit maximization to market-clearing mechanisms, this model offers valuable insights into the dynamics of competitive electricity markets. Further research should focus on enhancing the model's realism and exploring its applicability to scenarios involving renewable energy integration, grid constraints, and long-term market stability.

Formulation of the RD: The RD equations for the three-group $2 \times 2 \times 2$ evolutionary game model are formulated as follows:

$$\begin{cases} h_{\mathrm{SP}}(x) = \dfrac{\mathrm{d}x}{\mathrm{d}t} = x(E_{\mathrm{SP1}} - E_{\mathrm{SPav}}) \\ h_{\mathrm{MP}}(y) = \dfrac{\mathrm{d}y}{\mathrm{d}t} = y(E_{\mathrm{MP1}} - E_{\mathrm{MPav}}) \\ h_{\mathrm{LP}}(z) = \dfrac{\mathrm{d}z}{\mathrm{d}t} = z(E_{\mathrm{LP1}} - E_{\mathrm{LPav}}) \end{cases} \tag{22}$$

By substituting the payoff expressions, the RD can be expanded into the following form:

$$\begin{cases} h_{\mathrm{SP}}(x) = x(1-x)g_{\mathrm{SP}}(y,z) \\ h_{\mathrm{MP}}(y) = y(1-y)g_{\mathrm{MP}}(x,z) \\ h_{\mathrm{LP}}(z) = z(1-z)g_{\mathrm{LP}}(x,y) \end{cases} \tag{23}$$

$$\begin{cases} g_{\mathrm{SP}}(y,z) = (a_1 - b_1 - c_1 + d_1 - e_1 + f_1 + g_1 - h_1)yz + (b_1 - d_1 - f_1 + h_1)y + \\ \qquad (c_1 - d_1 - g_1 + h_1)z + d_1 - h_1 \\ g_{\mathrm{MP}}(x,z) = (a_2 - b_2 - c_2 + d_2 - e_2 + f_2 + g_2 - h_2)zx + (e_2 - f_2 - g_2 + h_2)z + \\ \qquad (b_2 - d_2 - f_2 + h_2)x + f_2 - h_2 \\ g_{\mathrm{LP}}(y,x) = (a_3 - b_3 - c_3 + d_3 - e_3 + f_3 + g_3 - h_3)xy + (c_3 - d_3 - g_3 + h_3)y + \\ \qquad (e_3 - f_3 - g_3 + h3)x + g_3 - h_3 \end{cases} \tag{24}$$

The RD can be expressed in matrix form for clarity:

$$\begin{bmatrix} \dfrac{\mathrm{d}x}{\mathrm{d}t} \\ \dfrac{\mathrm{d}y}{\mathrm{d}t} \\ \dfrac{\mathrm{d}z}{\mathrm{d}t} \end{bmatrix} = \begin{bmatrix} x(1-x)g_{\mathrm{SP}}(y,z) \\ y(1-y)g_{\mathrm{MP}}(x,z) \\ z(1-z)g_{\mathrm{LP}}(x,y) \end{bmatrix} \tag{25}$$

At equilibrium ($dx/dt = 0$, $dy/dt = 0$, $dz/dt = 0$), the system satisfies the following:

$$\begin{cases} x(1-x)g_{\mathrm{SP}}(y,z) = 0 \\ y(1-y)g_{\mathrm{MP}}(x,z) = 0 \\ z(1-z)g_{\mathrm{LP}}(x,y) = 0 \end{cases} \tag{26}$$

Stability Analysis: The Jacobian matrix (J) of the system corresponding to the RD Equation (25) derived earlier can be used to analyze the stability of the internal equilibrium points in the system by examining the positivity or negativity of the eigenvalues. The focus of this section is on the three groups SP, MP, and LP. Let J represent the associated three-by-three matrix, which has three eigenvalues denoted by λ_k ($k = 1, 2, 3$).

For simplification, let us assume the following:

$$\begin{cases} s_1 = a_1 - b_1 - c_1 + d_1 - e_1 + f_1 + g_1 - h_1 \\ s_2 = b_1 - d_1 - f_1 + h_1 \\ s_3 = c_1 - d_1 - g_1 + h_1 \\ s_4 = d_1 - h_1 \\ m_1 = a_2 - b_2 - c_2 + d_2 - e_2 + f_2 + g_2 - h_2 \\ m_2 = e_2 - f_2 - g_2 + h_2 \\ m_3 = b_2 - d_2 - f_2 + h_2 \\ m_4 = f_2 - h_2 \\ l_1 = a_3 - b_3 - c_3 + d_3 - e_3 + f_3 + g_3 - h_3 \\ l_2 = c_3 - d_3 - g_3 + h_3 \\ l_3 = e_3 - f_3 - g_3 + h_3 \\ l_4 = g_3 - h_3 \end{cases} \tag{27}$$

Then, Equation (24) can be simplified as follows:

$$\begin{cases} g_{\mathrm{SP}}(y,z) = s_1yz + s_2y + s_3z + s_4 \\ g_{\mathrm{MP}}(x,z) = m_1zx + m_2z + m_3x + m_4 \\ g_{\mathrm{LP}}(y,x) = l_1xy + l_2y + l_3x + l_4 \end{cases} \tag{28}$$

The corresponding Jacobian matrix (J) of the system is expressed as

$$\begin{aligned} J &= \begin{bmatrix} (1-2x)g_{\mathrm{SP}}(y,z) & x(1-x)\dfrac{\partial g_{\mathrm{SP}}(y,z)}{\partial y} & x(1-x)\dfrac{\partial g_{\mathrm{SP}}(y,z)}{\partial z} \\ y(1-y)\dfrac{\partial g_{\mathrm{MP}}(x,z)}{\partial x} & (1-2y)g_{\mathrm{MP}}(x,z) & y(1-y)\dfrac{\partial g_{\mathrm{MP}}(x,z)}{\partial z} \\ z(1-z)\dfrac{\partial g_{\mathrm{LP}}(x,y)}{\partial x} & z(1-z)\dfrac{\partial g_{\mathrm{LP}}(x,y)}{\partial y} & (1-2z)g_{\mathrm{LP}}(x,y) \end{bmatrix} \\ &= \begin{bmatrix} (1-2x)(s_1yz + s_2y + s_3z + s_4) & x(1-x)(s_1z + s_2) & x(1-x)(s_1z + s_3) \\ y(1-y)(m_1x + m_3) & (1-2y)(m_1zx + m_2z + m_3x + m_4) & y(1-y)(m_1x + m_2) \\ z(1-z)(l_1y + l_3) & z(1-z)(l_1x + l_2) & (1-2z)(l_1xy + l_2y + l_3x + l_4) \end{bmatrix} \end{aligned} \tag{29}$$

The set of all solutions formed by the internal equilibrium equations of the system is denoted as Ψ_{ESS}. These solutions can be categorized into four scenarios, as shown in Table 2.

Based on Table 2, we set the payoff distribution parameters as follows:

$$\begin{cases} a_1 = 24, b_1 = 24, c_1 = 24, d_1 = 2, e_1 = 24, f_1 = 24, g_1 = 24, h_1 = 5 \\ a_2 = 78, b_2 = 78, c_2 = 78, d_2 = 78, e_2 = 78, f_2 = 6, g_2 = 78, h_2 = 13 \\ a_3 = 120, b_3 = 120, c_3 = 120, d_3 = 120, e_3 = 120, f_3 = 120, g_3 = 4, h_3 = 8 \end{cases} \tag{30}$$

$$\begin{cases} s_1 = -3, s_2 = 3, s_3 = 3, s_4 = -3 \\ m_1 = -7, m_2 = 7, m_3 = 7, m_4 = -7 \\ l_1 = -4, l_2 = 4, l_3 = 4, l_4 = -4 \end{cases} \tag{31}$$

Table 2. Distribution of all the equilibrium points in Ψ_{ESS}.

Case 1	Case 2
$\begin{cases} x(1-x)=0 \\ y(1-y)=0 \\ z(1-z)=0 \end{cases}$	$\begin{cases} g_{\text{SP}}(y,z)=0 \\ g_{\text{MP}}(z,x)=0 \\ g_{\text{LP}}(y,x)=0 \end{cases}$
Case 3	
$\begin{cases} g_{\text{SP}}(y,z)=0 \\ y(1-y)=0 \\ z(1-z)=0 \end{cases}$ or $\begin{cases} x(1-x)=0 \\ g_{\text{MP}}(z,x)=0 \\ z(1-z)=0 \end{cases}$ or $\begin{cases} x(1-x)=0 \\ y(1-y)=0 \\ g_{\text{LP}}(y,x)=0 \end{cases}$	
Case 4	
$\begin{cases} x(1-x)=0 \\ g_{\text{MP}}(z,x)=0 \\ g_{\text{LP}}(y,x)=0 \end{cases}$ or $\begin{cases} g_{\text{SP}}(y,z)=0 \\ y(1-y)=0 \\ g_{\text{LP}}(y,x)=0 \end{cases}$ or $\begin{cases} g_{\text{SP}}(y,z)=0 \\ g_{\text{MP}}(z,x)=0 \\ z(1-z)=0 \end{cases}$	

Then, based on the parameter settings above, the evolutionary stability of each equilibrium point can be obtained, as summarized in Table 3.

Table 3. The evolutionary stability of each equilibrium point.

Ψ_{ESS}	Evolutionary Stability	Ψ_{ESS}	Evolutionary Stability
(0, 0, 0)	ESS	(1, 0, 1)	Unstable
(1, 0, 0)	Unstable	(1, 1, 0)	Unstable
(0, 1, 0)	Unstable	(0, 1, 1)	Unstable
(0, 0, 1)	Unstable	(1, 1, 1)	Unstable

Simulation results: The RD equations were implemented in MATLAB 2019b, with simulations performed at 0.2 s intervals over 50 iterations to analyze the evolutionary dynamics of a three-group system (SP, MP, LP). The results of the (x, y, z) phase trajectory are shown in Figure 5. The summary of results from Figure 5 is elaborated as follows.

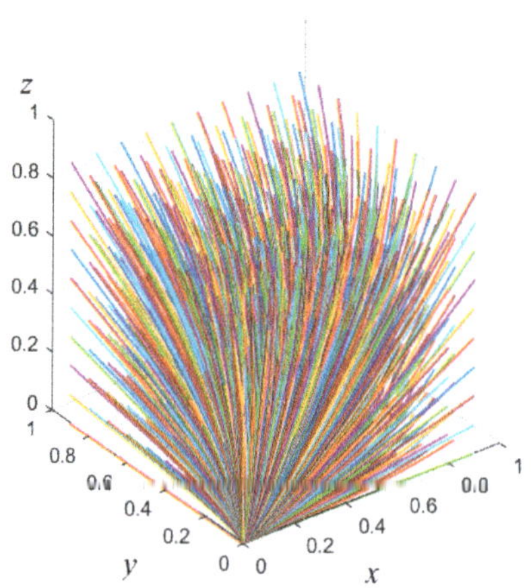

Figure 5. The phase trajectory simulation results of (x, y, z) for a typical asymmetric three-group two-strategy evolutionary game with a time step of 0.2 s.

(a) The system's phase trajectories converge only at (0, 0, 0): The simulation demonstrates that under the given parameter settings, the phase trajectories of the three groups converge exclusively at the equilibrium point (0, 0, 0). This equilibrium satisfies the conditions for an evolutionarily stable strategy (ESS), indicating that it is the only stable state in this specific scenario.

At (0, 0, 0), all three groups (SP, MP, LP) exclusively adopt the low-cost bidding strategy, leading to a uniform strategy profile. From a market perspective, this implies that all power producers opt for cost-efficient energy generation. However, this outcome may not be advantageous for producers who seek higher profits through aggressive bidding strategies.

As elaborated above, when high-price bidding strategies dominate (S_H), power producers achieve higher revenues at the expense of increased market prices. This scenario is suboptimal for consumers and policymakers, as it leads to higher electricity prices. Policymakers generally aim to align power generation costs with marginal prices to maintain market efficiency.

(b) Policy interventions to shift strategy adoption toward low-cost bidding: To encourage the adoption of low-cost strategies, regulators could implement penalty mechanisms for high-cost bidding strategies. For instance, modifying the payoff parameters to ensure $e_1<a_1$, $c_2<a_2$, $b_3<a_3$ would diminish the profits associated with high-cost strategies (S_H) while incentivizing low-cost bidding (S_L).

Specifically, penalty mechanisms can reduce the perceived benefits of aggressive bidding. This adjustment ensures that the system converges to the equilibrium point (0, 0, 0), where all groups consistently choose S_L.

Additionally, setting the coefficients s_4, m_4, l_4 to values below zero guarantees that the system stabilizes at (0, 0, 0). These parameters can be used as forecasting variables to assess whether the market is likely to exhibit aggressive competition among producers.

From a policy standpoint, regulators must pay close attention to the positivity of these penalty coefficients, as they directly influence the likelihood of achieving low-cost, cooperative equilibria. Monitoring these coefficients allows for more precise predictions of the market's evolution and potential strategic misalignments.

(c) The system exhibits 28 evolutionary scenarios across different parameters: The analysis reveals that the evolutionary stability of the system depends on the interplay between parameter settings and the eigenvalues associated with the Jacobian matrix (J). Across the 28 evolutionary scenarios considered, the eigenvalues' positivity or negativity determines whether a given equilibrium is evolutionarily stable.

Notably, the final stability of the system varies significantly with changes in payoff parameters. This variation underscores the importance of initial conditions and group-specific payoffs in determining the system's trajectory.

For example, an initial configuration favoring high-cost bidding may trap the system in suboptimal equilibria, necessitating targeted interventions to reorient the groups toward more cooperative outcomes. By carefully adjusting the initial payoff structure, it is possible to enhance the system's long-term stability while minimizing inefficiencies.

Overall, from the aspect of market dynamics and equilibria, the simulation highlights the sensitivity of equilibrium outcomes to changes in payoff parameters. By tuning these parameters, policymakers can effectively influence the strategic behaviors of power producers, ensuring that the system converges to desirable equilibria. For instance, the equilibrium at (0, 0, 0) corresponds to a highly cooperative market environment where all groups adopt low-cost strategies, minimizing overall electricity costs. From the aspect of policy design and implementation, introducing penalties for aggressive bidding strategies (S_H) can shift the equilibrium toward cooperative outcomes. However, such policies must be carefully calibrated to avoid excessive penalties that may disincentivize participation in the market altogether. Additionally, subsidies for low-cost strategies (S_L) could complement penalty mechanisms, further encouraging cost-efficient behavior among power producers. The simulation results from Figure 5 offer valuable insights into the dynamics of power generation bidding strategies within a three-group evolutionary game framework. By analyzing the conditions for evolutionary stability, this study provides a theoretical basis for designing market interventions that promote cooperative, cost-efficient equilibria. The findings underscore the importance of careful parameter calibration in achieving long-term market stability and efficiency, paving the way for future research on adaptive strategies and policy design in competitive electricity markets.

Future research directions: The findings suggest several avenues for future research. Incorporating stochastic elements into the RD could provide a more realistic representation of market volatility and uncertainty. Moreover, extending the model to include additional strategy options or heterogeneous player behaviors would enhance its applicability to complex electricity markets. Further analysis could also explore the long-term implications of different policy interventions on market stability, efficiency, and producer–consumer welfare.

3. Bidding Strategies for Power Generation Groups Based on Evolutionary Game Theory

3.1. Model Parameter Settings

This section utilizes the electricity market rules in East China as a case study to analyze the three-group evolutionary game model, particularly focusing on bidding mechanisms, transaction models, and carbon reduction strategies. The three groups—small producers (SP), medium producers (MP), and large producers (LP)—represent key players in the market, with generation capacities of 100 MW, 300 MW, and 500 MW, respectively. The model provides insights into how transaction models and bidding strategies play out among these groups. The parameters P_{SP}, P_{MP}, P_{LP} represent the bidding prices for each group, and the unit is measured in yuan per megawatt-hour (MWh).

The cost function for each group, along with bidding parameters, is illustrated in Table 4, showcasing the different strategies adopted by each group under varied market conditions. The unit costs are expressed as quadratic functions of the bidding volume. The generation groups submit bids across five capacity segments, with the minimum stable generation capacity for each group set as 50% of the rated capacity. The forecasted market-clearing price for electricity is assumed to be 480 yuan/MWh. Based on these assumptions and settings, Table 5 presents the calculated bidding strategies for small producers, medium producers, and large producers, which are influenced by transaction costs, market-clearing prices, and carbon reduction incentives.

Table 4. Bidding parameters for power generation groups.

Parameters	Group SP	Group MP	Group LP
Generation cost function (yuan/MWh)	$4800 + 170P_{SP} + 0.18P_{SP}^2$	$9600 + 136.8P_{MP} + 0.17P_{MP}^2$	$18760 + 118.6P_{LP} + 0.164P_{LP}^2$
Rated capacity (MW)	100	300	500
Minimum load capacity (MW)	50	150	250
Maximum controllable price (yuan/MWh)	480	480	480
Minimum controllable price (yuan/MWh)	275	244.9	234.6

The bidding strategies of the three groups—small producers (SP), medium producers (MP), and large producers (LP)—are influenced by several key factors such as market-clearing prices, generation costs, and capacity constraints. These strategies, both theoretical and practical, are outlined in Table 5 and analyzed in depth below:

1. Small Producers (SP)

Theoretical aspect: Small producers often face higher unit costs due to lower economies of scale, which directly influences their bidding strategy and their capacity to compete in carbon reduction efforts. Their bidding strategies focus on maintaining competitiveness at lower capacity levels by adopting more conservative pricing, which impacts both their market position and carbon reduction potential. The bidding strategy is influenced not only by the market-clearing price and production costs but also by the transaction models that account for carbon costs and environmental impact.

Practical aspect: In practice, small producers tend to align their bids with the market-clearing price, ensuring participation and minimizing risks associated with price volatility. This strategy also influences carbon reduction incentives as it ensures ongoing market participation, as seen in the price range for high-price bidding (SH) between 313 to 431 yuan/MWh and low-price bidding (SL) between 275 to 313 yuan/MWh. This strategy allows them to remain competitive while mitigating risks associated with price volatility.

Practical insight: As the market becomes more competitive, small producers may benefit from forming alliances with larger entities or adopting niche strategies focused on specific market segments or regional markets.

Table 5. Bidding strategies at 480 Yuan/MWh.

Group SP	Capacity Segment				
	(50 60]	(60 70]	(70 80]	(80 90]	(90 100]
High-price bidding strategy S_H	313	352	391	431	480
Low-price bidding strategy S_L	275	313	352	391	431
Group MP	**Capacity Segment**				
	(150 180]	**(180 210]**	**(210 240]**	**(240 270]**	**(270 300]**
High-price bidding strategy S_H	277	324	367	412	480
Low-price bidding strategy S_L	245	277	324	367	412
Group LP	**Capacity Segment**				
	(250 300]	**(300 350]**	**(350 400]**	**(400 450]**	**(450 500]**
High-price bidding strategy S_H	276	321	370	421	480
Low-price bidding strategy S_L	235	276	321	370	421

2. Medium Producers (MP)

Theoretical aspect: Medium producers strike a balance between cost efficiency, market positioning, and carbon reduction strategies, adapting their pricing in response to market shifts and regulatory frameworks. Their bidding strategy typically involves a wider range of pricing flexibility, allowing them to capture higher margins during favorable market conditions while remaining competitive in times of lower prices.

Practical aspect: The medium producers (MP) group exhibits versatility in their strategy, with high-price bids ranging from 277 to 480 yuan/MWh, reflecting a dynamic adjustment to both market conditions and carbon regulation incentives. The low-price bids from 244.9 to 367 yuan/MWh. Their ability to shift between these price ranges offers a strategic advantage, particularly when market conditions fluctuate.

Practical insight: Medium producers may adopt a dual strategy approach, utilizing both aggressive low-price bidding to gain market share and high-price bidding to optimize profit margins. Moreover, understanding demand fluctuations is crucial for adapting these strategies.

3. Large Producers (LP)

Theoretical aspect: Large producers (LP), benefiting from economies of scale, can offer lower prices at higher production capacities, ensuring they remain competitive in carbon-efficient bidding strategies. This strategic advantage is reflected in their ability to maintain competitive pricing even with large generation volumes. Their bidding strategy encompasses not only pricing but also market share maximization, with carbon reduction strategies often integrated into their long-term goals for profitability and market stability.

Practical aspect: Large producers exhibit a clear preference for lower-price bidding at higher capacities, with high-price bidding ranging from 292 to 528 yuan/MWh and low-price bidding from 210 to 292 yuan/MWh. Their ability to offer low prices at large capacities ensures they capture a larger share of the market.

Practical insight: Large producers are likely to dominate the market, but their strategic decisions, especially at high capacity levels, require careful balancing between maintaining competitiveness and optimizing revenue. As competition increases, LPs may be compelled to adopt more dynamic pricing models, especially with fluctuating market-clearing prices.

Analysis of Table 4: bidding parameters for power generation groups. Table 4 presents the cost functions, capacity constraints, and bidding limits for the three groups:

Small Producers (SP):

- Cost function: $4800 + 170P_{\mathrm{SP}} + 0.18P_{\mathrm{SP}}^2$;
- Rated capacity: 100 MW;
- Bidding price range: 275–431 yuan/MWh;
- Notably, small producers exhibit lower cost coefficients, making them more competitive at lower capacity levels.

Medium Producers (MP):

- Cost function: $9600 + 1136.8P_{\mathrm{MP}} + 0.17P_{\mathrm{MP}}^2$;
- Rated capacity: 300 MW;
- Bidding price range: 244.9–492 yuan/MWh;
- Medium producers balance between cost efficiency and scalability, providing flexible pricing strategies.

Large Producers (LP):

- Cost function: $18760 + 118.6P_{\mathrm{LP}} + 0.164P_{\mathrm{LP}}^2$;
- Rated capacity: 500 MW;
- Bidding price range: 210–528 yuan/MWh;
- Large producers demonstrate higher fixed costs but achieve economies of scale at higher capacities.

This setup highlights the heterogeneity of the groups, considering differences in generation scale, cost structure, and their respective bidding flexibility and carbon mitigation approaches.

Table 5 provides the detailed bidding strategies for the three groups when the market-clearing price is set at 480 yuan/MWh. Each group submits bids across five capacity segments, representing a range of potential production levels as follows.

Small Producers (SP):

- Capacity Segments: 50–60 MW, 60–70 MW, 70–80 MW, 80–90 MW, 90–100 MW.
- High-price bidding (S_{H}) ranges from 313 to 431 yuan/MWh, while low-price bidding (S_{L}) ranges from 275 to 313 yuan/MWh.
- Observations: Small producers tend to adopt conservative bidding strategies, aligning closely with the forecasted clearing price. Their bidding behavior indicates sensitivity to marginal cost increases at higher capacity levels.

Medium Producers (MP):

- Capacity Segments: 210–240 MW, 240–270 MW, 270–300 MW.
- High-price bidding (S_H) ranges from 367 to 492 yuan/MWh, while low-price bidding (S_L) ranges from 244.9 to 367 yuan/MWh.
- Observations: Medium producers exhibit a wider range of bidding prices compared to small producers. Their bidding strategies suggest flexibility in adapting to market conditions, balancing between competitiveness and profitability.

Large Producers (LP):

- Capacity Segments: 250–300 MW, 300–350 MW, 350–400 MW, 400–450 MW, 450–500 MW.
- High-price bidding (S_H) ranges from 292 to 528 yuan/MWh, while low-price bidding (S_L) ranges from 210 to 292 yuan/MWh.
- Observations: Large producers consistently offer lower prices at higher capacity levels due to economies of scale. Their strategic focus is on maintaining competitiveness while achieving high capacity utilization.

Based on the above, some critical insights and strategic implications are elaborated as follows.

Market dynamics: The bidding strategies reflect the interplay between cost structures, market positioning, and transaction models, which integrate carbon costs and regulatory factors that influence bidding behavior. Small producers focus on niche segments, medium producers leverage their flexibility in adapting bidding strategies, and large producers dominate due to economies of scale, with implications for carbon reduction and market fairness.

Policy implications: This analysis stresses the need for regulatory oversight, particularly to ensure fair competition and to foster bidding strategies that are both market-efficient and aligned with carbon reduction goals.

Future research directions: Incorporating stochastic elements, such as demand uncertainty and renewable energy variability, can enhance the model's realism, while also providing insights into carbon mitigation and evolving market dynamics. Analyzing the impact of policy mechanisms (e.g., carbon pricing, subsidies) on bidding strategies would provide valuable insights for sustainable market design.

Overall, the parameter settings and bidding strategies outlined in Tables 4 and 5 provide a comprehensive framework for understanding the strategic behaviors of heterogeneous power generation groups in a competitive electricity market. By linking cost structures to bidding strategies, this study demonstrates the importance of aligning market rules with the objectives of efficiency, fairness, and sustainability. Future research should focus on integrating real-world complexities to further refine the model's predictive accuracy and policy relevance.

3.2. *Bidding Strategies for Power Generation Groups*

3.2.1. Market Oversupply

1. **Parameter settings**

When the market demand is no less than the total rated generation capacity of the three groups, totaling 900 MW, each group is capable of selling its entire generation output. Under conditions without government regulation, the profit matrices of the generation groups are presented in Table 6.

Table 6. Benefit matrix for unregulated market conditions when the clearing price is set at 480 yuan/MWh.

Group SP and MP				LP	
				S_H	S_L
SP	S_H	MP	S_H	(24400, 78060, 120940)	(24400, 78060, 120940)
			S_L	(24400, 78060, 120940)	(24400, 78060, 120940)
	S_L		S_H	(24400, 78060, 120940)	(24400, 78060, 120940)
			S_L	(24400, 78060, 120940)	(5900, 13530, 8200)

Table 6 provides the bidding benefit matrix when the clearing price is set at 480 yuan/MWh under conditions of no government regulation, which is shown as follows.

Small producers (SP): Whether adopting high-cost bidding (S_H) or low-cost bidding (S_L), their payoffs are equal, demonstrating that all generation output is sold regardless of the bidding strategy.

Medium producers (MP): Similar to SP, the payoff for MP remains unchanged regardless of the bidding strategy. Their benefit is equally distributed across S_H and S_L, as demand is sufficient to absorb the total capacity of the market.

Large producers (LP): While high-cost bidding yields uniform benefits across scenarios, low-cost bidding (S_L) generates slightly reduced payoffs for LP due to its large-scale generation and cost inefficiencies at lower prices.

2. **RD equations**

These results highlight that, under unregulated conditions, there is no incentive for any group to alter their bidding behavior as the oversupply guarantees full dispatch of their capacities. Thus, the corresponding payment parameters under oversupply conditions are detailed as follows:

$$\begin{cases} a_1 = 24400, b_1 = 24400, c_1 = 24400, d_1 = 24400, \\ e_1 = 24400, f_1 = 24400, g_1 = 24400, h_1 = 5900, \\ a_2 = 78060, b_2 = 78060, c_2 = 78060, d_2 = 78060, \\ e_2 = 78060, f_2 = 78060, g_2 = 78060, h_2 = 13530, \\ a_3 = 120940, b_3 = 120940, c_3 = 120940, d_3 = 120940, \\ e_3 = 120940, f_3 = 120940, g_3 = 120940, h_3 = 8200, \end{cases} \tag{32}$$

By substituting these payment parameters into the RD equation, we derive the following:

$$\begin{cases} h_{SP}(x) = x(1-x)g_{SP}(y,z) \\ h_{MP}(y) = y(1-y)g_{MP}(x,z) \\ h_{LP}(z) = z(1-z)g_{LP}(x,y) \end{cases} \tag{33}$$

$$\begin{cases} g_{SP}(y,z) = 18500yz + (-18500)y + (-18500)z + 18500 \\ g_{MP}(x,z) = 64530zx + (-64530)z + (-64530)x + 64530 \\ g_{LP}(y,x) = 112740xy + (-112740)y + (-112740)x + 112740 \end{cases} \tag{34}$$

3. **Equilibrium stability analysis**

Using the parameters from Table 6, the stability of equilibrium points is analyzed based on evolutionary stability conditions. The results are summarized in Table 7, which shows the stability of all internal equilibrium points (Ψ_{ESS}): None of the eight equilibrium points, including (0, 0, 0), (1, 0, 0), (0, 1, 0), (1, 1, 0), etc., are evolutionarily stable under unregulated conditions. This indicates that the system lacks a stable evolutionary trajectory and is unable to reach a state of balance.

Table 7. Evolutionary stability analysis of each internal equilibrium point for unregulated market conditions when the clearing price is set at 480 yuan/MWh.

Ψ_{ESS}	Evolutionary Stability	Ψ_{ESS}	Evolutionary Stability
(0, 0, 0)	Unstable	(1, 0, 1)	Unstable
(1, 0, 0)	Unstable	(1, 1, 0)	Unstable
(0, 1, 0)	Unstable	(0, 1, 1)	Unstable
(0, 0, 1)	Unstable	(1, 1, 1)	Unstable

4. Simulation study

Figure 6 presents the results of the RD simulation conducted in MATLAB 2019b, with a time step of 0.1 s and 50 iterations. The x-y-z phase trajectories are visualized under the unregulated condition where the market-clearing price is set at 480 yuan/MWh. The results are analyzed as follows.

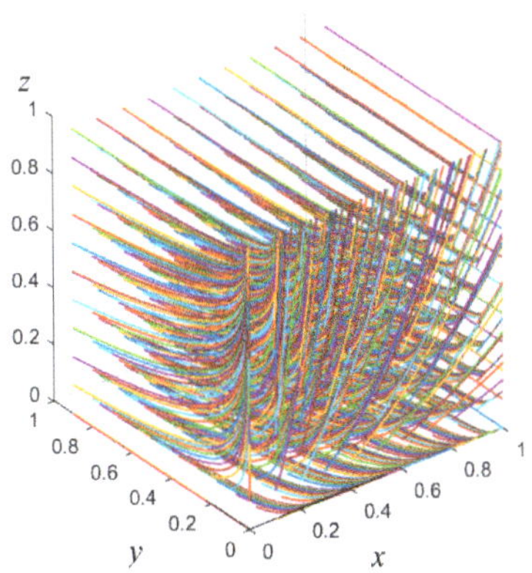

Figure 6. The RD simulation of the bidding evolution game under unregulated market conditions when the clearing price is set at 480 yuan/MWh.

Absence of stable equilibrium: The simulation demonstrates that in the absence of government regulation, the system fails to converge to any equilibrium. Instead, phase trajectories exhibit chaotic or divergent behavior.

Profit maximization through high-cost bidding: All power producers, regardless of group size, tend to pursue high-cost bidding (S_H) to maximize their individual payoffs. This is driven by the market-clearing price mechanism, which incentivizes aggressive bidding strategies.

Implications for arket dynamics: The lack of equilibrium reflects the inefficiencies of an unregulated market, where producers prioritize individual profits over collective market stability. This behavior leads to inflated electricity prices, reducing consumer welfare and undermining the efficiency of the electricity market.

Aiming at Figure 6, critical insights and policy recommendations are elaborated from several aspects as follows.

Market oversight is crucial: Figure 6 underscores the necessity of government regulation in electricity markets. Without oversight, aggressive bidding behavior dominates, leading to price distortion and market inefficiencies.

Role of penalty mechanisms: Introducing penalties for high-cost bidding or incentivizing low-cost bidding strategies could guide the system toward more stable equilibria. For example, adjusting payoff parameters such that $g_{SP}(y, z)$, $g_{MP}(x,z)$, $g_{LP}(x, y) \leq 0$ could shift the system dynamics toward cooperative outcomes.

Future research directions: The integration of renewable energy sources or stochastic demand variations can further enrich the model and provide deeper insights into how these dynamics affect carbon strategies and market performance. Additionally, analyzing the long-term effects of various regulatory policies would enhance the model's applicability to diverse market scenarios.

The findings from Figure 6 and the corresponding RD analysis highlight the limitations of unregulated electricity markets. Without intervention, the market fails to achieve equilibrium, and aggressive bidding strategies prevail, leading to higher electricity costs. To address these challenges, policymakers must design mechanisms that balance individual incentives with collective welfare, ensuring a stable and efficient electricity market.

As analyzed above, Figure 6 provides a dynamic simulation of the evolutionary game under government regulation in a power market with a clearing price of 480 Yuan/MWh. It clearly demonstrates that the introduction of government oversight fundamentally shifts the system's behavior. Unlike the unregulated scenario (Figure 6), which lacked equilibrium points and promoted bidding strategies aimed solely at maximizing profit by exploiting high-price strategies, government intervention leads to a stable evolutionary equilibrium. Under government regulation, the system converges to evolutionary stable equilibria (ESE) at (1, 1, 1), where all groups adopt low-price strategies, ensuring market fairness and efficiency while promoting carbon reduction. This result is significant because it indicates that all three market participants—SP (small producers), MP (medium producers), and LP (large producers)—eventually adopt the low-price strategy, achieving an optimal and stable outcome. The intervention ensures that the participants refrain from consistently engaging in high-price bidding, as the penalties and costs associated with such actions outweigh the potential benefits. This outcome aligns with the regulatory aim of ensuring market efficiency and discouraging manipulative bidding practices. It illustrates that when penalties are carefully calibrated, the inherent game-theoretic dynamics can steer market participants toward socially optimal strategies. Specifically, the penalties encourage producers to prioritize market stability and fairness over short-term profit maximization.

From Figure 6, it can be observed that in the absence of government supervision, there is no equilibrium point in the market. Each power generation company can sell all of its electricity at high prices. In practice, such a situation leads to power generation groups deliberately inflating electricity prices. Under the market-clearing price (MCP) mechanism, all companies intentionally bid high prices to raise the market-clearing price and maximize their profits. This outcome is a direct result of supply shortages and the lack of necessary oversight.

Obviously, based on the elaborations above, the absence of government regulation, as illustrated in Figure 6, is a critical issue that underscores a key flaw in the unregulated power market. The lack of a stable equilibrium when firms are free to manipulate their bidding strategies leads to a situation where all power generation groups push for higher prices, resulting in market inefficiency. This scenario exemplifies a classic problem in economic theory known as "market failure", where the absence of appropriate oversight allows companies to exploit the system for maximum profit, exacerbating the effects of supply shortages. Under these conditions, the MCP becomes excessively high, which, although benefiting large corporations in the short term, ultimately destabilizes the entire market.

Therefore, the introduction of government supervision alters the system's behavior, ensuring that market dynamics and bidding strategies are aligned with long-term sustainability and carbon reduction goals. The proposed penalty mechanism acts as a deterrent to excessive pricing behavior. By incorporating the costs of violations into the power generation companies' decision-making processes, the government forces companies to reassess the long-term profitability of their pricing strategies. When the financial gains from charging higher prices become outweighed by the penalties imposed for such actions, the companies are incentivized to lower their bids, thereby fostering more equitable market conditions.

Furthermore, the introduction of such regulatory measures ensures that the system tends toward a stable equilibrium, specifically at (1, 1, 1), where all participants are operating under reasonable pricing strategies. This stands in contrast to the unregulated market's tendency to spiral towards (0, 0, 0), an equilibrium point where no sustainable strategy exists, thus destabilizing the system in the long run.

The mathematical representation $e_1 = 0.5a_1$, $c_2 = 0.5a_2$, $b_3 = 0.5a_3$ simplifies the calculation of these equilibria, providing a convenient approximation to study the effects of regulatory oversight on the market. This adjustment also facilitates the understanding of how different payment parameters influence the dynamic interactions between firms under government supervision. The model suggests that with proper regulatory mechanisms, such as penalties for non-compliance, the market can stabilize and move toward a more efficient and sustainable equilibrium.

Based on the above, the practical implications and policy recommendations are summarized from several aspects as follows. This analysis highlights the importance of government intervention in preventing market manipulation and fostering a more competitive and fair energy market. From a policy perspective, the findings suggest several critical insights:

- Regulation and oversight: Government involvement is essential to curb the detrimental effects of unregulated high-price bidding strategies. Without oversight, power generation firms are likely to engage in price manipulation, harming consumers and destabilizing the market.
- Penalty mechanisms: Implementing a penalty system is a robust approach to discourage excessive pricing. When the cost of violating market norms is factored into decision-making, companies will naturally tend to adopt more reasonable, competitive pricing strategies.
- Stability through low-price strategies: The introduction of government regulation enables a shift toward low-price strategies, which benefits consumers and ensures market stability. This approach can help balance the interests of both large and small power generation firms, mitigating the risk of monopolistic behavior.
- Dynamic market adjustments: The mathematical modeling of this system, particularly through the payment parameters, provides a quantitative understanding of how regulatory measures affect market equilibrium. It also offers a foundation for further exploration into how specific parameters can be adjusted to optimize market performance.

In summary, the research in Section 3.2.1 underscores the need for proactive government regulation in energy markets, especially in scenarios where supply constraints lead to price inflation. By implementing well-designed penalty mechanisms and fostering a competitive bidding environment, the government can steer the market towards a stable and sustainable equilibrium, benefiting both producers and consumers in the long term. However, if government supervision is introduced, and a penalty mechanism is applied when malicious high pricing trends emerge, it forces the power generation groups to account for the costs of violations in their profit considerations. When the profit from high pricing is less than the profit from low pricing, power generation companies tend to adopt a low-price strategy. This government regulation ensures that the payment parameters meet the conditions $e_1 < a_1$, $c_2 < a_2$, and $b_3 < a_3$, and it is necessary to avoid situations where the warning parameters s_4, m_4, l_4 are negative. The system will ultimately converge to the stable equilibrium point (1, 1, 1), while it will be impossible to achieve long-term evolutionary stability at the equilibrium point (0, 0, 0). For the convenience of calculation, the model assumes the following relations for the payment parameters: $e_1 = 0.5a_1$, $c_2 = 0.5a_2$, and $b_3 = 0.5a_3$.

Therefore, based on the conditions above, new payoff parameters are calculated as follows:

$$\begin{cases} a_1 = 24400, b_1 = 24400, c_1 = 24400, d_1 = 24400, \\ e_1 = 12200, f_1 = 24400, g_1 = 24400, h_1 = 5900, \\ a_2 = 78060, b_2 = 78060, c_2 = 39030, d_2 = 78060, \\ e_2 = 78060, f_2 = 78060, g_2 = 78060, h_2 = 13530, \\ a_3 = 120940, b_3 = 60470, c_3 = 120940, d_3 = 120940, \\ e_3 = 120940, f_3 = 120940, g_3 = 120940, h_3 = 8200, \end{cases} \tag{35}$$

Based on Equation (35), the new RD equations are shown as

$$\begin{cases} h_{\mathrm{SP}}(x) = x(1-x)g_{\mathrm{SP}}(y,z) \\ h_{\mathrm{MP}}(y) = y(1-y)g_{\mathrm{MP}}(x,z) \\ h_{\mathrm{LP}}(z) = z(1-z)g_{\mathrm{LP}}(x,y) \end{cases} \tag{36}$$

$$\begin{cases} g_{\mathrm{SP}}(y,z) = 30700yz + (-18500)y + (-18500)z + 18500 \\ g_{\mathrm{MP}}(x,z) = 103560zx + (-64530)z + (-64530)x + 64530 \\ g_{\mathrm{LP}}(y,x) = 173210xy + (-112740)y + (-112740)x + 112740 \end{cases} \tag{37}$$

Then, we can obtain the stability of equilibrium points with government supervision and market-clearing price of 480 RMB/MWh, as shown in Table 8.

Table 8. The stability of equilibrium points with government supervision and market-clearing price of 480 RMB/MWh.

Ψ_{ESS}	Evolutionary Equilibrium Stability	Ψ_{ESS}	Evolutionary Equilibrium Stability
(0, 0, 0)	Unstable	(1, 0, 1)	Unstable
(1, 0, 0)	Unstable	(1, 1, 0)	Unstable
(0, 1, 0)	Unstable	(0, 1, 1)	Unstable
(0, 0, 1)	Unstable	(1, 1, 1)	ESS

Table 8 presents the stability of equilibrium points under the conditions where the market-clearing price (MCP) is set to 480 RMB/MWh, with the assumption that government supervision is in place. The equilibrium points (denoted as ESS) reflect the tendency of the system toward stability when the evolutionary game dynamics are in play. Notably, the table shows that, without government intervention, the system remains unstable in most scenarios, suggesting that power generation groups are likely to engage in aggressive pricing strategies that lead to competition that doesn't reach a stable equilibrium.

However, with the implementation of government supervision, an ESS at (1, 1, 1) emerges, which is an indicator that the system tends toward a balanced state. The ESS reflects a scenario where power generation companies would converge towards a mutually beneficial low-price strategy due to the enforcement of penalties for overpricing. The government's role here is pivotal in steering the system toward stability. Based on Table 8, we conduct a simulation study to verify this scenario, as illustrated in Figure 7. This figure demonstrates the evolutionary simulation results under government supervision with market-clearing price of 400 RMB/MWh. In this figure, we visualize the evolutionary simulation results under the condition where government supervision is introduced. The simulation results are based on a time interval of 0.1 s and are run for 50 iterations using MATLAB 2019b.

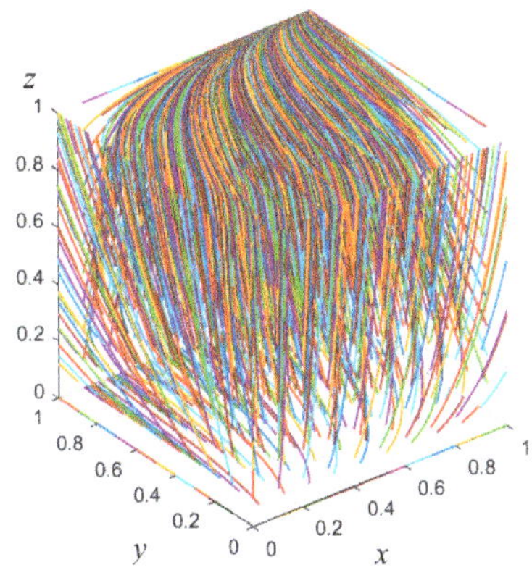

Figure 7. Evolutionary simulation results under government supervision with market-clearing price of 480 RMB/MWh.

From Figure 7, we also observe that the system, which was unstable under market-clearing conditions without government intervention, stabilizes once penalties are introduced for excessive pricing behavior. In particular, all three groups (Group SP, Group MP, and Group LP) tend toward the low-price strategy, reaching the ESS. This suggests that government regulation plays a critical role in ensuring fair competition in the electricity market. In the absence of such regulation, companies would likely engage in price manipulation, driving up the market-clearing price. The introduction of penalties for excessive pricing ensures that companies are incentivized to adopt strategies that benefit the market as a whole, rather than focusing on individual profits.

Based on Figure 7, we can further summarize some important findings, as discussed from three aspects as follows.

First, with respect to the evolutionary stability in power generation bidding, the concept of evolutionary stability is crucial when applying game theory to power generation bidding strategies. Under the assumption of competitive electricity markets, companies (or power generation groups) must continually adjust their strategies based on the actions of others, with the goal of maximizing profits. The evolutionary game framework used in this study accounts for the dynamic interactions between these players. In a deregulated electricity market, without supervision, companies have an incentive to bid aggressively, inflating prices and maximizing their revenue. However, this often leads to inefficiency, where the market price exceeds the marginal cost of electricity generation, leading to a socially suboptimal outcome. Figure 7 illustrate how, under government supervision, the system evolves towards a stable equilibrium where companies adopt lower bidding prices. This ensures a more competitive and fair pricing mechanism, ultimately benefiting consumers and promoting overall market efficiency.

Second, with respect to the impact of government supervision on market stability, from a broader perspective, the analysis demonstrates the importance of government supervision in maintaining market stability. In real-world electricity markets, when players (power generation companies) are left unchecked, they can collectively influence the market in ways that reduce overall welfare. The concept of the ESS in this context shows that the system can be stabilized through interventions such as penalties for overbidding or price manipulation. The critical finding here is that the implementation of government supervision not only ensures market efficiency but also facilitates a self-regulating mechanism among the firms. By penalizing firms that engage in excessive pricing strategies, the government forces them to internalize the costs of their actions. This leads to the adoption of lower price strategies, where firms realize that the long-term benefits of competitive pricing outweigh the short-term gains from overpricing.

Third, with respect to the implications and practical applications, the results of the evolutionary game analysis can be directly applied to the design of electricity market regulations. Specifically, regulators can implement policies that ensure firms adopt pricing strategies that align with the social welfare maximization objectives. This could include the following:

- Penalties for price manipulation: Ensuring that firms are not able to manipulate prices upwards without consequences.
- Incentives for low-price bidding: Encouraging firms to adopt low-price strategies that foster competition and consumer welfare.
- Market monitoring systems: Implementing systems that track and analyze the pricing behavior of firms to detect and prevent anti-competitive behavior.

The findings of this research also highlight the need for balancing market competition with support for smaller players in the market. As shown in the simulations, large-capacity firms (such as Group LP in this study) have a significant advantage due to their lower costs and larger market share. Without government intervention, smaller firms may struggle to compete effectively, leading to market monopolization. Therefore, policymakers should ensure that smaller firms receive the necessary support, possibly through subsidies or preferential treatment, to foster a more diverse and competitive market landscape.

Overall, the EGT approach to modeling the bidding strategies of power generation groups has provided valuable insights into the dynamics of electricity market competition. The findings from the simulations, especially the role of government supervision in stabilizing the market, emphasize the necessity of regulatory interventions to ensure that market behavior remains competitive and efficient. This research has practical implications for energy market regulators, who can use these insights to refine pricing mechanisms and competitive strategies. By leveraging evolutionary game theory, regulators can anticipate and mitigate market inefficiencies, fostering a fairer and more sustainable energy market for all stakeholders. Moreover, the findings underscore the importance of adaptive and flexible regulatory frameworks that can respond to the ever-changing dynamics of the power generation industry. Future research should continue to explore multi-strategy evolutionary models that can account for the complexities of real-world energy markets, including the role of renewable energy sources, technological innovations, and global energy trends.

3.2.2. Market Oversupply (Market Demand Decreases by 20%)

When market demand reduces to 720 MW, and there is no governmental oversight, the bid volumes and revenue volumes for each group are shown in Tables 9 and 10. From Table 10, the payment parameters when market demand is reduced to 720 MW are derived as follows:

$$\begin{cases} a_1 = 18448, b_1 = 10606, c1 = 15031, d_1 = 0, \\ e_1 = 23900, f_1 = 24400, g_1 = 22900, h_1 = 0, \\ a_2 = 61776, b_2 = 39100, c_2 = 73560, d_2 = 8154, \\ e_2 = 59712, f_2 = 26136, g_2 = 73560, h_2 = 8154, \\ a_3 = 97560, b_3 = 110940, c_3 = 82390, d_3 = 8390, \\ e_3 = 94834, f_3 = 110940, g_3 = 75294, h_3 = 8390, \end{cases} \quad (38)$$

Table 9. Cleared market price at 480 CNY/MWh - bid volumes.

Ψ_{ESS}	Bid Volume	Ψ_{ESS}	Bid Volume
(0, 0, 0)	(80, 240, 400)	(1, 0, 1)	(100, 120, 500)
(1, 0, 0)	(100, 232, 388)	(1, 1, 0)	(100, 300, 320)
(0, 1, 0)	(70, 300, 350)	(0, 1, 1)	(0, 220, 500)
(0, 0, 1)	(55, 165, 500)	(1, 1, 1)	(0, 220, 500)

Table 10. Revenue matrix for each group at a market-clearing price of 480 CNY/MWh.

Group SP and MP				LP	
				S_H	S_L
SP	S_H	MP	S_H	(18448, 61776, 97560)	(10606, 39100, 110940)
			S_L	(15031, 73560, 82390)	(0, 8154, 8390)
	S_L		S_H	(23900, 59712, 94834)	(22400, 26136, 110940)
			S_L	(22900, 73560, 75294)	(0, 8154, 8390)

By substituting these parameters into the RD equation, which is obtained as follows:

$$\begin{cases} h_{SP}(x) = x(1-x)g_{SP}(y,z) \\ h_{MP}(y) = y(1-y)g_{MP}(x,z) \\ h_{LP}(z) = z(1-z)g_{LP}(x,y) \end{cases} \tag{39}$$

$$\begin{cases} g_{SP}(y,z) = 16211yz + (-13794)y + (-7869)z \\ g_{MP}(x,z) = (-10900)zx + (-31830)z + 12964x + 17982 \\ g_{LP}(y,x) = (-4370)xy + 7096y + (-83010)x + 66904 \end{cases} \tag{40}$$

By incorporating these payment parameters into the evolutionary stability conditions, the stability of each equilibrium point is determined, as shown in Table 11.

Table 11. Stability of equilibrium points at a market-clearing price of 480 CNY/MWh.

Ψ_{ESS}	Evolutionary Stability	Ψ_{ESS}	Evolutionary Stability
(0, 0, 0)	Unstable	(1, 0, 1)	Unstable
(1, 0, 0)	Unstable	(1, 1, 0)	Unstable
(0, 1, 0)	ESS	(0, 1, 1)	Unstable
(0, 0, 1)	ESS	(1, 1, 1)	Unstable

By inputting the RD equation into MATLAB 2019b, and running 50 dynamic simulations with an interval of 0.1 s, the results are plotted in Figure 8. This figure demonstrates that the system reaches an ESS under two specific conditions: when the medium-capacity group adopts a low-price strategy while the other two groups adopt high-price strategies; and when the large-capacity group adopts a low-price strategy while the other groups adopt high-price strategies.

The equilibrium outcome is largely determined by the strategic choices of the large-capacity group. This is attributed to its advantages in capacity, lower operational costs, and significant market share. Without governmental regulation, such a dominant position can manipulate the bidding market, potentially leading to monopolistic practices and other forms of destructive competition. Based on this, from the simulation results presented in Figure 8, the following observations can be made:

(a) Influence of Large-Capacity and Medium-Capacity Groups on Market Stability

The findings highlight the critical role of the large-capacity (LP) group in determining market dynamics. As the primary market leader, the LP group possesses inherent advantages in volume, cost efficiency, and market share. These factors enable it to exert significant influence over the bidding outcomes. The MCP mechanism amplifies this advantage, as the

LP group can dictate price levels based on its strategic choices. Without external oversight, such dominance poses serious risks, including the potential for monopolistic behavior or aggressive predatory pricing strategies. This not only destabilizes the bidding environment but can also lead to long-term consequences, such as a reduction in market diversity and the marginalization of smaller participants. The study underscores the importance of designing regulatory frameworks to mitigate these risks, ensuring a more balanced and competitive market environment. Implications are as follows:

- Policy need: Regulatory bodies must monitor large-capacity groups to prevent exploitative behaviors. This may involve implementing price caps, market share limits, or stricter anti-monopoly measures.
- Strategic flexibility: The LP group's ability to switch between high-price and low-price strategies provides a significant edge. Encouraging transparency in strategy selection could reduce market manipulation risks.

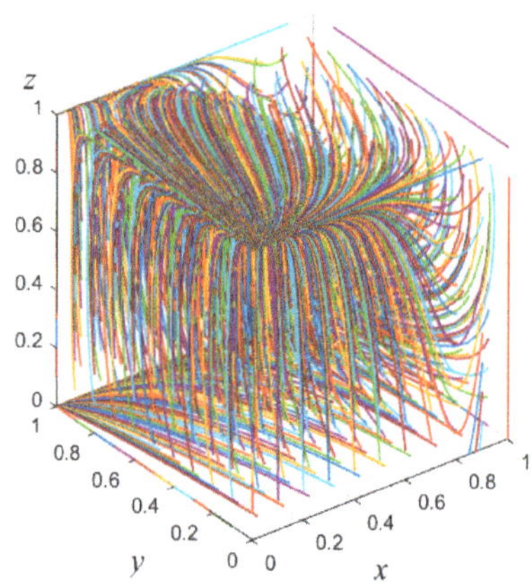

Figure 8. Evolutionary simulation results for three groups (Interval: 0.1s).

(b) Challenges Faced by Small-Capacity Groups in Competitive Markets

When the two larger-capacity groups simultaneously adopt low-price strategies, the small-capacity group faces insurmountable challenges. The fierce competition at lower price levels leaves the small-capacity group unable to secure contracts, primarily due to its higher generation costs. This dynamic illustrates the inherent structural disadvantage of smaller players in a deregulated electricity market.

The small-capacity group's exclusion from market participation has broader implications for market efficiency and equity. Over time, the lack of governmental support could lead to the elimination of smaller players, resulting in market consolidation. Such outcomes not only harm competition but may also reduce market resilience against supply disruptions. Recommendations for governmental intervention are elaborated as follows:

Support for small-capacity groups: Provide subsidies or tax incentives to reduce their operational costs and improve competitiveness. Establish quota mechanisms to ensure smaller participants secure a minimum share of contracts.

Balancing low-price competition: While promoting low-price strategies enhances consumer welfare and market efficiency, it is essential to implement safeguards that prevent the complete marginalization of smaller players. Introduce adaptive policies that adjust group revenue parameters dynamically based on market conditions, guiding the system toward more equitable outcomes.

Dynamic policy frameworks: Ensure that market adjustments are data-driven, using warning parameters such as s_4, m_4, l_4 as benchmarks to prevent imbalances. Allow for policy flexibility to address specific challenges faced by different groups while maintaining overall market stability.

Broader insights: Supporting smaller-capacity groups fosters market diversity and innovation, as they often bring unique approaches to energy generation. The absence of smaller players could reduce competition, leading to higher prices and less incentive for innovation in the long term.

(c) Critical Stability of the (1, 1, 1) Equilibrium

The (1, 1, 1) equilibrium is identified as a critically stable point, where the eigenvalues of the corresponding equilibrium are all zero. This critical stability highlights a key limitation of the current model: it does not achieve strict evolutionary stability. Consequently, the system is highly sensitive to external disturbances or perturbations, which could drive the market away from this equilibrium. This finding raises important questions about the robustness of the model and its applicability in real-world scenarios. In practice, market dynamics are often influenced by stochastic factors such as demand fluctuations, policy changes, and technological advancements. The absence of strict stability Implies that the market may oscillate or diverge from equilibrium in response to these factors. To this end, suggested model enhancements are elaborated as follows.

Incorporation of stochastic dynamics: Introduce random perturbations into the RD equation to simulate real-world uncertainties. Analyze how external shocks (e.g., sudden demand changes or policy shifts) impact the stability of the system.

Refinement of stability criteria: Use alternative stability concepts, such as asymptotic stability or Lyapunov functions, to provide a more comprehensive assessment of equilibrium behavior. Explore the use of non-linear dynamics to capture complex interactions between groups.

Simulation of real-world scenarios: Conduct simulations under varying market conditions (e.g., changes in demand, entry of new players) to evaluate the robustness of the (1, 1, 1) equilibrium. Assess the long-term behavior of the system under different policy interventions.

The analysis provides valuable insights into the dynamics of deregulated electricity markets. Key takeaways include the following:

Dominance of large-capacity groups: The LP group's strategic choices significantly influence market outcomes, necessitating robust regulatory oversight to prevent market distortions.

Vulnerability of small-capacity groups: Without targeted support, small-capacity groups are at risk of being excluded from the market, which could reduce competition and innovation.

Limitations of stability assumptions: The critical stability of the (1, 1, 1) equilibrium highlights the need for more sophisticated modeling approaches to capture real-world complexities.

Furthermore, when both the large-capacity and medium-capacity groups opt for low-price strategies, the small-capacity group faces severe disadvantages. Regardless of its strategy, the small group cannot secure contracts due to its higher production costs and inability to compete in an aggressive low-price environment. This dynamic highlights the vulnerability of smaller entities in deregulated markets.

(1) Role of Governmental Intervention

In such market conditions, governmental intervention becomes crucial. Specifically, Ppolicies should be enacted to support smaller-capacity generators, ensuring their survival in highly competitive low-price markets. Simultaneously, policies must encourage all groups to adopt low-price strategies to promote market efficiency. However, these interventions should be tailored to preserve diversity among generators and avoid excessive market dominance by any single group. One potential approach involves adjusting revenue parameters such that smaller-capacity groups can compete on a more equitable basis. For instance, subsidy mechanisms or preferential access to market opportunities could be explored.

(2) Stability of the Equilibria

The simulation identifies some equilibria as "critically stable" rather than strictly stable, due to the fact that the characteristic roots of the corresponding equilibrium points are zero. This does not satisfy the strict criteria for evolutionary stability as defined in game theory. As a result, the system is vulnerable to external perturbations and may oscillate rather than settle into a robust equilibrium. Enhancements to the model, such as incorporating stochastic dynamics or refining payoff structures, could provide a more realistic representation of market behavior.

(3) Research Implications and Practical Applications

This study yields several significant insights:

- Strategic behavior and market stability: The strategic choices of dominant players significantly influence market stability. In particular, large-capacity groups possess the leverage to dictate market outcomes, underscoring the importance of designing mechanisms to limit their undue influence.
- Policy recommendations: Policymakers should aim to: Support smaller-capacity generators to prevent market concentration. Use dynamic pricing regulations or incentives to discourage monopolistic behaviors. Adjust market parameters (e.g., subsidies, quotas) to guide the market toward more equitable and sustainable equilibria.
- Model refinements: The findings indicate the need for enhanced modeling approaches to capture real-world complexities. For example: Introducing stochastic elements into the RD equation can account for random fluctuations in market conditions. Analyzing the impact of external shocks, such as sudden demand surges or regulatory changes, could provide deeper insights into market dynamics.
- Broader applications: While this study focuses on electricity markets, its implications extend to other sectors characterized by competitive bidding dynamics, such as telecommunications or resource allocation in decentralized networks.

(4) Suggestions for Further Improvement

To advance the research further, the following avenues can be explored:

- Incorporating real-world data: Validate the model using empirical data from electricity markets to ensure its practical applicability and robustness.
- Expanding group dynamics: Extend the analysis to include more diverse groups with varying cost structures and market shares, enabling a more comprehensive understanding of market dynamics.
- Policy simulation: Conduct scenario-based simulations to evaluate the effectiveness of different regulatory interventions. For instance: Assess how subsidies for small-capacity groups impact overall market efficiency; and analyze the long-term implications of imposing price caps or quotas.
- Addressing critical stability: Refine the model to address the limitations of critically stable equilibria. Introducing non-linear dynamics or alternative stability criteria could yield more actionable insights.
- Environmental considerations: Incorporate environmental metrics, such as carbon emissions, into the payoff structure. This would align the study with broader sustainability objectives.

(5) Broader Implications and Recommendations

Policy implications: Design incentive mechanisms to level the playing field for small-capacity groups while maintaining overall market efficiency. Implement regulatory frameworks that discourage monopolistic practices and promote healthy competition.

Research directions: Extend the model to incorporate multi-period analysis, capturing the dynamic evolution of strategies over time. Explore the role of technological innovations (e.g., renewable energy sources) in reshaping market dynamics.

Practical applications: The findings can inform the design of bidding strategies for market participants, helping them optimize their outcomes under different market conditions. Policymakers can use the insights to develop targeted interventions that promote sustainability, equity, and efficiency in electricity markets.

By addressing the identified limitations and expanding the scope of analysis, future research can provide deeper insights into the interplay between strategic behavior, market stability, and policy interventions. These advancements will be instrumental in guiding the development of more resilient and sustainable energy markets. Overall, this study provides a rigorous analysis of the evolutionary dynamics in a deregulated electricity market with oversupply conditions. It underscores the need for targeted governmental interventions to balance market efficiency with equity. By refining the model and incorporating real-world data, future research can further enhance our understanding of competitive bidding strategies and their implications for market stability and sustainability.

3.3. *A Summary*

Based on Section 3.1, Section 3.2 employs the Chinese electricity market as a case study to examine the impact of governmental supervision, market-clearing prices, market demand, and generation group capacity sizes on their bidding strategies. The findings from simulations conducted under different conditions can be summarized as follows.

Firstly, our analysis shows that lowering the market-clearing price reduces the profitability of high-price strategies for enterprises, which incentivizes generation groups to adopt low-price strategies. However, a reduction in the market-clearing price narrows the competitive space for smaller-capacity groups, thereby intensifying the pressure to adopt low-price strategies.

Second, as market electricity demand decreases, the strategic choices of large-capacity generation groups increasingly influence the market. When the supply–demand balance reaches a critical point, large-capacity groups must adopt low-price strategies to remain competitive and sustain long-term market development. However, if large-capacity groups persist in aggressively adopting low-price strategies, smaller players are marginalized, diminishing their chances of securing contracts.

Third, effective government supervision of market competition is essential. When generation groups simultaneously adopt high-price strategies, government interventions, such as penalties or adjustments to subsidies, can encourage smaller-capacity groups to actively participate in the market through low-price strategies. This enables smaller-capacity groups to remain viable competitors in the market. Although the equilibrium point (1, 1, 1) represents a scenario where all groups adopt high-price strategies, it is neither optimal nor sustainable in the long term. Regulators should focus on developing policies that promote fair competition and encourage the majority of generation groups to adopt cost-aligned strategies, balancing both market efficiency and fairness.

In summary, Section 3.2 highlights the crucial balance between economic efficiency, fairness, and sustainability in competitive electricity markets. By tackling the challenges faced by smaller-capacity groups and fostering responsible behavior among large-capacity players, policymakers and market participants can contribute to a more resilient and equitable market. Future research should refine the models, explore new strategies, and integrate sustainability metrics to align market dynamics with broader policy objectives. These efforts will enhance market stability and also aid in transitioning toward a more sustainable and inclusive energy system.

4. The Evolutionary Game Between Local Governments and Power Producers Under Low-Carbon Mechanisms

4.1. Evolutionary Game Model Construction

4.1.1. Payoff Matrix Construction

Building upon the basic evolutionary game model and simulation study in Section 2, and compared with the scenario study in Ref. [39], we extend the analysis by focusing on the evolutionary game dynamics between local governments and power producers under low-carbon mechanisms, with an emphasis on bidding strategies and carbon reduction incentives. We first present the payoff matrix for both government and enterprise strategies in Table 12, integrating both transaction models and carbon trading mechanisms to evaluate the impact of regulatory interventions.

Table 12. The payoff matrix for government and power enterprise strategies.

Game Players		Power Producer Enterprise	
		Honest Strategy (S_{P1})	Dishonest Strategy (S_{P2})
Government	Supervision strategy (S_{G1})	$R + J - C_1 - P - S_E$ $H + B - J - C_2 - S_G$	$R' - C_1{}' - P' - F - S_E$ $H' + F - C_2 - L - S_P$
		Honest strategy (S_{P1})	Dishonest strategy (S_{P2})
	Non-Supervision strategy (S_{G2})	$R - C_1 - P - M_E$ $H - M_G$	$R' - C_1{}' - P' - M_E$ $H' - L - M_G$

Table 12 outlines the potential payoffs for both the government and enterprises, showcasing different strategic combinations and their implications for carbon reduction and market stability. Below, we provide a detailed analysis of the payoff outcomes for each player when specific strategies are selected, considering the probabilities of each strategy. The strategy combinations and payoffs are described as follows.

Case 1: When the government supervises and the enterprise acts honestly, their payoffs reflect the direct financial incentives and penalties related to carbon compliance and transaction costs. In this case, the payoffs are as follows.

Government's payoff: The government's payoff in this scenario (where supervision is chosen) is a combination of environmental benefits (H), societal benefits (B), and penalties for dishonesty, weighed against the costs of supervision (C_2) and reputational damage (S_G). The government chooses supervision strategy (denoted by strategy S_{G1}, with selection probability of x in each round of evolution game), and the power enterprise chooses honesty (denoted by strategy S_{P1}, with choosing probability of y in each round of evolution game), so that the government's payoff is shown as: $H + B - J - C_2 - S_G$.

Enterprise's payoff: The enterprise's payoff (under honest strategy) consists of market benefits I from carbon trading, plus rewards (J), minus compliance costs (C_1), carbon purchase costs (P), and reputational penalties (S_E). The enterprise's payoff in this scenario is $R + J - C_1 - P - S_E$.

Case 2: Government adopts the supervision strategy, and the enterprise adopts the dishonest strategy. In this case, the payoffs are as follows.

Government's payoff: When the government chooses supervision (S_{G1}, probability x) and the enterprise chooses dishonesty (probability $1 - y$), the government's payoff is as follows: $H' + F - C_2 - L - S_P$. Enterprise's payoff: The enterprise's payoff in this scenario is: $R' - C_1{}' - P' - F - S_E$.

Case 3: Government adopts the non-supervision strategy, and the enterprise adopts the honest strategy. In this case, the payoffs are as follows.

Government's payoff: When the government chooses non-supervision (probability $1 - x$) and the enterprise chooses honesty (probability y), the government's payoff is: $H - M_G$. Enterprise's payoff: The enterprise's payoff in this scenario is $R - C_1 - P - M_E$.

Case 4: Government adopts the non-supervision strategy, and the enterprise adopts the dishonest strategy. In this case, the payoffs are as follows.

Government's payoff: When the government chooses non-supervision (probability $1-x$) and the enterprise chooses dishonesty (probability $1 - y$), the government's payoff is: $H' - L - M_G$. Enterprise's payoff: The enterprise's payoff in this scenario is $R' - C_1' - P' - M_E$.

The variables related to both government and enterprise strategies are defined to capture the impact of transaction models, penalties, and carbon trading incentives on bidding behaviors and market outcomes.

- H: The environmental benefits from compliance (unit: yuan or equivalent environmental impact metric), including reductions in greenhouse gas emissions, mitigation of climate change, pollution alleviation, and societal energy savings, forming part of the carbon reduction strategy for enterprises.
- H': Reduced environmental benefits when enterprises act dishonestly (unit: yuan or equivalent environmental impact metric), reflecting the diminished environmental and societal gains due to non-compliance.
- B: Additional societal benefits from compliance (unit: yuan), representing co-benefits, such as health improvements from reduced pollution, enhanced renewable energy adoption, and long-term economic sustainability.
- J: Compliance rewards provided to enterprises (unit: yuan). Here, the financial incentives or subsidies granted by the government to promote honest carbon reduction behavior.
- F: Penalty for enterprise dishonesty (unit: yuan). Here, the monetary fines or punitive measures imposed on enterprises that fail to comply with carbon reduction targets.
- C_2: Supervision cost for the government (unit: yuan). Resources required for monitoring, auditing, and verifying enterprise carbon reduction activities.
- L: Environmental damage caused by dishonesty (unit: yuan or equivalent environmental impact metric). Quantifies the economic cost of environmental degradation, including loss of biodiversity, increased health burdens, and climate-related risks.
- S_G: Government reputational costs (unit: yuan). Reputational damage to the government resulting from perceived ineffectiveness or inability to enforce carbon reduction measures.
- M_G: Missed government environmental gains due to non-supervision (unit: yuan). Lost benefits resulting from a lack of monitoring, such as reduced emissions reductions and public dissatisfaction.

The variables related to enterprises are explained as follows:

- R: Market benefits from carbon trading under compliance (unit: yuan). The revenue gained by enterprises through carbon trading markets and improved market competitiveness.
- R': Reduced market benefits due to dishonesty (unit: yuan), reflecting reduced consumer willingness to pay, reputational losses, and diminished trading opportunities.
- C_1: Compliance costs for emissions reduction (unit: yuan). Direct investment costs for implementing carbon reduction technologies and meeting reduction targets.

- C_1': Reduced compliance costs from dishonest behavior (unit: yuan). The cost savings achieved by enterprises through non-compliance or avoiding emissions reduction investments.
- P: Cost of purchasing carbon quotas for compliance (unit: yuan). The expense incurred by enterprises in acquiring carbon credits to meet their reduction targets.
- P': Reduced trading costs from dishonesty (unit: yuan), reflecting reduced expenses when enterprises report lower-than-actual emissions or avoid purchasing quotas.
- S_E: Reputational penalties for enterprises (unit: yuan). The losses incurred by enterprises due to reputational damage caused by dishonesty, including reduced market share and public backlash.
- M_E: Missed enterprise market gains due to non-compliance (unit: yuan), representing forgone revenue and growth opportunities caused by a lack of regulatory alignment.
- S_P: Penalty-related social costs for enterprises (unit: yuan), which capture indirect penalties, such as strained relations with stakeholders, lawsuits, or exclusion from government incentives.

The discussion and practical implications for these payoff parameters are elaborated as follows.

1. Impacts of Supervision

When the government supervises, the payoff matrix reflects higher direct costs (C_2) but ensures better compliance, higher environmental benefits (H), and stronger societal gains (B). Enterprises acting dishonestly under supervision face substantial penalties (F), reputational costs (S_E), and additional social penalties (S_P), discouraging non-compliance.

2. Role of Rewards and Penalties

The inclusion of J (rewards) and F (penalties) creates a dynamic system that aligns financial incentives with carbon compliance goals, encouraging enterprises to adopt low-carbon strategies. The penalties (F) are dynamic and should scale with the severity of non-compliance to maximize their effectiveness.

3. Long-Term Implications

Honest enterprises benefit from increased market competitiveness and access to government incentives (J), which support sustainable carbon reduction strategies. This creates a virtuous cycle, aligning profitability with environmental responsibility. Dishonest enterprises might see short-term cost savings (C_1'), but long-term risks arise from penalties (F), reduced market benefits (R'), and reputational damage (S_E), which undermine sustainable carbon reduction strategies.

4. Strategic Choices for Non-Supervision

Non-supervision avoids immediate supervision costs (C_2) but leads to long-term environmental damage (L) and missed economic opportunities (M_G, M_E), while undermining the long-term sustainability of the low-carbon transition.

Design adaptive reward (J) and penalty (F) mechanisms that respond to enterprise performance, particularly emphasizing carbon reduction efforts and compliance behavior in the evolutionary game dynamics. For the enhanced monitoring technology, implement digital solutions (e.g., blockchain and IoT sensors) to reduce supervision costs (C_2) and enhance accuracy in monitoring carbon compliance, ensuring greater market stability and transactional transparency. For stakeholder engagement, promote consumer-driven incentives (R) by encouraging green branding and sustainability certifications, which support market competitiveness and carbon reduction behaviors. For integrated long-term planning, align enterprise compliance with broader sustainability goals, ensuring that emissions reduction integrates seamlessly with economic growth and energy transition strategies.

This payoff matrix in Table 12 provides a robust framework for analyzing government and enterprise interactions under carbon reduction policies. It integrates real-world complexities and offers actionable insights to guide policymakers and enterprises in achieving sustainable and equitable outcomes.

4.1.2. RD and Jacobian Matrix for the Enhanced Payoff Matrix

The RD framework models the evolutionary strategy dynamics between government and enterprise players, capturing how strategies evolve over time based on payoff outcomes and carbon incentives [40,41]. This mathematical framework is based on the principle that strategies with higher payoffs will increase in prevalence over time.

Step 1: Definitions and Setup

Let us define the following:

(1) Strategy probabilities:

- The probability that the government adopts the supervision strategy is x, and the probability of adopting non-supervision is $1-x$.
- The probability that the enterprise adopts the honest strategy is y, and the probability of adopting dishonesty is $1-y$.

(2) Payoffs from Table 12:

For the government, the payoffs for each strategy combination are:

$$\begin{cases} \pi_{\mathrm{G}}^{\mathrm{S,H}} = H + B - J - C_2 - S_{\mathrm{G}} \text{ (Supervision, Honest)} \\ \pi_{\mathrm{G}}^{\mathrm{S,D}} = H' + F - C_2 - L - S_{\mathrm{P}} \text{ (Supervision, Dishonest)} \\ \pi_{\mathrm{G}}^{\mathrm{NS,H}} = H - M_{\mathrm{G}} \text{ (Non-Supervision, Honest)} \\ \pi_{\mathrm{G}}^{\mathrm{NS,D}} = H' - L - M_{\mathrm{G}} \text{ (Non-Supervision, Dishonest)} \end{cases} \tag{41}$$

For the enterprise, the payoffs are as follows:

$$\begin{cases} \pi_{\mathrm{E}}^{\mathrm{S,H}} = R + J - C_1 - P - S_{\mathrm{E}} \text{ (Supervision, Honest)} \\ \pi_{\mathrm{E}}^{\mathrm{S,D}} = R' - C'_1 - P' - F - S_{\mathrm{E}} \text{ (Supervision, Dishonest)} \\ \pi_{\mathrm{E}}^{\mathrm{NS,H}} = R - C_1 - P - M_{\mathrm{E}} \text{ (Non-Supervision, Honest)} \\ \pi_{\mathrm{E}}^{\mathrm{NS,D}} = R' - C'_1 - P' - M_{\mathrm{E}} \text{ (Non-Supervision, Dishonest)} \end{cases} \tag{42}$$

(3) Average payoffs: The average payoff for each player is the weighted average based on the strategy probabilities.

For the government, average payoff is calculated as follows:

$$\begin{aligned} \overline{\pi}_{\mathrm{G}} &= x\left[y\pi_{\mathrm{G}}^{\mathrm{S,H}} + (1-y)\pi_{\mathrm{G}}^{\mathrm{S,D}}\right] + (1-x)\left[y\pi_{\mathrm{G}}^{\mathrm{NS,H}} + (1-y)\pi_{\mathrm{G}}^{\mathrm{NS,D}}\right] \\ &= xy(H + B - J - C_2 - S_{\mathrm{G}}) + x(1-y)(H' + F - C_2 - L - S_{\mathrm{P}}) + (1-x)y(H - M_{\mathrm{G}}) + (1-x)(1-y)(H' - L - M_{\mathrm{G}}) \end{aligned} \tag{43}$$

For the enterprise, average payoff is calculated as follows:

$$\begin{aligned} \overline{\pi}_{E} &= y\left[x\pi_{\mathrm{E}}^{\mathrm{S,H}} + (1-x)\pi_{\mathrm{E}}^{\mathrm{NS,H}}\right] + (1-y)\left[x\pi_{\mathrm{E}}^{\mathrm{S,D}} + (1-x)\pi_{\mathrm{E}}^{\mathrm{NS,D}}\right] \\ &= yx(R + J - C_1 - P - S_{\mathrm{E}}) + y(1-x)(R - C_1 - P - M_{\mathrm{E}}) + (1-y)x(R' - C'_1 - P' - F - S_{\mathrm{E}}) + (1-y)(1-x)(R' - C'_1 - P' - M_{\mathrm{E}}) \end{aligned} \tag{44}$$

Step 2: RD Equations

The RD describe how the probabilities x and y evolve over time. They are given as follows:

$$\begin{cases} \dot{x} = x(\pi_{\mathrm{G}}^{X} - \overline{\pi}_{\mathrm{G}}) \\ \dot{y} = y(\pi_{\mathrm{E}}^{Y} - \overline{\pi}_{\mathrm{E}}) \end{cases} \tag{45}$$

Step 3: Jacobian Matrix of the System

The Jacobian matrix is used to evaluate the stability of equilibrium points, providing insights into how government and enterprise strategies evolve in response to policy changes and carbon compliance incentives. The Jacobian matrix, J_{GP}, is derived from the partial derivatives of the replicator equations with respect to x and y.

$$J_{GP} = \begin{bmatrix} \frac{\partial \dot{x}}{\partial x} & \frac{\partial \dot{x}}{\partial y} \\ \frac{\partial \dot{y}}{\partial x} & \frac{\partial \dot{y}}{\partial y} \end{bmatrix} \tag{46}$$

where

$$\begin{cases} \frac{\partial \dot{x}}{\partial x} = \pi_G^S - \overline{\pi}_G + x\frac{\partial(\pi_G^S - \overline{\pi}_G)}{\partial x} \\ \frac{\partial \dot{x}}{\partial y} = x\frac{\partial(\pi_G^S - \overline{\pi}_G)}{\partial y} \\ \frac{\partial \dot{y}}{\partial x} = y\frac{\partial(\pi_E^H - \overline{\pi}_E)}{\partial x} \\ \frac{\partial \dot{y}}{\partial y} = \pi_E^H - \overline{\pi}_E + y\frac{\partial(\pi_E^H - \overline{\pi}_E)}{\partial y} \end{cases} \tag{47}$$

By substituting the payoff equations (π_G, π_E) into these expressions, the Jacobian matrix can be explicitly calculated. The stability of equilibrium points is determined by analyzing the eigenvalues of J_{GP}. The RD and Jacobian matrix derived from the enhanced payoff matrix provide a mathematical framework for understanding the evolution of government and enterprise strategies. The stability of equilibrium points hinges on payoff parameters (H, R, C_1, J, F, etc.), illustrating the interactions between government policies, enterprise behaviors, and carbon reduction strategies.

4.1.3. Evolutionary Stability Conditions

The evolutionary stability conditions for specific equilibria are summarized as follows:

1. Equilibrium $(x, y) = (0, 0)$:

   $x = 0$: Government always adopts non-supervision.
   $y = 0$: Enterprise always adopts dishonesty.
   Stability Condition: The government and enterprise payoffs under these strategies must make deviations $\pi_G^{S,D} < \pi_G^{NS,D}$ and $\pi_E^{S,D} < \pi_E^{NS,D}$.

2. Equilibrium $(x, y) = (0, 1)$:

   $x = 0$: Government always adopts non-supervision.
   $y = 1$: Enterprise always adopts honesty.
   Stability Condition: Deviations $\pi_G^{S,H} > \pi_G^{NS,H}$ and $\pi_E^{S,H} > \pi_E^{NS,H}$ must be unfavorable.

3. Equilibrium $(x, y) = (1, 0)$:

   $x = 1$: Government always adopts supervision.
   $y = 0$: Enterprise always adopts dishonesty.
   Stability Condition: $\pi_G^{S,D} > \pi_G^{NS,D}$ and $\pi_E^{S,D} > \pi_E^{NS,D}$ must be favorable for these strategies.

4. Equilibrium $(x, y) = (1, 1)$:

   $x = 1$: Government always adopts supervision.
   $y = 1$: Enterprise always adopts honesty.
   Stability Condition: $\pi_G^{S,H} > \pi_G^{NS,H}$ and $\pi_E^{S,H} > \pi_E^{NS,H}$.

5. Mixed Strategy Equilibrium

The mixed strategy equilibrium occurs when x and y lie strictly between 0 and 1 $(0 < x < 1, 0 < y < 1)$.

Mixed strategy probabilities are described as follows.

At the mixed strategy equilibrium:

(1) $\dot{x} = 0$: The government is indifferent between supervision and non-supervision, i.e., $\pi_G^S = \pi_G^{NS}$. Substituting payoffs:

$$y(H + B - J - C_2 - S_G) + (1 - y)(H' + F - C_2 - L - S_P) = y(H - M_G) + (1 - y)(H' - L - M_G) \quad (48)$$

(2) $\dot{y} = 0$: The enterprise is indifferent between honesty and dishonesty, i.e., $\pi_E^H = \pi_G^D$. Substituting payoffs:

$$x(R + J - C_1 - P - S_E) + (1 - x)(R - C_1 - P - M_E) = x(R' - C'_1 - P' - F - S_E) + (1 - x)(R' - C'_1 - P' - M_E) \quad (49)$$

Solving for x^* and y^*:

The solutions to the above equations provide the equilibrium probabilities x^* (probability of government supervision) and y^* (probability of enterprise honesty).

In conclusion, the equilibrium points are outlined as follows.

- 5 strategy equilibria as $(x, y) = (0, 0), (0, 1), (1, 0), (1, 1)$, and (x^*, y^*).
- Stability Conditions: Stability depends on payoff comparisons between strategies and eigenvalues of the Jacobian J_{GP}.
- Mixed Strategy Equilibrium: x^* and y^* are determined by the indifference conditions $\pi_G^S = \pi_G^{NS}$ and $\pi_E^H = \pi_E^D$, leading to probabilistic mixes of strategies.

4.2. Simulation Study and Analysis

Based on Section 4.1, we conduct a simulation study to verify the theoretical model constructed in Section 4.1. The simulation results are illustrated in Figure 9. Concretely, this simulation study investigates the evolutionary game between local governments and power producers under low-carbon mechanisms, emphasizing the impact of compliance costs (C_1) on bidding strategies and carbon reduction behaviors. The motivation for this research lies in understanding how regulatory strategies and compliance incentives interact dynamically to encourage power producers' honesty and governments' active supervision in carbon reduction efforts. The model uses RD to simulate the evolutionary process between these two players, with "government" (x) representing the probability of active supervision and "enterprise" (y) denoting the probability of honest compliance.

As shown in Figure 9, the key model features are summarized as follows:

- Payoff structure: The payoff matrix is derived from the enhanced Table 2, which includes compliance rewards, penalties, supervision costs, and market benefits for both players. This matrix captures the nuanced interactions of both sides under different strategies.
- Dynamic equations: The RD equations model how the strategy fractions evolve over time based on the relative payoffs of each strategy.
- Core innovation: The incorporation of realistic cost and benefit structures into the model highlights the interplay between government supervision probability and enterprise honesty probability under varying compliance costs.

Based on the above, the simulation parameters for Figure 9 are set as follows.

- Compliance costs for emissions reduction (C1): 500, 1000, and 2000 yuan.
- Simulation time: 50 time units, with 20 evaluation points.
- Initial conditions: Strategy fractions uniformly sampled in [0, 1].

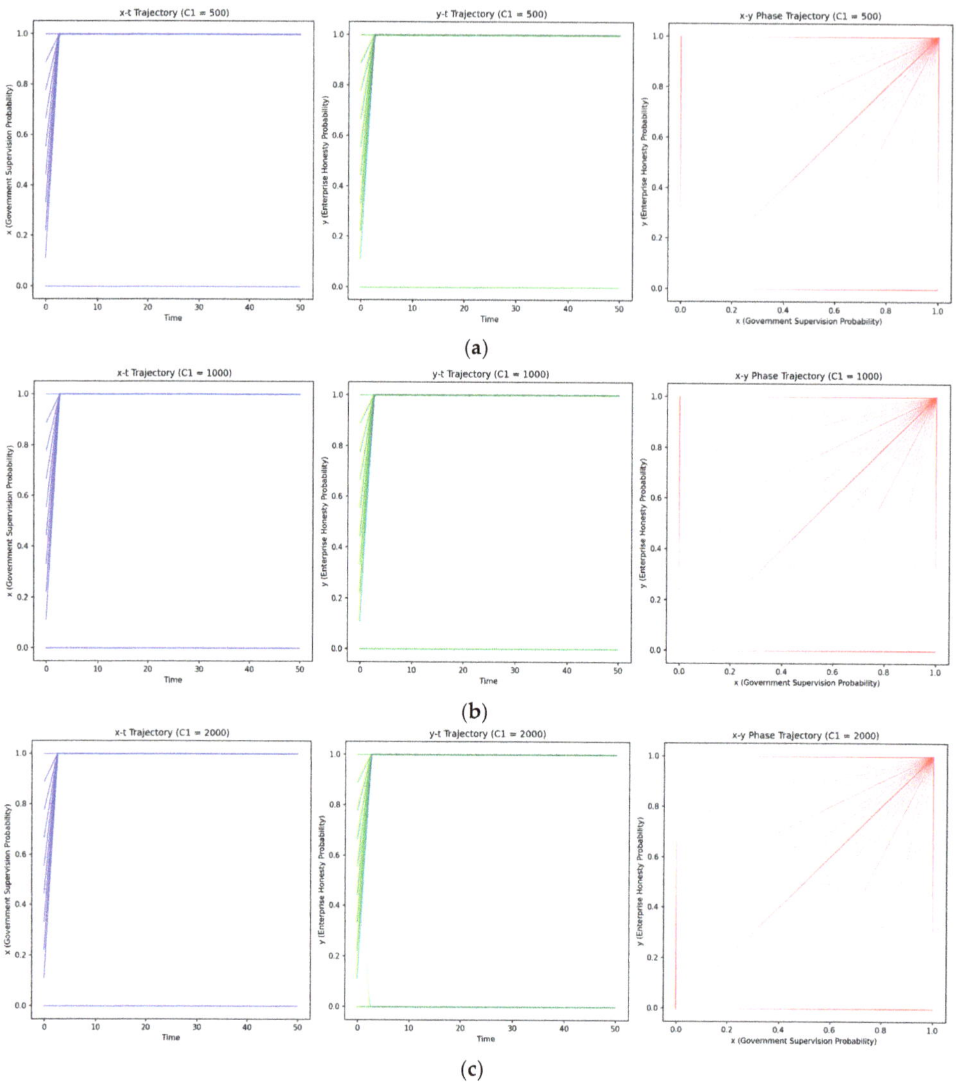

Figure 9. Simulation results for evolutionary game-theoretic analysis of government supervision and enterprise compliance strategies under low-carbon mechanisms. (**a**) The first subfigure shows the temporal evolution of the government's supervision probability (x) under compliance cost $C_1 = 500$. (**b**) The second subfigure demonstrates the temporal evolution of the enterprise's honesty probability (y) under compliance cost $C_1 = 1000$. (**c**) The third subfigure displays the phase trajectories of government supervision (x) and enterprise honesty (y) under compliance cost $C_1 = 2000$.

The simulation in Figure 9 is designed to observe how different compliance costs influence the stability and trajectory of strategies for both government and enterprises. Analysis of the simulation results is conducted as follows.

Figure 9a: Temporal Evolution of Government Supervision Probability ($C_1 = 500$)

Under lower compliance costs ($C_1 = 500$), the government supervision probability (x) converges rapidly to 1. This indicates that the government strongly favors active supervision when compliance costs are low, as the enterprises' honest strategies lead to higher payoffs for both players. This figure reveals that low compliance costs create a favorable environment for government supervision and enterprise cooperation, fostering a stable and effective low-carbon mechanism.

Figure 9b: Temporal Evolution of Enterprise Honesty Probability (C_1 = 1000)

With moderate compliance costs (C_1 = 1000), enterprises' honesty probability (y) also converges quickly to 1. The convergence is slightly slower compared to Figure 9a, indicating that higher compliance costs reduce the immediate benefits of honesty but still promote it over time. This figure indicates that moderate compliance costs strike a balance between affordability for enterprises and incentives for honest behavior, ensuring long-term cooperation.

Figure 9c: Phase Trajectories of Government and Enterprise Strategies (C_1 = 2000)

Under high compliance costs (C_1 = 2000), the phase trajectories still converge to the equilibrium point ($x = 1, y = 1$). However, the paths are more spread out, reflecting greater variability in the initial adjustments of strategies. This figure shows that even with higher compliance costs, the system converges to full supervision and honesty. However, the increased variability suggests that enterprises face greater challenges in adopting honest strategies when costs are prohibitive.

Overall, the simulations in Figure 9 validate the enhanced payoff matrix from Table 2, demonstrating that the proposed RD effectively capture the evolutionary interactions between local governments and enterprises under low-carbon mechanisms. The results highlight the following:

- Policy design: Lower compliance costs lead to faster convergence to cooperative strategies, emphasizing the importance of cost-effective carbon reduction policies.
- Robustness: The system converges to the equilibrium ($x = 1, y = 1$) across all tested compliance costs, showcasing the robustness of the model under varying conditions.
- Application: This research provides a quantitative framework for designing policies that balance government incentives with enterprise costs, fostering sustainable low-carbon development and enhancing regulatory compliance.

As demonstrated in Figure 10, the simulation study focuses on modeling the interaction between local governments and power producers under a low-carbon regulatory framework. The primary aim is to investigate the dynamic evolutionary behavior of government supervision strategies and enterprise honesty strategies, specifically under varying compliance cost scenarios. The study is motivated by the need to understand the incentives and trade-offs that shape cooperation between regulators and industry players in achieving low-carbon development goals. By employing evolutionary game theory, this research identifies conditions under which both parties converge to cooperative strategies, thus fostering sustainable carbon reduction.

The simulation results in Figure 10 validate the enhanced payoff matrix and demonstrate that the RD model effectively captures the interplay between government supervision and enterprise compliance under varying cost scenarios. Key conclusions include the following:

- Policy design: Low compliance costs promote rapid and stable cooperation, while moderate costs balance incentives and affordability. High compliance costs, though less effective, can still achieve cooperation with additional policy support.
- Model robustness: The system consistently converges to cooperative equilibria ($x = 1$, $y = 1$) across all tested scenarios, demonstrating the model's robustness.

  Based on these findings, future improvements are elaborated as follows.

- Dynamic costs: Introduce time-varying compliance costs to simulate real-world scenarios where policies evolve.

- Uncertainty analysis: Incorporate stochastic elements to study the impact of uncertainties in cost and benefits.
- Dynamic policy Adjustments: Introduce adaptive compliance costs to simulate real-world policy evolutions.
- Multi-agent interactions: Expand the model to include multiple governments and enterprises to study regional and sectoral dynamics.
- Multi-agent extension: Expand the model to include multiple governments and enterprises to explore regional and sectoral interactions in low-carbon mechanisms.
- Stochastic extensions: Incorporate uncertainty to reflect variability in costs and benefits, providing deeper insights into real-world applications.

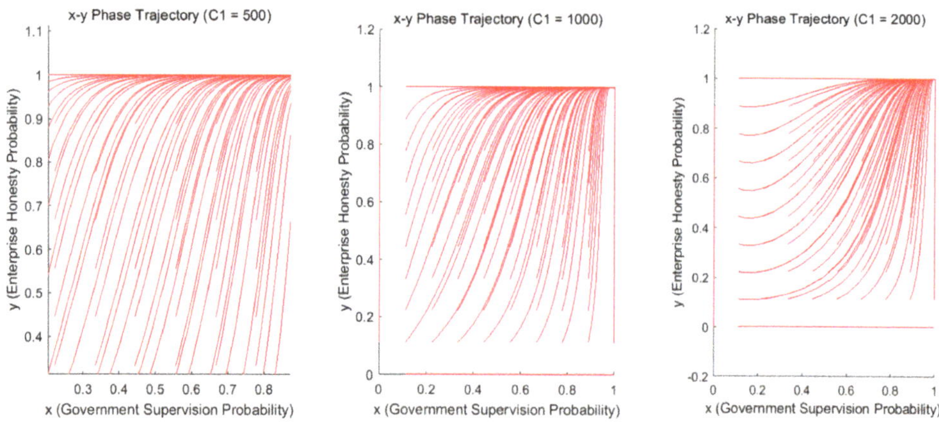

Figure 10. Simulation results for evolutionary game-theoretic analysis of government supervision and enterprise compliance strategies under low-carbon mechanisms. (**a**) The first subfigure shows the phase trajectories of government supervision probability (x) and enterprise honesty probability (y) under compliance cost $C_1 = 500$. (**b**) The first subfigure shows the phase trajectories of government supervision probability (x) and enterprise honesty probability (y) under compliance cost $C_1 = 1000$. (**c**) The first subfigure shows the phase trajectories of government supervision probability (x) and enterprise honesty probability (y) under compliance cost $C_1 = 2000$.

Overall, this study provides a rigorous framework for designing and evaluating low-carbon policies, offering actionable insights for fostering sustainable cooperation in carbon reduction initiatives. The evolutionary game-theoretic approach proves to be a powerful tool for modeling complex interactions and optimizing regulatory strategies. This study demonstrates the practical applicability of evolutionary game theory in low-carbon policy design, providing insights into the dynamic interplay between regulation and compliance. Further improvements can enhance the model's realism and policy relevance.

4.3. A Summary

1. **Findings from Simulation Study**

The simulation results reveal several key insights into evolutionary game dynamics, emphasizing the role of government supervision, penalties, and market incentives in shaping enterprise compliance and carbon reduction outcomes:

(i) Government supervision: As compliance costs decrease, the government is more likely to adopt a supervision strategy, promoting environmental compliance among power producers. This effect is most pronounced when penalties for dishonesty are high.

(ii) Power producer behavior: Power producers are more likely to adopt honest strategies when the costs of non-compliance are high, especially when regulatory supervision is strong. However, higher compliance costs lead to a trade-off where producers may be tempted to adopt dishonest strategies to reduce costs.
(iii) Stable equilibrium: The system converges to a stable equilibrium where both the government adopts supervision and power producers adopt honest strategies when regulatory incentives and penalties are effectively calibrated.

Overall, the findings offer actionable insights into the design of low-carbon policies and market mechanisms, demonstrating how evolutionary game theory can optimize regulatory interventions and bidding strategies to promote sustainability.

2. **Practical Implications**

The results of this study provide several policy implications for both local governments and power producers:

(i) Government regulation: The study underscores the importance of effective regulatory frameworks that incorporate appropriate incentives and penalties to ensure compliance with low-carbon initiatives. Government supervision strategies should be dynamically adjusted based on compliance costs and the level of dishonesty observed in the market.
(ii) Power producers' strategic adjustments: Power producers must carefully balance their strategies between profitability and environmental responsibility. The inclusion of carbon reduction rewards and penalties in market mechanisms can help align their incentives with societal goals.

3. **Recommendations for Future Research**

Future research should focus on the following:

(i) Incorporating stochastic elements: The current model assumes deterministic dynamics, but incorporating stochastic factors such as demand fluctuations and renewable energy variability can enhance the model's realism.
(ii) Exploring multi-agent interactions: Expanding the model to include multiple players, such as third-party intermediaries, can provide a more comprehensive understanding of the complex dynamics in low-carbon electricity markets.
(iii) Real-time policy adaptations: Developing real-time monitoring systems and adaptive policies can help governments respond dynamically to changes in market conditions and power producers' behaviors.

5. Discussions and Prospects

5.1. Discussions

This study explores the strategic dynamics of power producers within low-carbon mechanisms, employing EGT to incorporate carbon reduction strategies in competitive bidding. The findings offer insight into how governmental oversight, carbon pricing, and low-carbon policies shape the bidding behaviors and strategic interactions of power producers with diverse capacities. By incorporating RD equations, this study investigates stable equilibria across various scenarios, a key aspect in crafting practical and sustainable energy policies.

1. Core Contributions and Observations

(i) Model innovation: This study presents a three-group evolutionary game model for power producers, categorized into small, medium, and large producers, each with unique cost functions and capacity limitations. The model integrates regulatory policies and carbon incentives into the payoff structures, making it more applicable to real-world electricity markets.

(ii) Insights into equilibria: Simulations demonstrate that the system stabilizes at ESS under defined payoff conditions. For example, cooperative equilibria emerge when regulatory incentives are strong, encouraging groups to adopt low-cost bidding strategies. When carbon incentives are weak, competitive equilibria result in aggressive, high-cost bidding.

2. Policy Impacts

Table 13 illustrates the key findings from simulations across varying scenarios. Based on this, the policy impacts are summarized as follows.

(i) Penalties for dishonesty and subsidies for low-carbon compliance directly affect the evolution of strategies. For instance, a higher penalty for non-compliance shifts equilibrium points toward more cooperative and environmentally sustainable outcomes.

(ii) Rising compliance costs (such as carbon taxes) diminish the profitability of non-compliance, prompting shifts toward adopting low-emission technologies.

Table 13. Key findings from simulations across different scenarios.

Scenario	Parameters	Key Observations	Policy Implications
High Compliance Costs	$C_1 = 2000$, $C_2 = 800$	Low compliance probability ($x, y < 0.5$). Both groups adopt competitive, high-cost strategies.	Stronger subsidies or penalties are required to incentivize low-carbon behaviors.
Moderate Compliance Costs	$C_1 = 1000$, $C_2 = 500$	Stable equilibrium at cooperative strategies ($x, y > 0.7$).	Current regulatory policies are effective, but additional incentives can enhance stability further.
Low Compliance Costs	$C_1 = 500$, $C_2 = 300$	Both groups adopt low-cost, cooperative strategies ($x, y \approx 1$).	No additional intervention required; the market self-stabilizes.
Weak Carbon Penalties	$F = 500$, $P = 300$	High dishonesty levels ($x, y < 0.4$). Groups favor high-emission strategies.	Increase penalties for non-compliance to discourage dishonest behavior.
Strong Carbon Penalties	$F = 1500$, $P = 700$	High honesty levels ($x, y > 0.8$). Both groups favor low-emission, compliant strategies.	Carbon penalties are effective; maintain or increase support for low-carbon compliance.

Based on Table 13, the challenges and limitations are summarized as follows.

(i) Model simplification: Though the replicator dynamics equations capture strategic evolution, real-world complexities like stochastic fluctuations and information asymmetry are not fully accounted for.

(ii) Homogeneous groups: The model assumes uniform behavior within each group (small, medium, large producers), but this may not capture the individual heterogeneities seen in real-world markets.

(iii) Market constraints: Transmission and capacity constraints were not explicitly included, which limits the applicability of the findings to grid-constrained markets.

5.2. Prospects

The findings from this study provide a foundation for future research and policy formulation in low-carbon electricity markets. The evolutionary game-theoretic framework proves valuable for capturing multi-group interactions and offers promising avenues for further extension, as shown in Table 14, which summarizes the proposed extensions and their potential impacts. Based on this table, the future research directions are elaborated as follows:

(i) Integration of Stochastic Dynamics [41–43]:
- Rationale: Real-world electricity markets are subject to demand fluctuations, renewable energy intermittency, and policy uncertainties.
- Proposal: Incorporate stochastic elements into RD equations to account for these uncertainties.
- Expected outcome: Enhanced robustness and applicability of the model in predicting market behaviors under volatile conditions.

(ii) Expanded Strategy Spaces:
- Rationale: Current models restrict producers to binary strategies (compliance or non-compliance). Real markets involve a spectrum of strategies, including hybrid approaches.
- Proposal: Extend the model to include mixed strategies, such as partial compliance or adaptive bidding.
- Expected outcome: More nuanced insights into the interplay of multiple strategies in competitive environments.

(iii) Incorporating Grid Constraints:
- Rationale: Transmission and capacity constraints significantly impact bidding strategies and market outcomes.
- Proposal: Embed grid-related constraints into payoff functions and evolutionary dynamics.
- Expected outcome: Improved applicability of the model to grid-constrained electricity markets.

Table 14. Proposed extensions and their potential impacts.

Extension	Description	Expected Impact
Stochastic RD	Introduce randomness to capture demand volatility and renewable intermittency.	Better alignment with real-world market behaviors; improved predictive accuracy.
Multi-Dimensional Strategy Spaces	Allow mixed strategies (e.g., partial compliance) and adaptive behaviors.	Insights into hybrid strategies; applicability to complex market dynamics.
Inclusion of Transmission Constraints	Model transmission congestion and grid limitations in payoff calculations.	Increased relevance for electricity markets with grid reliability challenges.
Agent-Based Modeling	Replace deterministic equations with agent-based simulations for heterogeneous producers.	Improved representation of individual decision-making processes and market heterogeneity.
Real-Time Data Integration	Incorporate real-time market data, such as load forecasts and renewable generation levels.	Dynamic adaptation of strategies; potential for real-world policy implementation.

Based on Table 14, the policy recommendations are summarized as follows.

(i) Strengthening carbon pricing: Introduce dynamic carbon pricing mechanisms to penalize high-emission strategies while rewarding compliance.

(ii) Subsidy optimization: Design targeted subsidies to encourage low-carbon compliance among small and medium producers, leveling the competitive playing field.

(iii) Real-time monitoring: Establish mechanisms for real-time monitoring of market behaviors to enable adaptive policy interventions.

(iv) Broader implications: The integration of evolutionary game theory into electricity market analysis has implications beyond carbon reduction. It provides a framework for understanding the dynamics of renewable energy integration, grid stability, and market efficiency. By aligning market incentives with long-term sustainability goals, this approach can support the global transition to low-carbon energy systems.

In conclusion, this study highlights the effectiveness of evolutionary game-theoretic models in analyzing strategic interactions among power producers under low-carbon mechanisms. By addressing current limitations and pursuing the outlined prospects, future research can further enhance the theoretical and practical contributions of this field, paving the way for more sustainable and efficient electricity markets.

6. Conclusions

This study provides a comprehensive analysis of power producers' carbon emission reduction strategies and multi-group bidding dynamics in the low-carbon electricity market using EGT. The core contributions and findings of the study are as follows:

(1) **Development of an Evolutionary Game-Theoretic Framework:**

This research advances the application of EGT to model interactions among heterogeneous power producers. The framework incorporates compliance costs, carbon reduction strategies, and regulatory incentives, capturing the complex interplay between market forces and regulatory interventions.

(2) **Insights into Strategic Dynamics:**

(i) The study highlights how compliance costs influence bidding strategies. Higher compliance costs discourage cooperative behaviors, while lower costs promote stable cooperation and the adoption of low-carbon technologies.
(ii) Penalties for dishonesty and subsidies for compliance significantly affect the evolution of strategies. Strong penalties and subsidies drive the market toward cooperation and low-carbon objectives.
(iii) The model shows that well-calibrated regulatory policies can foster stable equilibria, aiding the transition to low-carbon energy systems.

(3) **Policy Implications:**

The findings highlight the need for targeted policies that balance compliance costs, penalties, and subsidies to facilitate a sustainable and equitable transition to a low-carbon electricity market.

The study identifies areas for future research, such as addressing real-world complexities, incorporating stochastic dynamics, and expanding the model to involve broader stakeholder participation. The transition to low-carbon electricity markets is a crucial aspect of global efforts to combat climate change. As the complexity of these markets grows with renewable energy integration and the implementation of carbon reduction policies, the demand for advanced analytical tools becomes more apparent. This study demonstrates the potential of evolutionary game theory to provide a robust framework for understanding and shaping market dynamics. By addressing the aforementioned limitations and pursuing the suggested research avenues, future work can refine this framework and contribute to the development of sustainable, efficient, and equitable electricity markets. Through these efforts, the insights derived from evolutionary game theory can significantly contribute to achieving global carbon neutrality goals.

Author Contributions: Conceptualization, J.T., B.Q., Y.L. and X.L.; methodology, J.T., Y.L., M.Z., F.Z. and H.W.; formal analysis, B.Q. and X.L.; investigation, J.T., Y.L., M.Z., F.Z. and H.W.; writing—original draft preparation, J.T., B.Q., Y.L., X.L., M.Z., F.Z. and H.W.; writing—review and editing, J.T., B.Q., Y.L., X.L., M.Z., F.Z. and H.W.; funding acquisition, J.T. All authors have read and agreed to the published version of the manuscript.

Funding: This research was funded by the Major Science and Technology Project of China Southern Power Grid Co., Ltd. under grant ZBKJXM20232456.

Data Availability Statement: The original contributions presented in this study are included in the article. Further inquiries can be directed to the corresponding author.

Acknowledgments: We would like to express our deepest gratitude to Lefeng Cheng and his team from the School of Mechanical and Electrical Engineering at Guangzhou University for their invaluable assistance, insights, and suggestions throughout the preparation of this paper. Further, we sincerely thank the associate editor and invited anonymous reviewers for their kind and helpful comments on our paper.

Conflicts of Interest: Authors Jianlin Tang, Bin Qian, Yi Luo, Xiaoming Lin, Mi Zhou, Fan Zhang and Haolin Wang are employed by Electric Power Research Institute of China Southern Power Grid and Guangdong Provincial Key Laboratory of Intelligent Measurement and Advanced Metering of Power Grid. The China Southern Power Grid Co., Ltd. had no role in the design of the study; in the collection, analyses, or interpretation of data; in the writing of the manuscript, or in the decision to publish the results.

Glossary

Term	Definition/Description
Agent-based Model (ABM)	A computational model for simulating the actions and interactions of autonomous agents to assess their effects on the system as a whole.
Bidding Dynamics	The competitive interactions between power producers in electricity markets as they strategize their bids to balance profitability and regulatory compliance.
Carbon Neutrality	A state where net carbon dioxide emissions are zero, achieved through a combination of emission reductions and offsetting measures such as carbon sequestration or trading.
Carbon Quotas	A limit set by regulatory agencies on the maximum amount of carbon emissions an entity is allowed to produce, often tied to carbon credit trading mechanisms.
Carbon Trading	A market-based mechanism designed to reduce greenhouse gas emissions by allowing entities to buy and sell carbon credits.
Clean Energy Transition	The shift from fossil fuel-based energy systems to renewable and low-carbon energy sources such as solar, wind, and hydropower.
Compliance Costs	The monetary expenditure incurred by firms to adhere to carbon emission regulations, including investments in clean technologies, penalties, and other operational adjustments.
Distributed Energy Resources (DERs)	Small-scale energy generation and storage technologies located close to the point of consumption, such as rooftop solar panels or home batteries.
Dynamic Pricing	A flexible pricing mechanism where energy prices fluctuate based on supply, demand, and time of use, incentivizing energy efficiency.
Enterprise Honesty Probability	The probability that enterprises will truthfully comply with carbon emission reduction regulations without engaging in dishonest behaviors such as falsifying emission data.
Evolutionary Game Theory (EGT)	A mathematical framework that studies the strategic interactions and behavioral evolution of competing groups or agents over time, particularly under changing environmental or regulatory conditions.
Government Supervision Probability	The likelihood that regulatory agencies will actively monitor and enforce compliance with carbon reduction policies among enterprises.

Low-Carbon Electricity Market	A regulated energy market that prioritizes the generation, trading, and consumption of electricity produced from low-carbon and renewable sources, incentivized through carbon pricing mechanisms and policies.
Low-Carbon Strategies	A set of operational, technological, and policy measures adopted by enterprises to minimize carbon emissions in compliance with environmental regulations.
Market Benefits	The economic advantages gained by enterprises from participating in low-carbon electricity markets, including increased market share, financial rewards, and improved reputation.
Penalty for Dishonesty	A financial or reputational penalty imposed on enterprises that fail to meet emission standards or falsify compliance data, designed to deter non-compliance.
Replicator Dynamics	A mathematical model used to describe the evolution of strategy proportions in a population, where strategies with higher-than-average payoffs increase in frequency over time.
Subsidy Allocation	Financial support provided by governments to enterprises to incentivize the adoption of low-carbon technologies and compliance with carbon emission reduction targets.

References

1. Liu, P.; Zhongfu, T. How to develop distributed generation in China: In the context of the reformation of electric power system. *Renew. Sustain. Energy Rev.* **2016**, *66*, 10–26.
2. Guo, H.; Davidson, M.R.; Chen, Q.; Zhang, D.; Jiang, N.; Xia, Q.; Zhang, X. Power market reform in China: Motivations, progress, and recommendations. *Energy Policy* **2020**, *145*, 111717.
3. Cheng, L.; Liu, G.; Huang, H.; Wang, X.; Chen, Y.; Zhang, J.; Yu, T. Equilibrium analysis of general N-population multi-strategy games for generation-side long-term bidding: An evolutionary game perspective. *J. Clean. Prod.* **2020**, *276*, 124123.
4. Cheng, L.; Chen, Y.; Liu, G. 2PnS-EG: A general two-population n-strategy evolutionary game for strategic long-term bidding in a deregulated market under different market clearing mechanisms. *Int. J. Electr. Power Energy Syst.* **2022**, *142*, 108182. [CrossRef]
5. Jamali, M.B.; Rasti-Barzoki, M.; Khosroshahi, H.; Altmann, J. An evolutionary game-theoretic approach to study the technological transformation of the industrial sector toward renewable electricity procurement: A case study. *Appl. Energy* **2022**, *315*, 118968. [CrossRef]
6. Jamali, M.B.; Rasti-Barzoki, M.; Altmann, J. An evolutionary game-theoretic approach for investigating the long-term behavior of the industry sector for purchasing renewable and non-renewable energy: A case study. *Energy* **2023**, *269*, 126751. [CrossRef]
7. Sun, J.; Wu, F.; Shi, M.; Yuan, X. Coordination of renewable energy integration and peak shaving through evolutionary game theory. *Processes* **2024**, *12*, 267. [CrossRef]
8. Huang, F.; Fan, H.; Shang, Y.; Wei, Y.; Almutairi, S.Z.; Alharbi, A.M.; Ma, H.; Wang, H. Research on renewable energy trading strategies based on evolutionary game theory. *Sustainability* **2024**, *16*, 2671. [CrossRef]
9. Zhang, X.; Guo, X.; Zhang, X. Bidding modes for renewable energy considering electricity-carbon integrated market mechanism based on multi-agent hybrid game. *Energy* **2023**, *270*, 120940. [CrossRef]
10. Perera, R.S. An evolutionary game theory strategy for carbon emission reduction in the electricity market. *Int. Game Theory Rev.* **2018**, *20*, 1850006. [CrossRef]
11. Wang, R.; Li, Y.; Gao, B. Evolutionary game-based optimization of green certificate-carbon emission right-electricity joint market for thermal-wind-photovoltaic power system. *Glob. Energy Interconnect.* **2023**, *6*, 92–102. [CrossRef]
12. Peng, C.; Lin, Z. Optimal strategy for generating and bidding of generating side in electricity market. *Electr. Power Autom. Equip.* **2005**, *25*, 1–5.
13. Zhao, Q.; Guo, R.; Zeng, M.; Qi, X. Application of bidding strategy based on genetic algorithm for power generation enterprises. *Electr. Power* **2002**, *35*, 44–46.
14. Conejos, A.J.; Nogales, F.J.; Arroyo, J.M. Price-taker bidding strategy under price uncertainty. *IEEE Trans. Power Syst.* **2002**, *17*, 1081–1088.
15. Wang, X. The research on Gencos' bidding strategy in regional electricity market based on evolutionary game theory. Master's Thesis, North China Electric Power University, Baoding, China, 2006.
16. Ma, L.; Wen, F.; Xu, N. Fuzzy set theory based bidding strategies for generation companies in electricity market environment. *Power Syst. Technol.* **2003**, *27*, 10–14.

17. Liu, Y.; Tan, Z.; Liu, M.; Yang, L.; Wang, C. Optimal bidding strategy based on Markov decision process for generation companies. *Trans. China Electrotech. Soc.* **2005**, *20*, 36–42.
18. Gountis, V.P.; Bakirtzis, A.G. Bidding strategies for electricity producers in a competitive electricity marketplace. *IEEE Trans. Power Syst.* **2004**, *19*, 356–365.
19. Zhang, S.; Li, Q.; Fu, C. Game analysis of power generation companies and grid companies in tariff. *J. Shanghai Univ. Electr. Power* **2015**, *31*, 389–392.
20. Zhang, D.; Wang, Y.; Luh, P.B. Optimization-based bidding strategies in the deregulated market. *IEEE Trans. Power Syst.* **2000**, *15*, 981–986.
21. Wen, F.; David, A.K. Optimal bidding strategies and modeling of imperfect information among competitive generators. *IEEE Trans. Power Syst.* **2001**, *16*, 15–21.
22. Guo, J.; He, G.; Wang, Z.; Chen, X.; Zhou, F. Behavior analysis on the bidding strategy of generation companies. *Autom. Electr. Power Syst.* **2002**, *5*, 5–8.
23. Wang, Y.; Yu, J.; Qiu, W.; Shen, H.; Cheng, X.; Lin, C. Evolutionary game model and analysis methods for network group behavior. *Chin. J. Comput.* **2015**, *38*, 282–300.
24. Wang, Y.H.; Liu, X.Y. Dynamics and optimization of control networked evolutionary games with local information. *Control Theory Appl.* **2019**, *36*, 279–285.
25. Cheng, L.; Pan, P.; Lu, W.; Huang, P.; Chen, Y. Study of flexibility transformation in thermal power enterprises under multi-factor drivers: Application of complex-network evolutionary game theory. *Mathematics* **2024**, *12*, 2537. [CrossRef]
26. Zhu, Z.; Cheng, L.; Shen, T. Spontaneous formation of evolutionary game strategies for long-term carbon emission reduction based on low-carbon trading mechanism. *Mathematics* **2024**, *12*, 3109. [CrossRef]
27. Wang, G.; Chao, Y.; Cao, Y.; Jiang, T.; Han, W.; Chen, Z. A comprehensive review of research works based on evolutionary game theory for sustainable energy development. *Energy Rep.* **2022**, *8*, 114–136.
28. Dong, Z.; Yu, X.; Chang, C.T.; Zhou, D.; Sang, X. How does feed-in tariff and renewable portfolio standard evolve synergistically? An integrated approach of tripartite evolutionary game and system dynamics. *Renew. Energy* **2022**, *186*, 864–877.
29. Cheng, L.; Zhang, M.; Huang, P.; Lu, W. Game-theoretic approaches for power-generation companies' decision-making in the emerging green certificate market. *Sustainability* **2025**, *17*, 71. [CrossRef]
30. Ning, J.; Xiong, L. Analysis of the dynamic evolution process of the digital transformation of renewable energy enterprises based on the cooperative and evolutionary game model. *Energy* **2024**, *288*, 129758.
31. Li, X.; Chen, L.; Sun, F.; Hao, Y.; Du, X.; Mei, S. Share or not share, the analysis of energy storage interaction of multiple renewable energy stations based on the evolution game. *Renew. Energy* **2023**, *208*, 679–692. [CrossRef]
32. Cheng, L.F.; Yin, L.F.; Wang, J.H.; Shen, T.; Chen, Y.; Liu, G.Y.; Yu, T. Behavioral decision-making in power demand-side response management: A multi-population evolutionary game dynamics perspective. *Int. J. Electr. Power Energy Syst.* **2021**, *129*, 106743. [CrossRef]
33. Lv, H.; Wu, Q.; Ren, H.; Li, Q.; Zhou, W. Collaborative strategy towards a resilient urban energy system: Evidence from a tripartite evolutionary game model. *Energy* **2024**, *313*, 133818. [CrossRef]
34. Xiao, X.; Yin, J.; Chen, L.; Wang, M.; Zhao, Y.; Li, Z. Evolutionary game-theoretic modeling of massive distributed renewable energy deployment towards low-carbon distribution networks. *J. Mod. Power Syst. Clean Energy* **2023**, *11*, 1519–1528. [CrossRef]
35. Cheng, L.F.; Yu, T. Nash equilibrium-based asymptotic stability analysis of multi-group asymmetric evolutionary games in typical scenarios of electricity market. *IEEE Access* **2018**, *6*, 32064–32086. [CrossRef]
36. Wang, J.; Zhou, Z.; Botterud, A. An evolutionary game approach to analyzing bidding strategies in electricity markets with elastic demand. *Energy* **2011**, *36*, 3459–3467. [CrossRef]
37. Cheng, L.F.; Zhang, J.; Yin, L.F.; Chen, Y.; Wang, J.H.; Liu, G.Y.; Wang, X.G.; Zhang, D.X. General three-population multi-strategy evolutionary games for long-term on-grid bidding of generation-side electricity market. *IEEE Access* **2021**, *9*, 5177–5198. [CrossRef]
38. Han, O.; Ding, T.; Bai, L.; He, Y.; Li, F.; Shahidehpour, M. Evolutionary game-based demand response bidding strategy for end-users using Q-learning and compound differential evolution. *IEEE Trans. Cloud Comput.* **2021**, *10*, 97–110. [CrossRef]
39. Jiao, J.; Chen, J.; Li, L.; Li, F. A study of local governments' and enterprises' actions in the carbon emission mechanism of subsidy or punishment based on the evolutionary game. *Chin. J. Manag. Sci.* **2017**, *25*, 140–150. [CrossRef]
40. Cheng, L.F.; Huang, P.; Zou, T.; Zhang, M.Y.; Peng, P.; Lu, W.T. Evolutionary game-theoretical approaches for long-term strategic bidding among diverse stakeholders in large-scale and local power markets: Basic concept, modelling review, and future vision. *Int. J. Electr. Power Energy Syst.* **2025**, *166*, 110589. [CrossRef]
41. Cheng, L.F.; Peng, P.; Lu, W.T.; Sun, J.; Wu, F.; Shi, M.M.; Yuan, X.D.; Chen, Y. The evolutionary game equilibrium theory on power market bidding involving renewable energy companies. *Int. J. Electr. Power Energy Syst.* **2025**, *167*, 110588. [CrossRef]

42. Wijk, L. On Stochastic Control Theory for Dynamic Carbon Emission Reduction. Master's Thesis, Utrecht University, Utrecht, The Netherlands, 2024.
43. Huang, W.; Liang, J.; Dong, Y. Optimal Stochastic Control Problem for a Carbon Emission Reduction Process. *SIAM J. Appl. Math.* **2023**, *83*, 1272–1295. [CrossRef]

processes

MDPI

Article

Process Improvement and Economic and Environmental Evaluation of Bio-Hydrogenated Diesel Production from Refined Bleached Deodorized Palm Oil

Amata Anantpinijwatna [1], Lida Simasatitkul [2,*], Kanokporn Yooyen [2], Suksun Amornraksa [3], Suttichai Assabumrungrat [4,5] and Karittha Im-orb [6]

1 Department of Chemical Engineering, School of Engineering, King Mongkut's Institute of Technology Ladkrabang, Bangkok 10520, Thailand; amata.an@kmitl.ac.th
2 Department of Industrial Chemistry, Faculty of Applied Science, King Mongkut's University of Technology North Bangkok, Bangkok 10800, Thailand
3 Chemical and Process Engineering Program, The Sirindhorn International Thai-German Graduate School of Engineering, King Mongkut's University of Technology North Bangkok, Bangkok 10800, Thailand; suksun.a@tggs.kmutnb.ac.th
4 Center of Excellence in Catalysis and Catalytic Reaction Engineering, Department of Chemical Engineering, Faculty of Engineering, Chulalongkorn University, Bangkok 10330, Thailand; suttichai.a@chula.ac.th
5 Bio-Circular-Green-Economy Technology & Engineering Center (BCGeTEC), Faculty of Engineering, Chulalongkorn University, Bangkok 10330, Thailand
6 Program in Food Process Engineering, School of Food Industry, King Mongkut's Institute of Technology Ladkrabang, Bangkok 10520, Thailand; karittha.im@kmitl.ac.th
* Correspondence: lida.s@sci.kmutnb.ac.th

Abstract: The co-production of BHD with other renewable fuels (i.e., using a novel process involving carbon dioxide utilization to achieve the global sustainability goal) is presented. The three configurations of BHD production from refined bleached deodorized palm oil (RBDPO), including (1) the conventional BHD process with hydrogen recovery (BHD process), (2) the BHD process coupled with the Fischer–Tropsch process (BHD-FT process), and (3) the BHD process coupled with the bio-jet fuel and methanol processes (BHD-BIOJET-MEOH process) are investigated using the process model developed in Aspen Plus. The effect of the operating parameters is studied, and the condition of each process offering the highest BHD yield is proposed. Then, the pinch analysis and heat exchanger network (HEN) design of each proposed process are performed to find the highest energy-efficient configuration. The economic and environmental analysis is later performed to investigate the sustainability performance of each configuration. The conventional BHD process requires less hydrogen and consumes less energy than the others. The BHD-BIOJET-MEOH process is the most economically feasible, offering the highest net present value (NPV) of USD 7.93 million and the shortest payback period of 3 years and 1 month. However, it offers the highest carbon footprint of 0.820 kgCO_2 eq./kg of BHD, and it presented the highest potential environmental impact (PEI) in all categories.

Keywords: bio-hydrogenated diesel; process improvement; hydrodeoxygenation; economic analysis; carbon footprint

1. Introduction

Rising energy demands across the transportation, building, and all industrial sectors lead to a high rate of fossil fuel consumption, which releases large amounts of greenhouse gases during combustion, contributing to air pollution and global warming issues. As a result, several countries agreed to set a target of net-zero carbon emissions by 2050 [1]. The European Union (EU) has expressed its commitment to reducing greenhouse gas emissions by at least 55% by 2030 [2]. To achieve carbon neutrality, the energy strategy encourages the government to pursue technology for renewable energy production. For example, the Neste company produces renewable diesel and renewable bio-jet fuel to reduce greenhouse gas emissions in accordance with the greenhouse gas protocol by 2035 [3]. Using biomass derived from agricultural crops and residues as an energy source for the production of biofuels and biochemicals is a suitable method for agricultural countries.

For green diesel production, vegetable oils such as canola oil, rapeseed oil, flaxseed oil, soybean oil, and sunflower oil are used as raw materials for hydroprocessing vegetable oils in Canada [4]. In Asia, palm fatty acid distillate, palm oil, and palm kernel oil are more popular due to their large-scale cultivation [5]. Waste cooking oil can also be used in hydroprocessing vegetable oil because the price is less expensive [6,7]. Thailand is an agricultural country with various raw materials that are suitable for the production of bio-hydrogenated diesel (BHD) or green diesel, which mainly contains hexadecane. BHD is a renewable fuel that can be produced from the hydrodeoxygenation of edible and non-edible vegetable oils, animal fats, and used oils. According to the Ministry of Agriculture and Cooperatives [8], animal fats and used oils do not have the potential to be commercially viable for BHD production because the amount that can be collected each year is relatively small. According to business trends and the oil palm industry in 2020–2022, Thailand has approximately 500,000 tons per year of palm oil reserved for consumption and biodiesel production [9]. The oil palm has the highest potential for oil production per area and, thus, the lowest production cost compared with other oil crops [9]. Therefore, it has great potential to be used as a raw material source for BHD production.

In the production of BHD, vegetable oil reacts with hydrogen under certain conditions. The triglycerides (TG) and fatty acids (FA) are transformed into hydrocarbon compounds by the removal of oxygen atoms in the molecules. The main reactions of BHD production include the exothermic cracking of TG, reduction of FA, and hydrogenation reactions for the breaking of a double bond into a single bond. Hydrodeoxygenation (HDO) is the reaction to remove oxygen from the molecule in the form of water in the presence of hydrogen. Decarbonylation (DCO) and decarboxylation (DCO_2) release carbon monoxide and carbon dioxide, respectively. HDO and DCO_2 are exothermic reactions, while DCO is an endothermic reaction [10]. Refined bleached deodorized palm oil (RBDPO) is obtained from refining, bleaching, and deodorizing crude palm oil. It can be used as a raw material for fatty acid methyl ester (FAME) and BHD production. BHD contains hydrocarbon compounds in the same range of FA contents. RBDPO is a commonly used feedstock due to its availability, relatively low cost, and favorable properties for conversion into BHD [10].

A large amount of hydrogen and a high operating pressure are generally required for the reactions to obtain high yield [10–12]. The main products produced are long-chain hydrocarbons (n-alkane), typically in the range of diesel fuel, and thus, they are referred to as BHD. Propane (C_3H_8) and carbon monoxide are also produced as byproducts. The properties of the BHD are similar to petroleum-based diesel, and thus, it can be used in conventional combustion engines without any modifications. In addition, BHD has better properties compared to biodiesel in many aspects [13]. The current technology for BHD production is carried out in a fixed bed reactor with a commercial catalyst (i.e., $NiMo/Al_2O_3$ [14,15]) under high temperature and high pressure. For example, Kaewtrakulchai et al. [16] produced BHD from palm oil under 340–420 °C. They found that a high BHD yield and 100% conversion were achieved [16]. To enhance the hydrogen dispersion in raw material, the solvent, such as dodecane, decane, heptadecane, and mesitylene, was incorporated throughout the reaction to decrease mass transfer resistance [17–19].

The slurry phase reactor was utilized for the hydroprocessing of cottonseed acidic oil and offered a high production rate of BHD [20]. Previously, several studies carried out experiments on BHD production via the hydroprocessing of palm oil and waste cooking oil using a batch continuous fixed-bed reactor [7,21]. Additionally, BHD production was examined using process models developed in commercial software in order to study the effect of operating parameters on process performance [22–24]. As there are some excess hydrogen and gaseous products released from the reactions, conversion of these gases to high-value products not only improves the economic performance but also the environmental one.

The management of byproducts, or wastes, is crucial to the sustainability of any production process. Specifically, it is more preferable if the carbon monoxide byproduct from the BHD production process can be converted into a more valuable product. One possible option is to mix it with hydrogen and utilize it as a syngas feedstock in the Fischer–Tropsch (FT) process to produce synthetic hydrocarbon fuels [25]. The produced fuels are renewable and clean fuels. They can be burnt in existing combustion engines and release less harmful emissions due to their very low sulfur and aromatic contents compared to fossil fuels [26]. Another possible and attractive option to utilize carbon monoxide is to use it as a feedstock to produce a primary chemical such as methanol. Methanol is an important industrial chemical because it is a base substance used to produce many other valuable chemicals, such as formaldehyde, acetic acid, methyl tertiary butyl ether (MTBE), and more [27]. Methanol can be synthesized via the hydrogenation of carbon monoxide. Phichisurathaworn et al. [28] pointed out that the co-production of BHD with methanol could offer more benefits in terms of economics compared with the stand-alone BHD process.

Bio-jet fuel, or sustainable aviation fuel, is another potential product that can be produced from the catalytic cracking of BHD. Bio-jet fuel is a renewable aviation fuel that has been recently developed and used for aircraft. Its use in air transportation has been greatly promoted to mitigate carbon emissions and reach carbon neutrality. It is targeted that, in 2050, 70% of bio-jet fuel will be incorporated into normal aviation fuel consumption [29]. Bio-jet fuel can be produced from renewable feedstocks, such as biomass or vegetable oils, via hydrotreating TG or FA with hydrogen under a high-pressure condition [30]. This hydrodeoxygenation reaction is strongly exothermic and can cause an uncontrollable rise in temperature in the BHD production system [31]. Moreover, the FT reaction to synthesize paraffinic kerosene can also be carried out [32]. Among the studies, in bio-jet fuel production, a hydroprocessing reactor was designed as reactive distillation (RD) to reduce the energy requirement [33,34]. Carrasco-Suárez et al. [35] proposed a bio-jet fuel production process from waste cooking oil, consisting of the esterification process, the transesterification process, and the hydrotreating of biodiesel. Naphtha, BHD, bio-jet fuel, and glycerol were obtained. The co-production of BHD and bio-jet fuel offered an economical process in case of large plant size [36].

To investigate the economic viability of the BHD process, the economic analysis using the payback period, NPV, and return on investment as indicators should be performed. The BHD production from palm fatty acid distillate, a byproduct from palm oil refineries, through the conventional BHD process and the one with heat integration can make a profit and give a payback period of 3 years [37]. However, the stand-alone BHD production was not a feasible process comparable with the conventional biodiesel production process [22]. The processes offering the co-production of BHD with bio-jet fuel and with methanol were more economically feasible than a stand-alone BHD production process [28,36].

For the environmental aspect, a life cycle assessment (LCA) is used to evaluate the environmental impacts under system boundaries such as gate-to-gate, cradle-to-gate, and cradle-to-grave. The inventory data collected from mass balance and energy balance is used to evaluate the environmental impact. The LCA of BHD production from palm fatty acid distillate and FAME was analyzed in terms of a cradle-to-gate boundary. The result showed that the cultivation and feedstock production stages significantly contributed to environmental impacts in the categories of greenhouse gas emission, human toxicity, freshwater eutrophication, ecotoxicity, natural land transformation for BHD production, and BHD production from the co-products generated during palm oil production [38,39]. Therefore, process improvement based on the bio-circular economy concept is a promising strategy for developing a more sustainable BHD process.

Although BHD production from palm oil has been widely studied, the co-production of BHD with other renewable fuels using a novel process involving carbon dioxide utilization has not been discussed. Therefore, this work aims to enhance the BHD production process in terms of technical, economic, and environmental perspectives. Three different scenarios of the improved process were designed and assessed. The first process is a base case of BHD production, with hydrogen recovery to reduce hydrogen consumption (BHD process). The second process is BHD production coupled with synthesis fuel production via FT synthesis (BHD-FT process). The last process is BHD production coupled with bio-jet fuel and methanol production (BHD-BIOJET-MEOH process). First, the technical performance of the three processes is studied by investigating the effect of the operating conditions. Pinch analysis and heat exchanger network (HEN) design are performed for each configuration that offers the highest BHD yield to minimize energy consumption and improve energy efficiency. The economic analysis is performed using NPV, payback period, and internal rate of return (IRR) as indicators. Then, the LCA is used to evaluate the environmental performance of the three processes. Finally, the process offering the best performance is presented.

2. Methodology

A flow diagram showing the methodology used in this work is illustrated in Figure 1. First, the models of three BHD processes (i.e., the BHD process, BHD-FT process, and BHD-BIOJET-MEOH process) are developed in Aspen Plus V12.1 based on the reaction kinetics of BHD reactions. Then, the effect of the operating parameters is investigated in order to find the suitable condition offering the highest BHD yield. After that, the pinch analysis and heat exchanger network (HEN) design are performed to optimize the use of external utilities and improve the energy efficiency of the process, as well as to maximize the heat recovery of the selected BHD processes. Finally, the selected processes were evaluated by an economic assessment and a life cycle assessment (LCA).

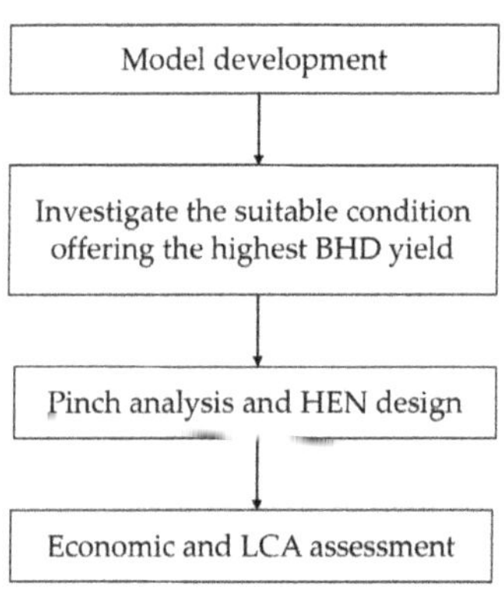

Figure 1. Methodology of the process design.

2.1. Process Description

Normally, the raw materials for BHD production are refined bleached deodorized palm oil (RBDPO) and hydrogen. However, since RBDPO contains triglycerides with a wide variety of FA compositions, tripalmitin is used to represent RBDPO in this study in order to simplify the calculation. The reactions involved in BHD production are shown in Equations (1) to (6). There are two reaction pathways involved the BHD production. The first pathway (RP-I) is the decarboxylation of aldehydes, giving pentadecane as the final product. The second pathway (RP-II), the reaction 1 and reaction 2 are similar to the RP-I as shown in Equations (1) and (2). Then other reactions are the reduction of aldehyde and the dehydration of alcohol, giving hexadecane as shown in Equations (4) to (6) as the final product.

RP-I

$$\text{Reaction 1 Tripalmitin} + 3H_2 \rightarrow \text{Palmitic acid} + C_3H_8 \quad (1)$$

$$\text{Reaction 2 Palmitic acid} + H_2 \xrightarrow{k_1} \text{Hexadecanal} + H_2O \quad (2)$$

$$\text{Reaction 3 Hexadecanal} \xrightarrow{k_2} \text{Pentadecane} + CO \quad (3)$$

RP-II

$$\text{Reaction 4 Hexadecanal} \overset{K_3}{\leftrightarrow} \text{Hexadecanol} \quad (4)$$

$$\text{Reaction 5 Hexadecanol} \xrightarrow{k_4} \text{Hexadecene} + H_2O \quad (5)$$

$$\text{Reaction 6 Hexadecene} + H_2 \xrightarrow{k_5} \text{Hexadecane} \quad (6)$$

The block flow diagrams of three configurations of BHD production were preliminarily generated prior to the rigorous design. The isothermal and adiabatic conditions of the plug-flow reactor (PFR) were assumed in this study. Three BHD production processes with different downstream managements were created and simulated using Aspen Plus, namely (a) a BHD process, (b) a BHD-FT process, and (c) a BHD-BIOJET-MEOH process, as shown in Figure 2.

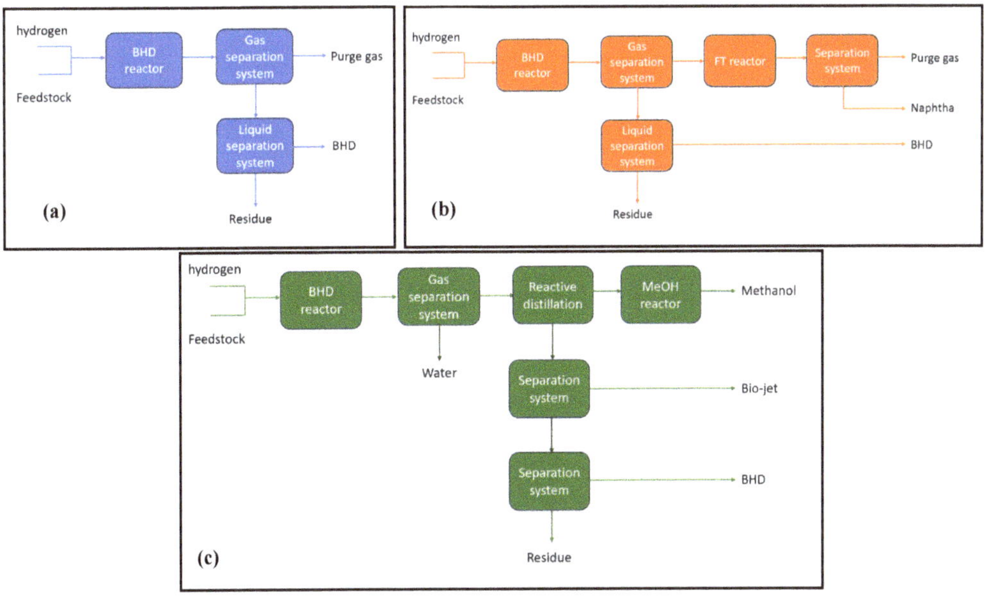

Figure 2. Block flow diagrams of (**a**) BHD process, (**b**) BHD-FT process, and (**c**) BHD-BIOJET-MEOH process.

Mass balances and energy balances for the three processes were performed by Aspen Plus. The tripalmitin is represented by RBDPO. The nonrandom two liquids (NRTL) activity coefficient model was used to calculate the vapor–liquid equilibrium in the BHD reactor, RD, and distillation column because the mixtures consist of ester, alcohol, aldehyde, and water. The Peng–Robinson (PENG-ROB) model, which was a cubic equation of state, was used for nonpolar or mildly polar mixtures, such as hydrocarbon in the FT reactor [40]. The Soave–Redlich–Kwong (SRK) model, which was a cubic equation of state, was applied in the methanol reactor because the mixtures consisted of wet synthesis gas and methanol [28,41,42]. The UNIFAC model was an activity coefficient model, which was used for liquid systems [43]. After all process flowsheets were generated, a pinch analysis and heat integration were applied to the HEN to minimize energy consumption and improve the process's energy efficiency. High-pressure steam and cooling water were used as hot and cold utilities, respectively. Then, the Aspen Energy Analyzer was used to analyze the energy consumption and design the HEN.

2.2. Economic Evaluation

Economic evaluations of three BHD processes were evaluated and compared. The economic calculation followed the method reported by R. Turton et al. [44]. The assumptions for the economic feasibility assessment of this research are as follows:

(a) operation of 8 h per day, and 340 working days per year;
(b) project lifetime is 15 years;
(c) costs of raw materials and products are based on 2022;
(d) income and expenses are constant throughout the project's lifetime;
(e) the manufacturing cost does not include the waste treatment cost;
(f) the deactivated catalyst and catalyst regeneration are neglected.

The total investment cost consisted of fixed capital investment cost (FCI) and the working capital cost. The operating cost consisted of the raw materials cost and the utility cost. NPV, payback period, and IRR were used as the economic indicators.

2.3. Environmental Evaluation

For the environmental aspect, the LCA of the BHD processes was analyzed on the basis of cradle-to-gate using LCSoft software V6.2 from PSE for the SPEED company [45] and normalized by the ILCD 2011 method. The database of RBDPO in LCSoft included refining, bleaching, and deodorizing. The environmental impacts of cultivation and transportation were constant for all cases. The effects of catalyst regeneration and disposal on the environment were not considered. The carbon footprint was the total amount of GHGs, which were presented as kilograms of CO_2 equivalent and could be calculated from Equation (7).

$$CO_{2,eq} = \frac{(m_{GHG} \times CF_{GHG})}{m_{product}} \quad (7)$$

where m_{GHG} was the mass flow rate of the GHG (kg h^{-1}), and CF_{GHG} was the characterization factor of the GWP.

3. Model Development

After the preliminary flowsheets were generated, the models of three BHD processes were developed in Aspen Plus based on the kinetic data of BHD reactions obtained from Yenumala et al. [11], in which the reactions were the first order of concentration in the liquid phase, and the rate constant (k) of each reaction followed the Arrhenius equation, as shown in Equation (8). To perform the feasibility study of BHD production using various process configurations, an RBDPO of 22,000 kg/day, which was a pilot scale capacity of a refinery plant, was selected as model input. The model development of three BHD processes was explained in the following sections.

$$k = A \exp(-E_a/RT) \quad (8)$$

3.1. BHD Process

For the BHD process, the NRTL activity coefficient model was used to calculate the vapor–liquid equilibrium in the BHD reactor. And the isothermal and adiabatic conditions of a PFR were maintained. The Aspen Plus model flowsheet of the BHD process is shown in Figure 3. Tripalmitin was fed to a pump (P-01) and a heater (E-03) to increase the pressure to 30 bar with temperatures in the range of 330–460 °C, respectively. The hydrogen gas was fed to multistage compressors (C-01 and C-02) and coolers (E-01 and E-02) until the pressure of 30 bar and a temperature in the range of 330–460 °C were achieved. Hydrogen gas and tripalmitin were mixed at a mixer (M-01) before entering the adiabatic BHD reactor (RE-101) in which Ni/Al_2O_3 was selected as the catalyst. The main products were pentadecane (n-C_{15}) and hexadecane (n-C_{16}), while propane, hexadecanol, water, and carbon monoxide were side products. All products, including hydrogen gas and palmitic acid residue from the reactions, were sent to a chiller (E-04) and a flash tower (F-01) to separate the gaseous products (propane, carbon monoxide, and hydrogen) from the liquid products (pentadecane, hexadecane, hexadecanol, water, and palmitic acid). The gas stream (S12) was sent to pressure swing adsorption (PSA) to adsorb carbon monoxide and propane at the condition reported in previous work [46]. The unadsorbed gas, mainly hydrogen, was then partially purged into the environment. The remaining hydrogen was adjusted to the same pressure and temperature by the compressor (C-03) and heater (E-05) before re-circulation back with the feed stream. The liquid product was sent to a decanter (DE-01) to remove water. A rigorous design of the distillation column by using the RadFrac model, in which the NRTL activity coefficient model was used to calculate the vapor–liquid equilibrium and the UNIFAC model was used for the liquid system [43], was applied for the distillation column. Finally, the propane and BHD were purified to 96.0% and 99.4%, respectively.

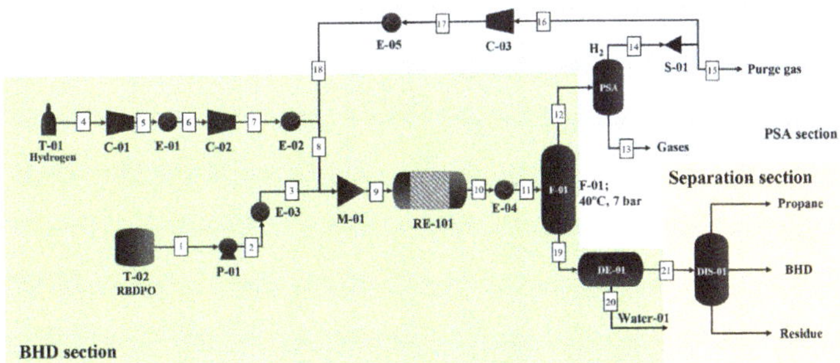

Figure 3. Aspen Plus model flowsheet of BHD process.

3.2. BHD-FT Process

The BHD process in the BHD-FT process was similar to the BHD process explained in Section 3.1, but the FT process was integrated for the production of paraffin from the carbon monoxide and hydrogen gas obtained from the BHD process, as shown in Figure 4. The PENG-ROB model was used to estimate the product composition of the FT reactor [40]. Thus, the product from the BHD reactor (RE-101) was cooled to 100 °C before entering the flash tower (F-01) to separate the gaseous compound from liquids. A gas stream (S12), consisting of hydrogen, carbon monoxide, and propane, was preheated by the heater (E-05) to increase its temperature in the range of 210–270 °C before entering the FT reactor (RE-201) in which the Co-based catalyst was used. This process could produce naphtha (n-C_5 to n-C_{10}) and a small amount of kerosene (n-C_{11} to n-C_{14}). The reactor product (S14) was sent to a three-phase flash tower (F-02) to separate the unreacted gases (carbon monoxide and hydrogen) and water from liquid hydrocarbons and propane. The hydrocarbon stream (S16) was then mixed with the treated bottom product (S22) from the first flash tower (F-01) and then sent to the separation section. The gaseous products (carbon monoxide, hydrogen gas, and propane) were separated from naphtha by the first distillation column (D-01). Then, propane and naphtha were purified using the distillation column (D-02). Finally, BHD, consisting of kerosene and diesel, was separated from the palmitic acid and hexadecanol (residue) using the distillation column (D-03).

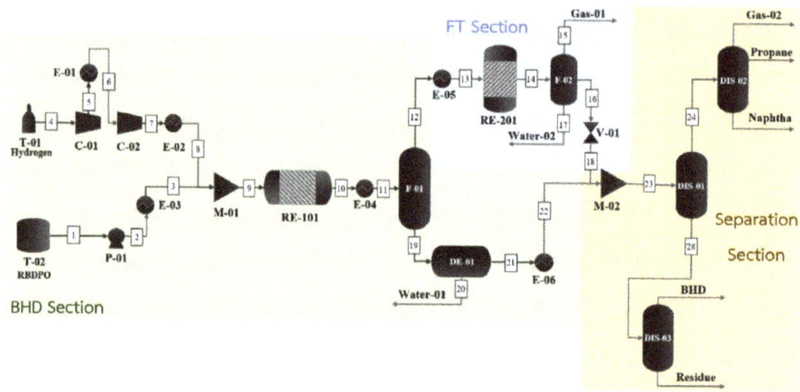

Figure 4. Aspen Plus model flowsheet of BHD-FT process.

3.3. BHD-BIOJET-MEOH Process

The Aspen Plus model flowsheet of the BHD-BIOJET-MEOH process is shown in Figure 5. The BHD process was similar to the BHD process explained in Section 3.1. However, the products from the RE-101 reactor were cooled to 35 °C using a heat exchanger (E-04) and sent to a three-phase flash separation (F-01) to remove water residue. Then, gaseous and liquid products from F-01 were separately heated to 240 °C by using heat exchangers (E-05 and E-06) before feeding to the RD using a Co-Mo/Al_2O_3 catalyst in which the NRTL activity coefficient model was used to calculate the production of jet fuel. This research assumed the cracking of BHD molecules of 10%. A gaseous product from RD (S17), containing synthesis gases (hydrogen gas and carbon monoxide), was heated by the heater (E-07) before entering the MEOH section at which the equilibrium reactor (RE-301) was used to produce methanol from synthesis gas under a Cu/Zn catalyst. As the gaseous product of the RD had low H_2/CO, most of the H_2 was consumed during methanol synthesis. As a result, recirculation of the gas product leaving the methanol reactor to improve methanol yield was not cost-effective to carry out. The SRK model was applied to calculate the product composition of this unit [10,20,21]. The top liquid product from RD (S22), containing jet fuel (n-C8 to n-C16) and propane, was sent to a valve (V-01) to reduce the pressure to 1 bar. The bottom product from RD (S26), containing BHD, hexadecanol, and palmitic acid residue, was sent to a valve (V-02) and the heat exchanger (E-08) to reduce the pressure and temperature to 1 bar and 340 °C, respectively. In the separation section, methanol was purified using the distillation column (DIS-01) with a partial condenser, resulting in highly purified methanol (99.6% by volume) as the bottom product. Crude propane and crude bio-jet fuel from the V-01 valve were purified by a flash separation unit (F-02). Highly purified jet fuel (99.9% by volume) was obtained as the bottom product, while propane was derived as the overhead product. BHD in the treated bottom product (S28) was further purified using the distillation column (DIS-02) to obtain the BHD as the overhead product and the other residue at the bottom.

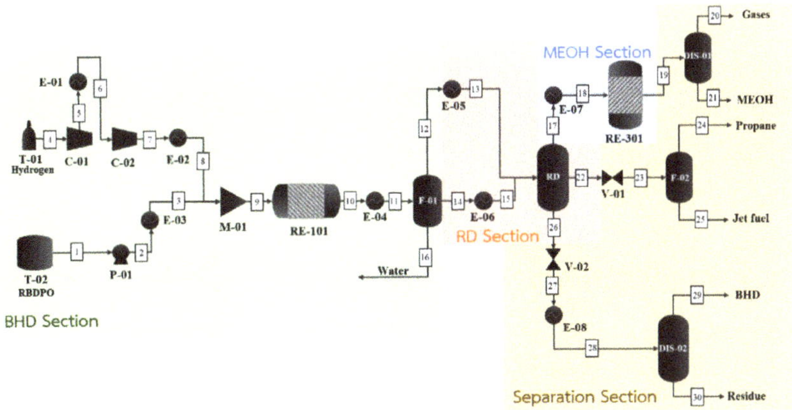

Figure 5. Aspen Plus model flowsheet of BHD-BIOJET-MEOH process.

4. Results and Discussion

The process design and optimization of the three processes were performed. When the optimal operating conditions are found, a HEN was designed to improve energy consumption. Finally, the three configurations of the BHD process were evaluated in terms of economic and environmental aspects.

4.1. Technical Performance

The technical performance of the three BHD processes is performed to investigate the effect of the operating conditions offering the highest BHD yield. The optimal process conditions of BHD, BHD-FT, and BHD-BIOJET-MEOH processes were summarized in Tables 1–3, respectively. And the technical performance of each BHD process is shown in Table 4. It was found that the BHD process offered the highest BHD of 1966.6 kg/h, followed by the BHD-FT process (1961.08 kg/h) and the BHD-BIOJET-MEOH process (1745.96 kg/h), respectively. However, the BHD-FT process offered 60.86 kg/h of naphtha as a co-product. For the BHD-BIOJET-MEOH process, the methanol of 99.08 kg/h and bio-jet of 189.25 kg/h were produced as additional products. Regarding energy consumption, the BHD process released the highest amount of energy, followed by the BHD-FT and BHD-BIOJET-MEOH processes, respectively.

Table 1. Optimal operating conditions of BHD process.

Operating Unit	Parameter	Condition
RE-101	Temperature (°C)	340
	Pressure (bar)	30
	H_2/TP ratio	6.4
	L:D reactor	5:1

Table 1. *Cont.*

Operating Unit	Parameter	Condition
RE-102	Temperature (°C)	430
	Pressure (bar)	30
	H_2/TP ratio	11.83
	L:D reactor	6:1
F-01	Temperature (°C)	40
	Pressure (bar)	7
DE-01	Temperature (°C)	35
	Pressure (bar)	7
PSA	Pressure (bar)	7
Pressure (bar)	Number of stages	18
	Feed stage	9
	Reflux ratio	0.864
	Distillate-to-feed ratio	0.915
	Distillate vapor fraction	0.1
	Pressure (bar)	7

Table 2. Optimal operating conditions of BHD-FT process.

Operating Unit	Parameter	Condition
RE-201	Temperature (°C)	130
	Pressure (bar)	30
	H_2/CO ratio	1.19
	L:D reactor	0.185:0.046
F-01	Temperature (°C)	100
	Pressure (bar)	30
F-02	Temperature (°C)	25
	Pressure (bar)	30
DE-01	Temperature (°C)	100
	Pressure (bar)	12
DIS-01	Number of stages	11
	Feed stage	6
	Reflux ratio	1.045
	Distillate-to-feed ratio	0.2595
	Pressure (bar)	12
DIS-02	Number of stages	7
	Feed stage	4
	Reflux ratio	3
	Distillate-to-feed ratio	0.76
	Distillate vapor fraction	0.7
	Pressure (bar)	12
DIS-03	Number of stages	21
	Feed stage	11
	Reflux ratio	2
	Distillate to feed ratio	0.906
	Pressure (bar)	1

Table 3. Optimal operating conditions of BHD-BIOJET-MEOH process.

Operating Unit	Parameter	Condition
RE-301	Temperature (°C)	210
	Pressure (bar)	30
	H_2/CO ratio	1.02
	L:D reactor	1.15:0.288
F-01	Temperature (°C)	35
	Pressure (bar)	30
F-02	Temperature (°C)	25
	Pressure (bar)	1
RD	Number of total stages	7
	Feed stage	4
	Reactive stage	4
	Reflux ratio	0.075
	Distillate rate (kmol/h)	22
	Distillate vapor fraction	0.75
	Pressure (bar)	30
DIS-01	Number of stages	5
	Feed stage	3
	Reflux ratio	0.6
	Distillate-to-feed ratio	0.702
	Pressure (bar)	1
DIS-02	Number of stages	9
	Feed stage	5
	Reflux ratio	2
	Distillate-to-feed ratio	0.899
	Pressure (bar)	1

Table 4. Technical performance of the three BHD processes.

Technical Performance	BHD Process	BHD-FT Process	BHD-BIOJET-MEOH Process
BHD (kg/h)	1966.60	1961.08	1745.96
Propane (kg/h)	45.26	35.75	184.17
Naphtha (kg/h)	-	60.86	-
Methanol (kg/h)	-	-	99.08
Bio-jet (kg/h)	-	-	189.25
Overall energy consumption (MW)	−0.52	−0.46	−0.22

4.2. Energy Analysis

An energy analysis was performed using an Aspen Energy Analyzer in Aspen Plus in order to find the energy-efficient conditions of the selected BHD processes. The pinch analysis provided the composite curve with a temperature difference of 10 °C, which is depicted in Figure 6. The high-pressure steam (HP steam) and cooling water were used as hot and cold utilities for all BHD processes. It was found that the BHD process required external hot and cold utilities of 1.85 MW and 2.37 MW, respectively. The BHD-BIOJET-MEOH process showed the same trend, as it required both external hot and cold utilities of 1.35 MW and 1.57 MW, respectively. Specifically, only external hot utilities of 2.30 MW were required for the BHD-FT process.

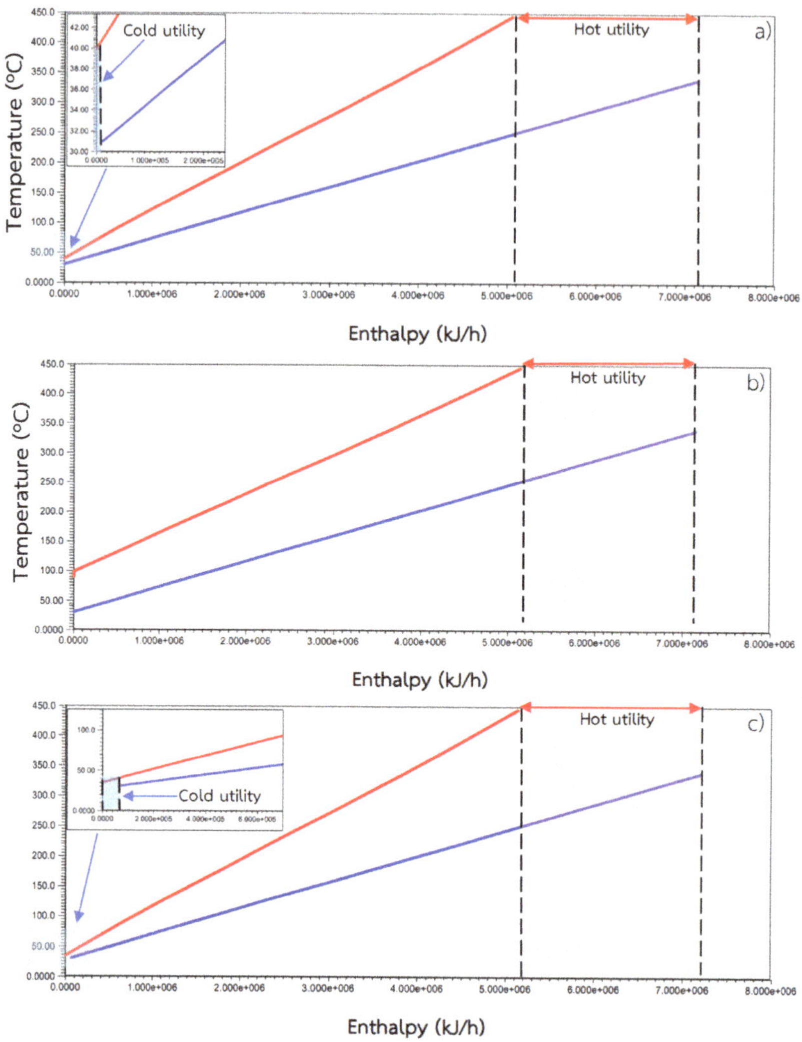

Figure 6. Composite curves of hot and cold utility of (**a**) BHD process, (**b**) BHD-FT process, and (**c**) BHD-BIOJET-MEOH process when red line is hot utility and blue line is cold utility.

The HEN design was later performed to maximize the heat recovery of the selected BHD processes. The HEN design diagram of the BHD process is shown in Figure 7. The red arrows are hot streams, while the blue arrows are cold streams. A HEN analysis indicated that the BHD process required two heaters (i.e., HEAT-01 and HEAT-02), which supply the external heat source, as shown in Figure 8. The HEN design diagram of the BHD-FT process is shown in Figure 9. The newly designed BHD-FT process including HEN required three coolers, namely COOL-01, COOL-02, and COOL-03, as shown in Figure 10. The HEN design diagram of the BHD-BIOJET-MEOH process is shown in Figure 11. After the HEN analysis, one heater and one cooler were required, namely HEAT-01 and COOL-01, as shown in Figure 12.

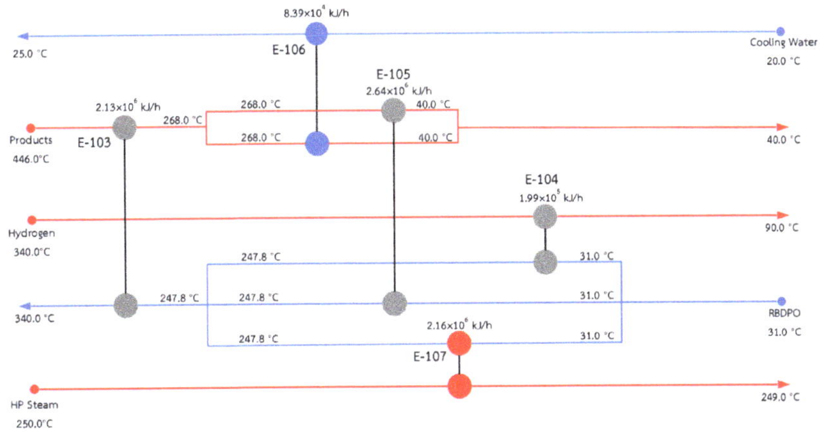

Figure 7. HEN design diagram of BHD process.

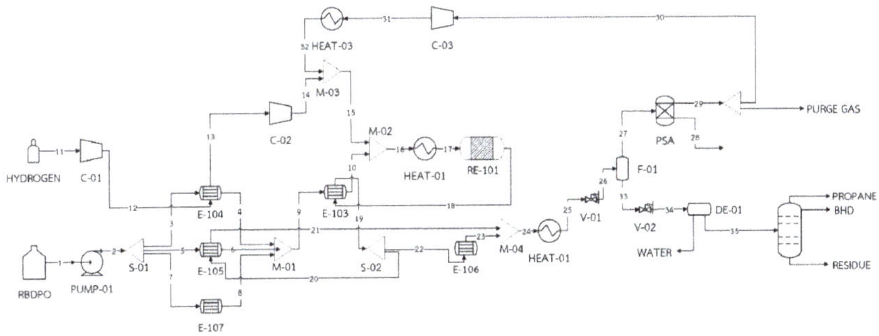

Figure 8. Newly designed BHD process including HEN.

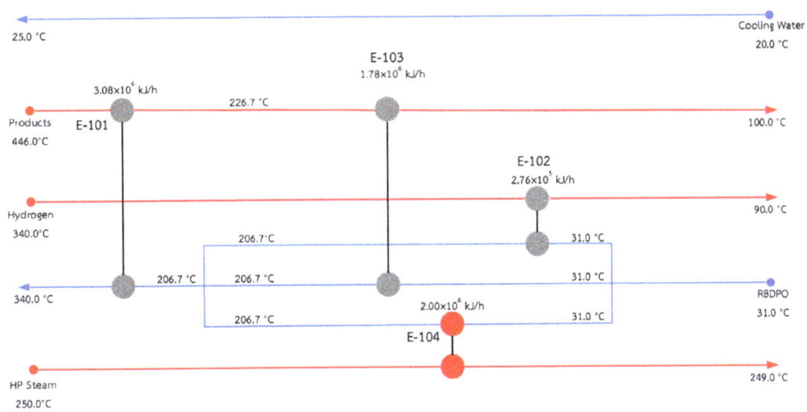

Figure 9. HEN design diagram of BHD-FT process.

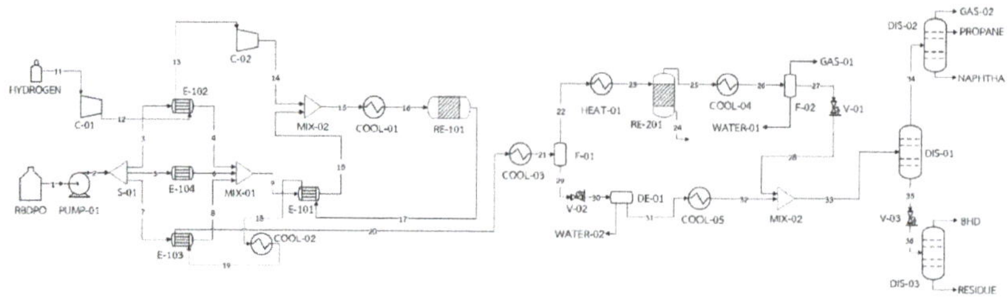

Figure 10. Newly designed BHD-FT process including HEN.

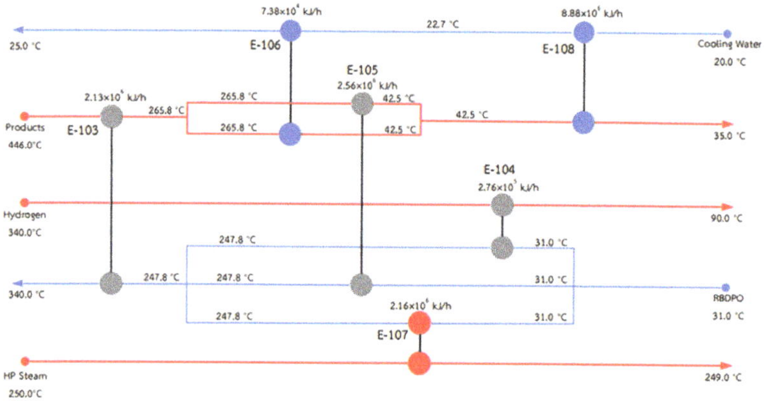

Figure 11. HEN design diagram of BHD-BIOJET-MEOH process.

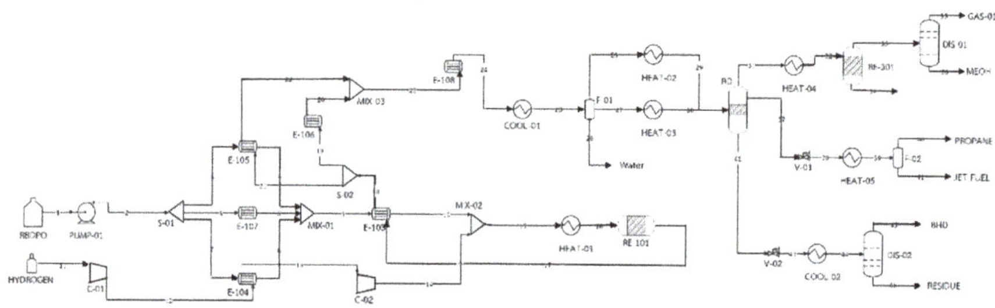

Figure 12. Newly designed BHD-BIOJET-MEOH process including HEN.

After performing the heat integration, the comparison revealed that the energy requirements for heating and cooling in the BHD-FT process were slightly higher than those of the BHD and BHD-BIOJET-MEOH processes due to a specific emphasis on the energy required for the FT reactor and the distillation columns. Notably, the BHD-BIOJET-MEOH process demonstrated the lowest energy demands for heating and cooling due to the simultaneous utilization of energy for reaction and separation in RD. The comparison of energy utilization before and after heat integration for the three BHD processes was illustrated in Figure 13. For the BHD process, the HEN design reduced the hot utility consumption of the process by 31.32% and reduced the cold utility consumption of the process by 26.85%. For the BHD-FT process, the hot and cold utility consumption could be reduced by 34.64% and 29.42%, respectively, after performing the HEN design. While the BHD-BIOJET-MEOH process showed the same trend, the requirement of the hot and cold utilities could be reduced by 29.16% and 26.14%, respectively.

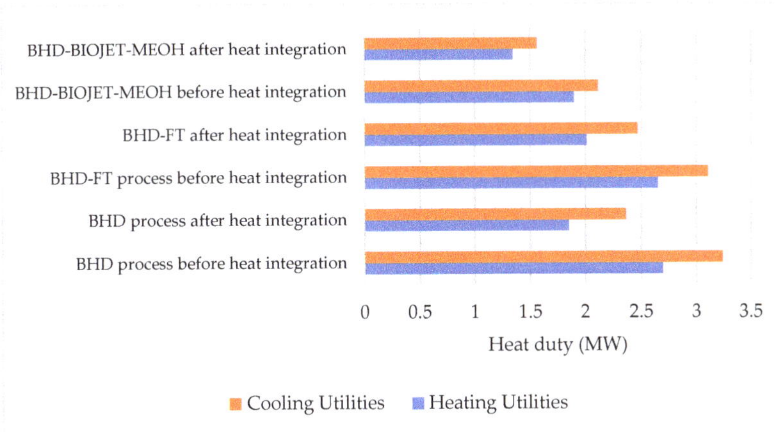

Figure 13. Hot and cold utility requirements before and after heat integration.

4.3. Economic Analysis

The economic analysis was performed for three BHD processes based on the same RBDPO feed rate of 22,000 kg/day. Raw material, product, and catalyst costs are shown in Table 5. The catalyst costs depended on reactor sizing and inlet flow rate. However, these costs were low compared with the cost of raw materials and equipment. Therefore, the catalyst costs had no influence on the total production cost. The economic feasibility assessment indicated that the BHD-FT process offered the highest FCI due to the high costs of the compressor, distillation, and flash separator, as shown in Table 6. The compressor cost and distillation cost accounted for 70% and 30% of the FCI, respectively. The total investment cost of the BHD-FT process was higher than that of the BHD process by approximately 6.93%. Considering the BHD-BIOJET-MEOH process, the additional heaters that provided the external heat source caused a higher cost compared with the BHD process. Thus, the BHD process offered the lowest FCI due to a smaller number of operating units. The total production costs of the three BHD processes are shown in Table 7. The total production costs of the BHD-FT process and the BHD-BIOJET-MEOH process were quite similar and higher than those of the BHD process due to the higher cost of raw materials. It was noted that the BHD process was still preferable in terms of total investment cost and total production cost.

Table 5. Raw material and product price.

Chemicals	Price (USD/kg)	References
Hydrogen	1.15	[47]
RBDPO	0.78	[48]
Ni/γ-Al_2O_3	10	[49]
Co-Mo/Al_2O_3	7.8	[49]
Cobalt catalyst	751.5	[50]
Cu/ZnO/Al_2O_3	300	[49]
Propane	0.725	[48]
Diesel	1.46	[51]
Naphtha	1.24	[48]
Jet-fuel	1.49	[51]
Methanol	0.48	[48]

Table 6. The summary of total investment cost.

Description	BHD Process (Millions USD)	BHD-FT Process (Millions USD)	BHD-BIOJET-MEOH Process (Millions USD)
Reactor	0.13	0.13	0.13
PSA	0.09	0	0
Flash	0.09	0.19	0.18
Pump	0.04	0.04	0.04
Compressor	1.04	1.11	1.11
Heat exchanger	0.03	0.02	0.03
Heater	0.02	0.01	0.14
Cooler	0.00	0.12	0.10
Decanter	0.08	0.06	0.00
Distillation	0.61	0.59	0.52
Fixed capital investment cost	3.51	3.76	3.74
Working capital cost	0.53	0.56	0.56
Total investment cost	4.04	4.32	4.30

Table 7. The summary of total production cost.

Description	BHD Process (Millions USD)	BHD-FT Process (Millions USD)	BHD-BIOJET-MEOH Process (Millions USD)
Raw material	5.93	5.93	5.98
Utilities	0.05	0.05	0.05
Catalyst	0.002	0.0023	0.03
Labor	0.05	0.05	0.05
Maintenance	0.18	0.19	0.19
Depreciation	0.35	0.38	0.37
Local tax	0.04	0.05	0.04
Insurance	0.18	0.19	0.19
Total production cost	6.72	6.95	7.93

The BHD-BIOJET-MEOH process offered more benefits than the other BHD processes in terms of NPV and IRR. The cumulative cash flow of each process is shown in Figure 14. It was found that the cumulative cash flow continuously increased when the project lifetime increased. And the payback periods of the BHD, BHD-FT, and BHD-BIOJET-MEOH processes were 3 years and 4 months, 3 years and 4 months, and 3 years and 1 month, respectively. Among the three BHD processes, the BHD-BIOJET-MEOH process was the most economically feasible process because it offered the highest NPV and IRR, followed by the BHD-FT and BHD processes, as shown in Table 8.

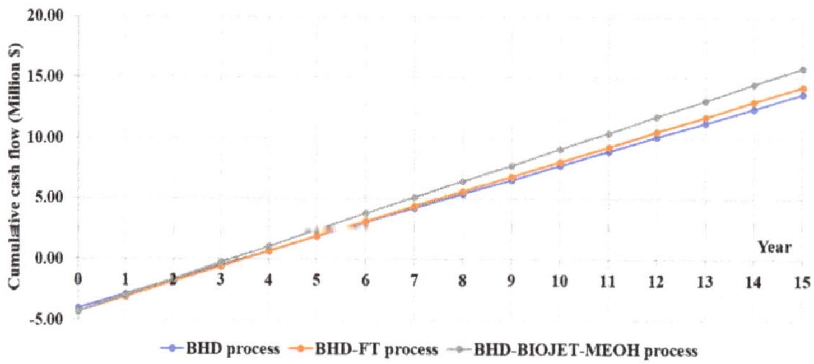

Figure 14. Cumulative cash flow of three BHD processes.

Table 8. The NPV and IRR of three BHD processes.

Parameter	BHD Process	BHD-FT Process	BHD-BIOJET-MEOH Process
NPV (Million USD)	6.69	6.95	7.93
Payback period	3 Y 4 M	3 Y 4 M	3 Y 1 M
IRR (%)	28.48	27.92	30.75

Considering the sensitivity analysis of only the BHD-BIOJET-MEOH process, in case the product selling price changes +30% and −30% of the base case, it was found that the NPV was negative when the selling price of the products decreased, as shown in Table 9. According to the literature [52,53], the price of diesel and jet fuel fluctuated according to the market price. Therefore, the BHD-BIOJET-MEOH process was the most attractive process because it could adjust the operating parameters of the RD, such as the distillate-to-feed ratio, the reflux ratio, etc., to maximize the profit.

Table 9. The sensitivity analysis of the BHD-BIOJET-MEOH process in case the product selling price changes.

Sensitivity (%)	Price List (USD/kg)				NPV (Million USD)
	Propane	BHD	Bio-Jet Fuel	MEOH	
−30	0.51	1.02	1.04	0.34	−14.53
−20	0.58	1.17	1.19	0.38	−6.92
−10	0.65	1.31	1.34	0.43	0.28
0	0.73	1.46	1.49	0.48	7.93
10	0.80	1.61	1.64	0.53	15.59
20	0.87	1.75	1.79	0.58	22.79
30	0.94	1.90	1.94	0.62	30.40

4.4. Life Cycle Assessment

In this section, environmental impacts including carbon footprint, human toxicity non-cancer (HTNC), human toxic cancer (HTC), human toxicity by exposure (HTPE), human toxicity by ingestion (HTPI), human toxicity carcinogenics (HTC), photochemical ozone formation (POCP), photochemical oxidation potential (PCOP), ozone depletion potential (ODP), global warming potential (GWP), terrestrial eutrophication, marine eutrophication potential (MEP), acidification potential (AP), terrestrial toxicity potential (TTP), aquatic toxicity (ATP), and freshwater ecotoxicity (ET) are evaluated for all BHD processes. The inputs were raw materials, chemicals, and fuels. The outputs of the process were effluent gas, wastewater, and solid waste.

The carbon footprint of the raw materials (including H_2 and RBDPO) and that of each operating unit of the three BHD processes are shown in Figure 15. Although all processes consumed the same amount of RBDPO, the hydrogen consumption of the BHD process was 28% less than the BHD-FT process and the BHD-BIOJET-MEOH process, resulting in a lower carbon footprint of the raw materials. As the BHD-BIOJET-MEOH process involved a large number of operating units, it offered the highest carbon footprint, followed by the BHD process and the BHD-FT process, respectively. The carbon footprints of each operating unit in the BHD, BHD-FT, and BHD-BIOJET-MEOH processes are shown in Figure 16a–c, respectively. It was found that the distillation columns showed the highest carbon footprint because they consumed a large amount of energy.

The analysis of potential environmental impacts (PEIs) of all BHD processes by normalization was performed using the ILCD 2011 software. The PEIs in the categories of human health and ecosystem are shown in Figure 17a,b, respectively. For the category of human health, the PEI of the BHD-BIOJET-MEOH process was the one that most contributed to

HTNC, while the HTC was comparable among all processes. Regarding the category of the ecosystem, the PEIs in the categories of PCOP and AP were the two biggest impacts for all processes. And the BHD-BIOJET-MEOH process offered the highest effect on the ecosystem.

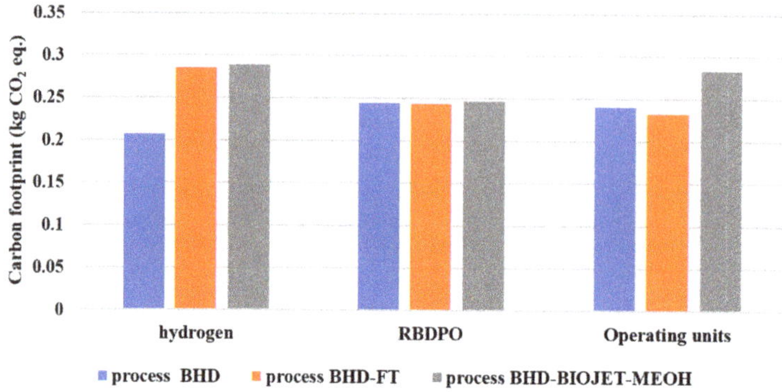

Figure 15. Carbon footprint of operating units and raw materials of three BHD processes.

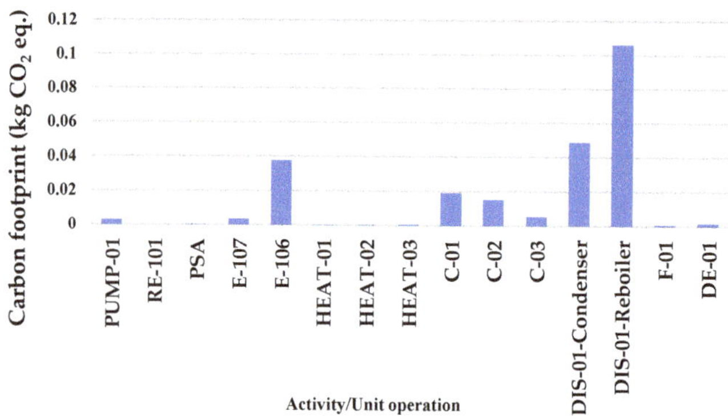

(a)

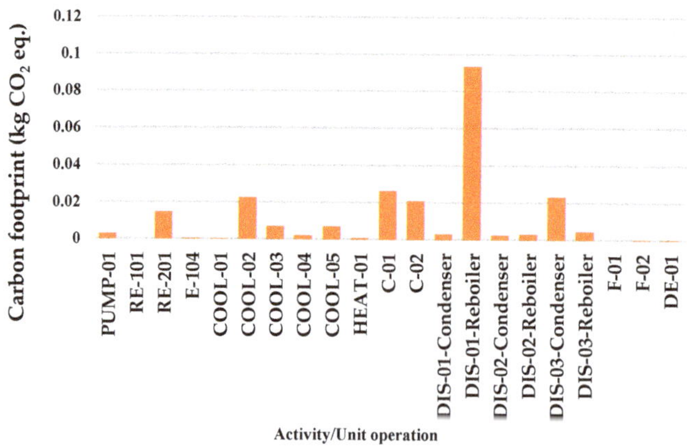

(b)

Figure 16. *Cont.*

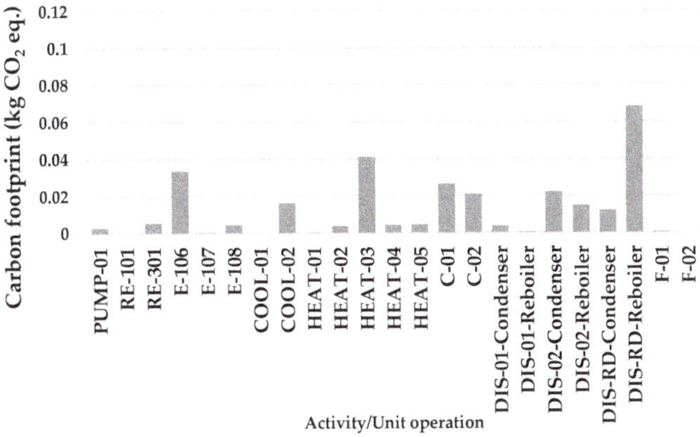

(c)

Figure 16. Carbon footprint of each operating unit of (**a**) BHD process, (**b**) BHD-FT process, and (**c**) BHD-BIOJET-MEOH process.

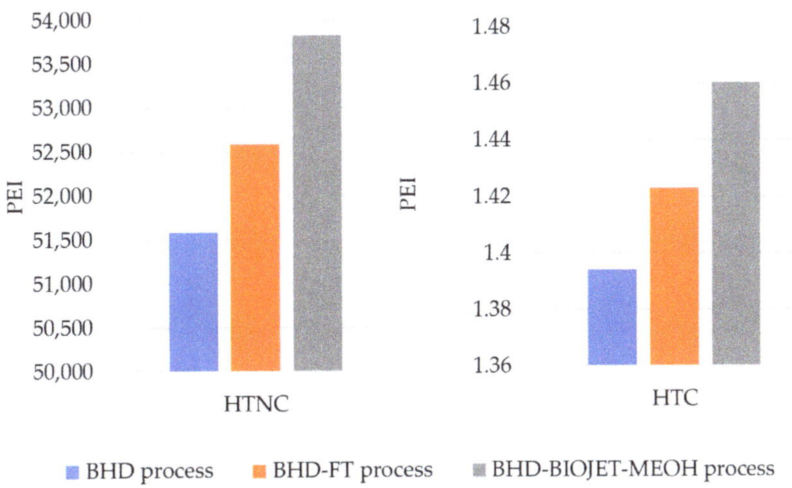

(a)

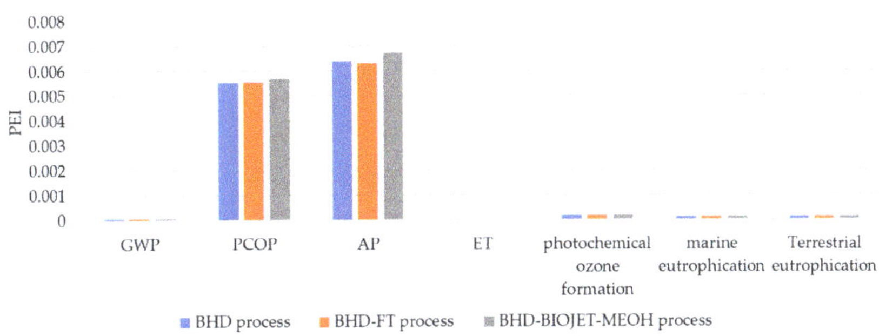

(b)

Figure 17. The environmental impact of various BHD processes in categories of (**a**) human health and (**b**) ecosystem.

5. Conclusions

In order to enhance process performance and achieve a sustainable objective, the co-production of BHD with other renewable fuels using a unique process including carbon dioxide usage is investigated. Three alternative processes of BHD production from RBDPO, including (1) the conventional BHD process with hydrogen recovery (BHD process), (2) the BHD process coupled with the Fischer–Tropsch process (BHD-FT process), (3) the BHD process coupled with the bio-jet fuel and methanol processes (BHD-BIOJET-MEOH process) are studied and compared with the conventional BHD process, in terms of technical, economic, and environmental aspects. The conventional BHD process requires less hydrogen and consumes less energy than the others. The pinch analysis and HEN design were performed to find the energy-efficient condition of the selected BHD process. The BHD-BIOJET-MEOH process could reduce hot utilities by 50% and cold utilities by 52% compared with the BHD process. The economic analysis indicates that the BHD-BIOJET-MEOH process is the most economically feasible, as it has the shortest payback period of 3 years and 1 month, the highest NPV of USD 7.93 million, and the highest IRR of 30.75%. Regarding environmental performance, the BHD process has the lowest carbon footprint of 0.692 kgCO_2eq./ kg of BHD. The BHD-BIOJET-MEOH process offered the highest PEI in the categories of human health and ecosystem. To exactly justify which process is the most suitable process offering the highest performance regarding the technical, economic, and environmental viewpoints, a multi-objective optimization that combines all criteria should be further performed.

Author Contributions: Conceptualization, L.S., S.A. (Suksun Amornraksa), A.A.; methodology, L.S., S.A. (Suksun Amornraksa), A.A.; software, L.S., S.A. (Suksun Amornraksa); validation, K.Y.; investigation, L.S., S.A. (Suksun Amornraksa), A.A.; resources, K.Y; writing—original draft preparation, L.S., S.A. (Suksun Amornraksa), K.Y.; writing—review and editing, A.A., S.A. (Suttichai Assabumrungrat); visualization, S.A.; supervision, S.A.; writing—review and editing, K.I.-o. All authors have read and agreed to the published version of the manuscript.

Funding: This work was funded by King Mongkut's University of Technology North Bangkok, Contract no. KMUTNB-67-BASIC-05.

Data Availability Statement: Data are contained within the article.

Acknowledgments: This work was supported and funded by King Mongkut's University of Technology North Bangkok, Contract no. KMUTNB-67-BASIC-05. The authors would like to acknowledge the support of Aspen Plus from TGGS- Thai—German Graduate School of Engineering King Mongkut's University of Technology North Bangkok and the support of LCSoft from PSE for SPEED Company Limited.

Conflicts of Interest: The authors declare no conflicts of interest.

References

1. Renewables 2023 Global Status Report. "Renewable in Energy Demand: Global Trends". Available online: https://www.ren21.net/gsr-2023/modules/energy_demand/ (accessed on 4 December 2022).
2. European Commission. Renewable Energy Targets. Available online: https://energy.ec.europa.eu/topics/renewable-energy/renewable-energy-directive-targets-and-rules/renewable-energy-targets_en (accessed on 4 December 2022).
3. Environmental + Energy Leader. Renewable Diesel Maker Neste Targets Carbon Neutral Production. Available online: https://www.environmentenergyleader.com/stories/renewable-diesel-maker-neste-targets-carbon-neutral-production,5985 (accessed on 4 December 2022).
4. Stumborg, M.; Wong, A.; Hogan, E. Hydroprocessed vegetable oils for diesel fuel improvement. *Bioresour. Technol.* **1996**, *56*, 13–18. [CrossRef]
5. Sumathi, S.; Chai, S.; Mohamed, A. Utilization of oil palm as a source of renewable energy in Malaysia. *Renew. Sustain. Energy Rev.* **2008**, *12*, 2404–2421. [CrossRef]

6. Mohammed, S.T.; Gheni, S.A.; Aqar, D.Y.; Hamad, K.I.; Ahmed, S.M.; Mahmood, M.A.; Abdullah, G.H.; Ali, M.K. Evaluation and optimal design of a high stability hydrothermal deoxygenation process for production of green diesel fuel via deoxygenation of waste cooking oil. *Process. Saf. Environ. Prot.* **2022**, *159*, 489–499. [CrossRef]
7. Munoz-Arjona, A.; Ayala-Cort´es, A.; Di Stasi, C.; Torres, D.; Pinilla, J.L.; Suelves, I. Catalytic hydrodeoxygenation of waste cooking oil into green diesel range hydrocarbons: From batch to continuous processing. *Chem. Eng. J.* **2025**, *503*, 158303. [CrossRef]
8. Department of Alternative Energy Development and Efficiency. Project to Study the Feasibility of Establishing a Commercial BHD (Bio-Hydrogenated Diesel) Plant. Available online: https://www.dede.go.th (accessed on 4 November 2022).
9. Sowcharoensuk, C. Palm Oil Industry. Available online: https://www.krungsri.com/en/research/industry/industry-outlook/agriculture/sugar-(1)/IO/io-oil-palm-20-th (accessed on 4 March 2022).
10. Srifa, A.; Faungnawakij, K.; Itthibencharong, V.; Viriya-empikul, N.; Charinpanitkul, T.; Assabumrungrat, S. Production of Bio-hydrogenated Diesel by Catalytic Hydrotreating of Palm Oil Over $NiMoS_2/\gamma$-Al_2O_3 Catalyst. *Bioresour. Technol.* **2014**, *158*, 81–90. [CrossRef]
11. Yenumala, S.R.; Maity, S.K.; Shee, D. Reaction mechanism and kinetic modeling for the hydrodeoxygenation of triglycerides over alumina supported nickel catalyst. *React. Kinet. Catal. Lett.* **2016**, *120*, 109–128. [CrossRef]
12. Noriega, A.K.; Tirado, A.; Méndez, C.; Marroquín, G.; Ancheyta, J. Hydrodeoxygenation of vegetable oil in batch reactor: Experimental considerations. *Chin. J. Chem. Eng.* **2020**, *28*, 1670–1683. [CrossRef]
13. Amin, A. Review of diesel production from renewable resources: Catalysis, process kinetics and technologies. *Ain Shams Eng. J.* **2019**, *10*, 821–839. [CrossRef]
14. Mikkonen, S. Second-generation renewable diesel offers advantages: Hydrotreated vegetable oils are more compatible with engine design and operations: Clean Fuels. *Hydrocarb. Process.* **2008**, *87*, 63–66.
15. Du, Y.; Wang, F.; Xia, X.; Zhu, H.; Zhang, Z.; You, C.; Jiang, X.; Jiang, J.; Li, C. MOF-derived Co nanoparticle on nitrogen-rich carbon for fatty acid hydrotreatment into green diesel. *Renew. Energy* **2022**, *198*, 246–253. [CrossRef]
16. Kaewtrakulchai, N.; Kaewmeesri, R.; Itthibenchapong, V.; Eiad-Ua, A.; Faungnawakij, K. Palm oil conversion to bio-jet and green diesel fuels over cobalt phosphide on porous carbons derived from palm male flowers. *Catalysts* **2020**, *10*, 694. [CrossRef]
17. Zhou, Y.; Liu, X.; Yu, P.; Hu, C. Temperature-tuned selectivity to alkanes or alcohol from ethyl palmitate deoxygenation over zirconia-supported cobalt catalyst. *Fuel* **2020**, *278*, 118295. [CrossRef]
18. Immer, J.G.; Kelly, M.J.; Lamb, H.H. Catalytic reaction pathways in liquid-phase deoxygenation of C18 free fatty acids. *Appl. Catal. A Gen.* **2010**, *375*, 134–139. [CrossRef]
19. Mäki-Arvela, P.; Snåre, M.; Eränen, K.; Myllyoja, J.; Murzin, D.Y. Continuous decarboxylation of lauric acid over Pd/C catalyst. *Fuel* **2008**, *87*, 3543–3549. [CrossRef]
20. Luo, H.; Yang, H.; Deng, W.; Li, C.; Du, F.; Li, S. Slurry-phase hydrotreating of waste oil to bio-hydrogenated diesel using in situ oil—Soluble MoS_2 particle. *Renew. Energy* **2023**, *19*, 119506. [CrossRef]
21. Kaewchada, A.; Akkarawatkhoosith, N.; Bunpim, D.; Bangjang, T.; Ngamcharussrivichai, C.; Jaree, A. Production of bio-hydrogenated diesel from palm oil using Rh/HZSM-5 in a continuous mini fixed-bed reactor. *Chem. Eng. Process.-Process. Intensif.* **2021**, *168*, 108586. [CrossRef]
22. Glisic, S.B.; Pajnik, J.M.; Orlović, A.M. Process and techno-economic analysis of green diesel production from waste vegetable oil and the comparison with ester type biodiesel production. *Appl. Energy* **2016**, *170*, 176–185. [CrossRef]
23. Cheah, K.W.; Yusup, S.; Singh, H.K.G.; Uemura, Y.; Lam, H.L. Process simulation and techno economic analysis of renewable diesel production via catalytic decarboxylation of rubber seed oil – A case study in Malaysia. *J. Environ. Manage.* **2017**, *203*, 950–961. [CrossRef]
24. Seibel, J.; Wancura, J.H.C.; Mayer, F.D. Process simulation and technology prospection to the hydrotreating of vegetable oils and animal fats. *Energy Convers. Manag.* **2024**, *315*, 118811. [CrossRef]
25. Mahmoudi, H.; Mahmoudi, M.; Doustdar, O.; Jahangiri, H.; Tsolakis, A.; Gu, S.; LechWyszynski, M. A review of Fischer Tropsch synthesis process, mechanism, surface chemistry and catalyst formulation. *Biofuels Eng.* **2017**, *2*, 11–31. [CrossRef]
26. Schulz, H. Short history and present trends of Fischer–Tropsch synthesis. *Appl. Catal. A Gen.* **1999**, *186*, 3–12. [CrossRef]
27. Van-Dal, S.; Bouallou, C. Design and simulation of a methanol production plant from CO_2 hydrogenation. *J. Clean. Prod.* **2013**, *57*, 38–45. [CrossRef]
28. Phichitsurathaworn, N.; Simasatitkul, L.; Amornraksa, S.; Anantpinijwatna, A.; Charoensuppanimit, P.; Assabumrungrat, S. Techno-economic analysis of co-production of bio-hydrogenated diesel from palm oil and methanol. *Energy Convers. Manag.* **2021**, *244*, 114464. [CrossRef]
29. RefuelEU Aviation Initiative: Council Adopts New Law to Decarbonise the Aviation Sector—Consilium. Available online: https://www.consilium.europa.eu/en/press/press-releases/2023/10/09/refueleu-aviation-initiative-council-adopts-new-law-to-decarbonise-the-aviation-sector/ (accessed on 23 May 2024).

30. Zhang, Y.; Bi, P.; Wang, J.; Jiang, P.; Wu, X.; Xue, H.; Liu, J.; Zhou, X.; Li, Q. Production of jet and diesel biofuels from renewable lignocellulosic biomass. *Appl. Energy* **2015**, *150*, 128–137. [CrossRef]
31. Tirado, A.; Trejo, F.; Ancheyta, J. Simulation of bench-scale hydrotreating of vegetable oil reactor under non-isothermal conditions. *Fuel* **2020**, *275*, 117960. [CrossRef]
32. Meurer, A.; Kern, J. Fischer–Tropsch Synthesis as the Key for Decentralized Sustainable Kerosene Production. *Energies* **2021**, *14*, 1836. [CrossRef]
33. Gutiérrez-Antonio, C.; Ornelas, M.L.S.; Gómez-Castro, F.I.; Hernández, S. Intensification of the hydrotreating process to produce renewable aviation fuel through reactive distillation. *Chem. Eng. Process.-Process. Intensif.* **2018**, *124*, 122–130. [CrossRef]
34. Gutiérrez-Antonio, C.; Romero-Izquierdo, A.G.; Israel Gomez-Castro, F.; Hernández, S. Energy integration of a hydrotreatment process for sustainable biojet fuel production. *Ind. Eng. Chem. Res.* **2016**, *55*, 8165–8175. [CrossRef]
35. Carrasco-Suárez, M.T.; Romero-Izquierdo, A.G.; Gutiérrez-Antonio, C.; Gómez-Castro, F.I.; Hernández, S. Production of renewable aviation fuel by waste cooking oil processing in a biorefinery scheme: Intensification of the purification zone. *Chem. Eng. Process.-Process. Intensif.* **2022**, *181*, 109103. [CrossRef]
36. Martinez-Hernandez, E.; Ramírez-Verduzco, L.F.; Amezcua-Allieri, M.A.; Aburto, J. Process simulation and techno-economic analysis of bio-jet fuel and green diesel production–Minimum selling prices. *Chem. Eng. Res. Des.* **2019**, *146*, 60–70. [CrossRef]
37. Kantama, A.; Narataruksa, P.; Hunpinyo, P.; Prapainainar, C. Techno-economic assessment of a heat-integrated process for hydrogenated renewable diesel production from palm fatty acid distillate. *Biomass-Bioenergy* **2015**, *83*, 448–459. [CrossRef]
38. Boonrod, B.; Prapainainar, C.; Narataruksa, P.; Kantama, A.; Saibautrong, W.; Sudsakorn, K.; Mungcharoen, T.; Prapainainar, P. Evaluating the environmental impacts of bio-hydrogenated diesel production from palm oil and fatty acid methyl ester through life cycle assessment. *J. Clean. Prod.* **2017**, *142*, 1210–1221. [CrossRef]
39. Boonrod, B.; Prapainainar, P.; Varabuntoonvit, V.; Sudsakorn, K.; Prapainainar, C. Environmental impact assessment of bio-hydrogenated diesel from hydrogen and co-product of palm oil industry. *Int. J. Hydrogen Energy* **2020**, *46*, 10570–10585. [CrossRef]
40. Forero, L.A.; Velasquez, J.A. A generalized cubic equation of state for non-polar and polar substances. *Fluid Ph. Equilib.* **2016**, *418*, 74–87. [CrossRef]
41. Iyer, S.S.; Renganathan, T.; Pushpavanam, S.; Kumar, M.V.; Kaisare, N. Generalized thermodynamic analysis of methanol synthesis: Effect of feed composition. *J. CO2 Util.* **2015**, *10*, 95–104. [CrossRef]
42. Bisotti, F.; Fedeli, M.; Prifti, K.; Galeazzi, A.; Dell'angelo, A.; Manenti, F. Impact of Kinetic Models on Methanol Synthesis Reactor Predictions: In Silico Assessment and Comparison with Industrial Data. *Ind. Eng. Chem. Res.* **2022**, *61*, 2206–2226. [CrossRef]
43. Fredenslund, A. UNIFAC and related group-contribution models for phase equilibria. *Fluid Phase Equilibria* **1989**, *52*, 135–150. [CrossRef]
44. Turton, R.; Bailie, R.C.; Whiting, W.B.; Shaeiwitz, J.; Bhattacharyya, D. *Analysis, Synthesis and Design of Chemical Processes*, 3rd ed.; Pearson Education: Toronto, ON, Canada, 2008.
45. PSEforSPEED Company. Available online: https://www.pseforspeed.com/product/lcsoft-and-econ/ (accessed on 4 November 2022).
46. Pratiwi, V.D.; Juwari, J.; Handogo, R. Simulation of Hydrogen Purification using Two Bed System Pressure Swing Adsorption. *Iptek J. Proc. Ser.* **2017**, *3*, 14–18. [CrossRef]
47. International Energy Agency. Available online: https://www.iea.org/ (accessed on 4 November 2022).
48. Trading Economics. Available online: https://tradingeconomics.com/ (accessed on 4 November 2022).
49. Alibaba. Available online: https://www.alibaba.com/ (accessed on 4 November 2022).
50. Merck. Available online: https://www.sigmaaldrich.com/TH/en (accessed on 4 November 2022).
51. U.S. Energy Information Administration. Available online: https://www.eia.gov/ (accessed on 4 November 2022).
52. U.S. Energy Information Administration. Gasoline and Diesel Fuel Update. Available online: https://www.eia.gov/petroleum/gasdiesel/ (accessed on 4 January 2023).
53. U.S. Energy Information Administration. U.S. Gulf Coast Kerosene-Type Jet Fuel Spot Price FOB (Dollars per Gallon). Available online: https://www.eia.gov/dnav/pet/hist/eer_epjk_pf4_rgc_dpgD.htm (accessed on 4 January 2023).

processes

MDPI

Article

Forecasting CO_2 Emissions in India: A Time Series Analysis Using ARIMA

Hrithik P. M. [1], Mohd Ziaur Rehman [2], Amir Ahmad Dar [1,*] and Tashi Wangmo A. [1]

1 School of Chemical Engineering and Physical Sciences, Lovely Professional University, Phagwara 144411, Punjab, India; hrithik.sarh@gmail.com (H.P.M.); tassymo506@education.gov.bt (T.W.A.)
2 Department of Finance, College of Business Administration, King Saud University, P.O. Box 71115, Riyadh 11587, Saudi Arabia; zmuhammad@ksu.edu.sa
* Correspondence: sagaramir200@gmail.com

Abstract: This study evaluates the capability of the ARIMA (Auto Regressive Integrated Moving Average) to predict CO_2 emissions in India using data from 1990 to 2023, addressing a critical need for accurate forecasting amid various economic and environmental uncertainties. It is observed that ARIMA yields high accuracy with respect to the prediction, and hence, it is reliable for environmental forecasting. These predictions give policymakers evidence-based information to aid in implementing sustainable climate policies within India. To ensure reliable predictions, the study methodology utilizes the Box–Jenkins approach, which encompasses model identification, estimation, and diagnostic checking. The initial step in the study is the Augmented Dickey–Fuller (ADF) test, which assesses data stationarity as a prerequisite for precise time series forecasting. Model selection is guided by the Akaike Information Criterion (AIC), which balances prediction accuracy with model complexity. The efficiency of the ARIMA model is assessed by comparing the actual observed values to the predicted CO_2 emissions and the results demonstrate ARIMA's effectiveness in forecasting India's CO_2 emissions, validated by statistical measures that confirm the model's robustness. The value of the present study lies in its focused assessment of the relevance of the ARIMA model to the specific environmental and economic context of India, with actionable insight for policymakers. This study enhances prior research by incorporating a focused approach to data-driven policy formulation that increases climate resilience. The establishment of a reliable model for the forecasting of CO_2 will aspire to support informed decision making in environmental policy and help India move forward toward sustainable climate goals. This study only serves to highlight the applicability of ARIMA in terms of environment-based forecasting and permits further emphasis on how much this method can be a useful data-based tool in climate planning.

Keywords: CO_2 emissions forecasting; ARIMA time series model; environmental policy analysis; India climate policy; climate change forecasting

1. Introduction

The increasing environmental and economic challenges associated with climate change underscore the importance of robust predictive models to assess and mitigate its impact. Predicting CO_2 emissions is essential for comprehending the course of climate change and for developing strategies for sustainable development. Making successful environmental plans requires precise forecasting, especially considering India's increasing emission levels, which are a significant contribution to global greenhouse gas emissions [1–3]. The ARIMA (Auto Regressive Integrated Moving Average) model is one of the advanced time series forecasting techniques used in this study to examine the trends in India's CO_2 emissions. This study attempts to offer insightful information that will help policymakers navigate the many problems of climate action by analyzing historical trends in India's emissions [4]. Comprehending the variations, enduring patterns, and possible seasonality in CO_2 emissions might enable interested parties to make informed choices for sustainable growth [5–7].

The continuously increasing rate of CO_2 emission is highly threatening the global climatic condition and affecting almost every section of the world. According to Raihan et al. (2022), addressing the challenges of these areas with respect to a reduction in CO_2 emission and increasing environmental quality has made it a crucial priority at the global level towards the attainment of sustainable growth and mitigation of adverse climatic conditions [8]. India is an emerging and developing country that happens to be the seventh most vulnerable to climate change according to the Global Climate Risk Index 2020 [9]. For a better contrast, India is among the fastest-growing economies of the world. The World Bank adds that fast economic and population growth highly relies on fossil fuels, providing about 74% of the energy needs for India [10]. The unprecedented increase in this energy source and economic growth has therefore caused a massive increase in the country's CO_2 emissions. In being the third largest emitter of carbon in the world, the country contributed 7% of the world's emissions in 2020. CO_2 emissions in the country increased by 4.8% in 2018, the increase equally distributed within the power sector, transport and other industries [11]. The high consumption of fossil fuel and ever-increasing energy demand have brought severe environmental and ecological consequences to India, threatening its sustainable development goals. It raises critical questions about whether economic growth is combined with improved environmental sustainability.

Precise CO_2 forecasting plays a pivotal role in aiding policymakers to identify high-priority sectors requiring immediate emission reductions, optimize resource allocation, and track progress toward achieving sustainable development goals. Such forecasting equips decision-makers with actionable insights, enabling targeted interventions to mitigate climate impacts effectively. For environmental stakeholders, it offers a data-driven foundation to monitor progress, advocate for stringent climate policies, and ensure accountability in meeting global climate commitments. Given India's significant contribution to global emissions, its CO_2 trajectory directly influences global climate dynamics. Successful emission reductions in India can substantially enhance international efforts to combat climate change and limit global warming.

Because it offers insights into patterns and trends throughout time, time series data analysis is extensively useful across disciplines like meteorology, epidemiology, finance, and environmental research. To predict CO_2 emissions in India, we use fundamental time series concepts such as stationarity, ACF, partial autocorrelation function (PACF), and the ARIMA model. A key component of time series data is stationarity, which is the ability of a series to maintain its statistical characteristics across time, including mean, variance, and autocorrelation. In the absence of trends or seasonality, ensuring stationarity makes modelling easier and produces more accurate forecasts [12].

The classical machine learning methods have developed the ability to learn complex patterns in data, which involves minimal human interference [13]. Classic machine learning models have been widely utilized for the forecast of CO_2 emissions, in which the most general techniques include linear regression analysis [14,15], support vector machines [16], random forest [17], neural network autoregressive model [14] and feed-forward neural networks [18]. However, a very important bottleneck for these machine learning techniques regarding CO_2 forecasting is that, unlike statistical methods for time series analysis, these machine learning techniques are not inherently capable of dealing with time series. First, machine learning models consider all the data points to be independent and identically distributed, whereas in the case of a time series, the observations will be highly related and have temporal dependencies. Another important aspect is that time-series data often contain an underlying trend and seasonal variation, hard to capture with classic machine learning algorithms without serious preprocessing [19].

Emission reduction and environmental quality are considered global priorities in relation to sustainable growth, fighting against the dangers of climatic change. This research contributes to investigating the impact of economic growth, renewable energy use, urbanization, industrialization, tourism, agricultural productivity, and forest area on CO_2 emissions in India between 1990 and 2020 using the Dynamic Ordinary Least Squares method [9].

In fact, all other statistical time series methods are designed to work with datasets that are sequential in nature—like for the recordings of CO_2 measurement consistently over time. Methods such as autoregressive, moving average, exponential smoothing, and structural models are all naturally suited to handle this sequential data structure. Among these methods, the ARIMA model has emerged as one of the most applied for CO_2 emissions forecasts [18].

Among the many approaches used in the literature on time series analysis and environmental forecasting is the ARIMA model, which is well known for its convenience of use and effectiveness in identifying linear patterns in time series data. Testing for stationarity with visual evaluations, the Augmented Dickey–Fuller (ADF) test, and the Kwiatkowski–Phillips–Schmidt–Shin (KPSS) test are crucial prior to implementing time series models. These tests establish if there is a regular, predictable trend in the time series across the examined period [20,21].

ACF gives these correlations a visual representation, while autocorrelation quantifies the link between a time series and its lagged values. Seasonality may be suggested by a sharp drop in autocorrelation at a certain lag, while non-stationarity is shown by a gradual decay. On the other hand, when calculating the correlation between a data point and its lagged values, the PACF plot separates the impact of intermediate lags, which is very useful for figuring out the sequence of an autoregressive process.

One of the most used methods for time series forecasting is the ARIMA model, which consists of three primary parts: Moving Average (MA), Integrated (I), and Auto-Regressive (AR). Features of the time series are captured by each component [1,22–25]:

Auto-Regressive (AR): The link between the current observation and its historical values is captured by auto-regressive (AR) analysis. The AR component's order, represented by "p", indicates how many lags were considered.

Integrated (I): The term "integrated" (I) describes the process of differencing time series data to attain stationarity. The "d", which stands for differencing order, denotes how frequently differencing is used.

Moving Average (MA): Takes into consideration the effect of previous white noise or random shocks. "q", the order of the MA component, indicates how many lag forecast mistakes are incorporated into the model.

The AIC, a frequently used metric in model selection, evaluates the level of fitness of a statistical model while taking complexity into account. The AIC value is very helpful in ARIMA modelling since it helps balance the trade-off between model complexity and performance. Better-fitting models with fewer parameters are shown by lower AIC values, which provide an objective means of comparing different parameter combinations (p, d, and q) and choosing the best model for precise forecasting. As a result, AIC is a crucial tool for determining the best ARIMA model for CO_2 emissions forecasting that balances predictability and simplicity [26,27].

The current work of research on time series analysis in environmental science is expanded by using ARIMA to forecast CO_2 emissions in India. To build models that account for long-term trends and possible seasonality, this study makes use of historical data on CO_2 emissions from India. Since it can offer insights that direct the creation of policies and the distribution of resources in attempts to slow down climate change, accurate emissions data forecasting is essential. This method is in line with earlier studies on emissions forecasting, which show how important historical data analysis is for strategic planning and environmental management (Figure 1) [1].

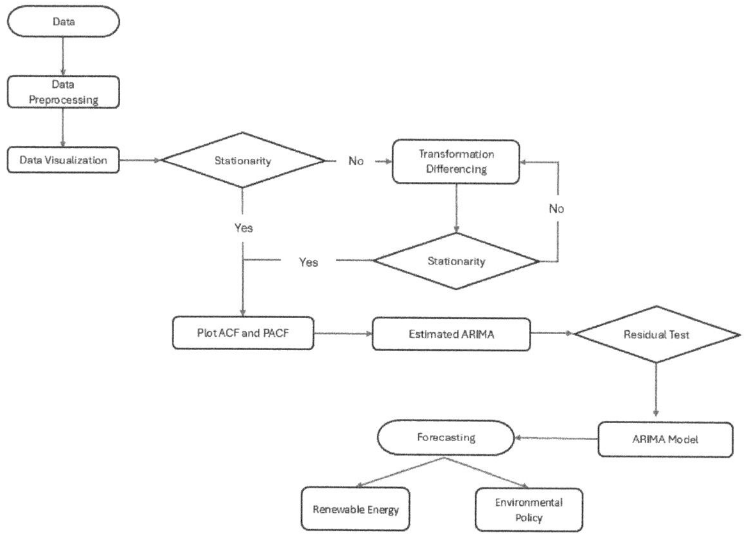

Figure 1. Box–Jenkins methodology.

The main tool for this study's methodological implementation is R software (RStudio 2024.09.0). R, which is well known for its statistical computing and data processing skills, provides the adaptability and transparency needed to create and assess intricate forecasting models like ARIMA. India's environmental policies and sustainable development goals can be supported by data-driven decision making, made possible by R's vast library of statistical tools, which enable accurate forecasting [28,29].

Because it uses differencing to handle nonstationary data, the ARIMA model is very useful for time series forecasting. This makes it appropriate for environmental datasets, which frequently show trends because of ongoing shifts in industrialization, energy use, and human activities. Although ARIMA has demonstrated potential in emissions forecasting, research has pointed out that it is not very good at taking sudden shocks or abrupt changes in environmental data into account. Nevertheless, ARIMA can provide competitive accuracy for short- to medium-term emissions forecasting with the right parameter tweaking.

For predicting complex, nonlinear interactions in environmental data, machine learning approaches have recently gained popularity as alternatives to conventional statistical models. ARIMA, being simpler and consuming fewer computational resources, is theoretically well suited and outcompetes LSTM on handling small datasets and time series with a linear trend. Furthermore, ARIMA does not require extensive training or tuning, in contrast to LSTM, and hence, it remains far more accessible and efficient for simple stationary series analyses [30].

ARIMA models can be used for a variety of emissions forecasting scenarios. Research has extended the use of ARIMA to several contaminants and environmental indicators in several different industries. One study, for example, used ARIMA to anticipate urban $PM_{2.5}$ and PM_{10} pollution levels. This showed that ARIMA was useful for creating baseline projections, but further modelling was needed to take abrupt changes in environmental regulations or industrial operations into consideration. For similar applications in agricultural emissions, where seasonal patterns are prominent, weather-related factors had to be added to improve the accuracy of the model [31–33].

The aim of this study is to provide the fundamental ideas of time series analysis for environmental forecasting, such as the significance of stationarity and the functions of ARIMA, PACF, and ACF in emissions data prediction. By using these techniques, scientists and environmental analysts can learn about the trends in CO_2 emissions over time, which helps them make well-informed decisions on things like sustainability planning and climate policy.

2. Methodology

2.1. Data Preparation

Data Cleaning and Preprocessing: Missing values or outliers are checked in the data, which contain values. To preserve the quality of the data, all variations are handled carefully.

2.2. Stationarity Testing Using ADF

Stationarity Check: Since stationarity suggests that the data's statistical characteristics (such mean and variance) remain constant over time, it is a crucial prerequisite for time series modelling using ARIMA. For the model to correctly detect and forecast recurring patterns in the data, trends or seasonality must be eliminated for CO_2 emissions in India. The ARIMA model may produce skewed or unstable forecasts if the series is non-stationary because it may mistakenly interpret trends as continuous patterns. By ensuring that the model concentrates on the actual underlying relationships, stationarity produces forecasts that are more accurate and dependable. The detection of unit roots, which signify non-stationarity in the data, is performed using the Augmented Dickey–Fuller (ADF) test [34].

Differencing for Stationarity: Differencing is used if it is determined that the data are not stationary. This involves lowering the emissions value from the prior year from the emissions value from the current year. The results of the ADF test and a visual examination of the time series plot are used to calculate the necessary degree of differencing. For example, if it stabilizes the mean and enables a more accurate ARIMA model, a first-order difference (i.e., one level of differencing) would be adequate.

2.3. ACF and PACF Analysis for Model Identification

Autocorrelation and Partial Autocorrelation: ACF gives these correlations a visual representation, while autocorrelation quantifies the link between a time series and its lagged values. Seasonality may be suggested by a sharp drop in autocorrelation at a certain lag, while non-stationarity is shown by a gradual decay. On the other hand, when calculating the correlation between a data point and its lagged values, the PACF plot separates the impact of intermediate lags, which is very useful for figuring out the sequence of an autoregressive process. When choosing the right model parameters for ARIMA modelling, ACF and PACF charts offer crucial insights into the structure of time series data.

Plots of the ACF and PACF are essential instruments for figuring out the sequence of the ARIMA model's elements. While the PACF plot aids in separating the direct impacts of each lag, the ACF plot shows correlations between the series and its lagged values [35].

Order Selection: We can determine possible orders for the autoregressive (AR) and moving average (MA) components by looking at graphs. Significant spikes in the ACF plot at lags, for instance, would point to an MA component, whereas spikes in the PACF plot might point to an AR component. The choice of the (p, d, q) parameters in ARIMA is guided by this analysis, where:

- p is the order of the AR term;
- d is the number of differencing required to make the series stationary;
- q is the order of the MA term.

2.4. Model Selection Using AIC

AIC Calculation and Model Evaluation: The AIC is used to assess various model configurations following the identification of possible orders for the ARIMA model. By penalizing extra parameters, the AIC reduces the chance of overfitting and helps strike a balance between the model's complexity and goodness of fit.

Optimal Model Selection: The model that best fits the data is the one with the lowest AIC score. This model should ensure accurate projections without needless complexity by offering a reasonable balance between accuracy and simplicity. CO_2 emissions forecasts for the future are based on the chosen ARIMA model [36].

2.5. Forecasting CO_2 Emissions

Generating Forecasts: After determining the best ARIMA model, we create projections for CO_2 emissions. The output of the model provides a range of expected future emissions, including both point projections and confidence ranges. This stage offers useful forecasts that help direct environmental regulations.

Validation and Comparison: By contrasting the predicted emissions with actual emissions data (when values become available), the accuracy of the model is assessed. This comparison enables us to evaluate the prediction ability of the model and, if required, modify it for upcoming projections. If the Mean Absolute Percentage Error (MAPE) value is approximately close to 10, then it indicates high model accuracy. For instance, extra variables like economic indicators or policy changes may need to be considered if the predicted value deviates greatly from the observed values [37].

3. Result and Analysis

Understanding Emission Trends: The final study reveals underlying trends in CO_2 emissions throughout time in addition to forecasts. Context for the projections is provided by discussing insights into the growth rate, stability, and any notable changes in emission levels.

Implications for Policy: This study intends to assist decision-makers in developing environmental policy by precisely forecasting future emissions. The findings can be used to track India's progress toward its emission reduction goals and predict future difficulties in limiting CO_2 emissions.

3.1. Stationarity Testing Using the ADF Test

The Augmented Dickey–Fuller (ADF) test was used to ascertain whether the CO_2 emissions data are stationary. The ADF test's null hypothesis (H_0) suggests that there may be a unit root, which would suggest non-stationarity. A p-value of 0.01 and a Dickey–Fuller test statistic of -4.4891 were obtained from the ADF test. We reject the null hypothesis, since the p-value is less than the significance level of 0.05, which shows that the CO_2 emissions data are stable. This is a crucial discovery since stationarity keeps trends from skewing the data, guaranteeing the accuracy of ARIMA-based forecasting.

A differencing value of d = 2 was selected based on the dataset, as the first series showed significant non-stationarity, most likely because of a clear trend. Persistent patterns in the ACF and PACF plots and the stationarity test demonstrated that the series still showed evidence of non-stationarity after applying initial differencing (d = 1). However, the dataset reached stationarity, with the mean and variance stabilizing and no significant trends remaining after applying second differencing (d = 2). To convert the data into a stable series suited for ARIMA modelling, d = 2 is appropriate, as demonstrated by the ACF plot, which shows no significant autocorrelations at higher lags, and the PACF plot, which only identifies significant correlations at the first few lags.

3.2. ACF and PACF Analysis for Model Identification

We examined the ACF and PACF plots following the procedure of second-order differencing (d = 2) to determine the proper values of p (autoregressive order) and q (moving average order) for the ARIMA model. The selection of the AR and MA components is aided by these graphs, which show any connections or dependencies in the data.

3.3. ACF Analysis

With autocorrelation values near zero at all lags, the ACF plot for the differenced series showed no notable spikes. This implies that there are no notable autocorrelations in the data, and therefore, an autoregressive component is not helpful. Because of this, we set the autoregressive order p to 0, which means that, in the context of this dataset, past values of the time series have little predictive value for current values.

3.4. PACF Analysis

On the other hand, lag 4 displayed a significant spike in the PACF plot, whereas the remaining lags were quite modest and lacked significant associations. This spike at lag 4 indicates that the data have a substantial moving average relationship, meaning that the last four years' worth of random shocks or error terms can accurately estimate the current value of CO_2 emissions. The moving average order q is thus set to 4.

Figure 2 below shows the plot for ACF and PACF.

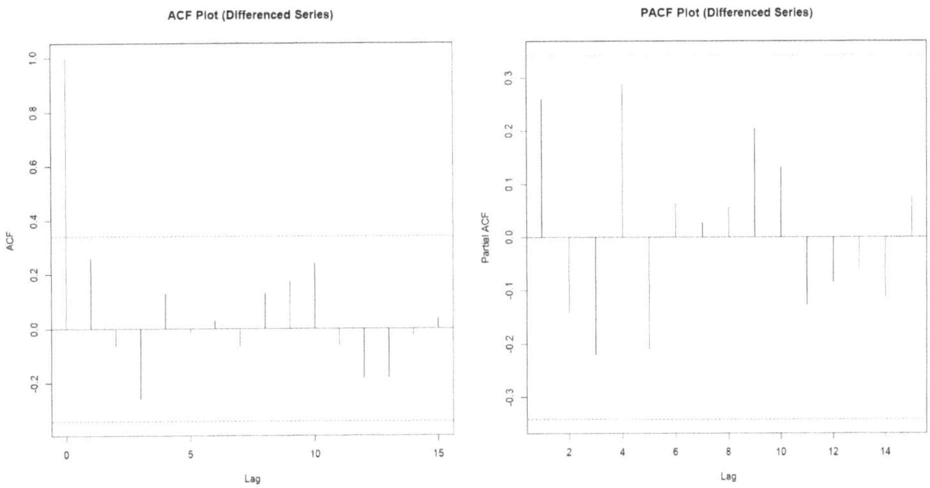

Figure 2. Plot for ACF and PACF.

3.5. ARIMA Model Selection

There were no notable lags in the ACF plot, and all autocorrelation values stayed within the confidence intervals. The absence of notable spikes suggests that there is no need for a moving average (MA) component because the series' current values do not exhibit a strong correlation with lagged values. To simplify the model and make sure it accurately depicts the underlying structure of the data, the moving average order p was set to 0. On the other hand, no significant correlations were found at any of the other lags, and the PACF plot showed a clear spike at lag 4. After taking into consideration the impacts of intermediate lags, this single spike suggests that there is a significant correlation between the series' present value and its value four steps earlier. The addition of a fourth-order autoregressive component (q = 4) to the model is supported by this finding.

ARIMA (0, 2, 4) is the most suitable model structure for the CO_2 emissions data, according to the ACF and PACF plot evaluations. This model setup consists of the following:

$p = 0$: The absence of significant autocorrelation in the ACF plot indicates that there is no autoregressive component ($p = 0$).

d = 2: The Augmented Dickey–Fuller test verified that stationarity was achieved using second-order differencing.

q = 4: A moving average component of the fourth order derived from the PACF plot's spike.

Based on ACF and PACF analysis, which indicates strong autocorrelations at higher lags and the requirement for two differencing steps to ensure stationarity, ARIMA (0, 2, 4) was chosen to model CO_2 emissions in India. Additionally, the model minimizes the AIC, suggesting that fitness and complexity are well balanced. But it is important to prevent overfitting, which can happen if the model becomes too complicated and captures both the underlying pattern and noise, making it difficult to generalize to fresh data. ARIMA (0, 2, 4) achieves a balance between a strong fit to historical data and preserving generalizability for future predictions by employing a very simple model with few AR components and maximizing the MA terms.

3.6. Residual Analysis

We carried out an in-depth residual study to guarantee the ARIMA (0, 2, 4) model's dependability. To verify that the residuals, which are the variations between the model's fitted values and the observed values, complied with the presumptions of the ARIMA framework, they were analyzed, as shown in Figure 3.

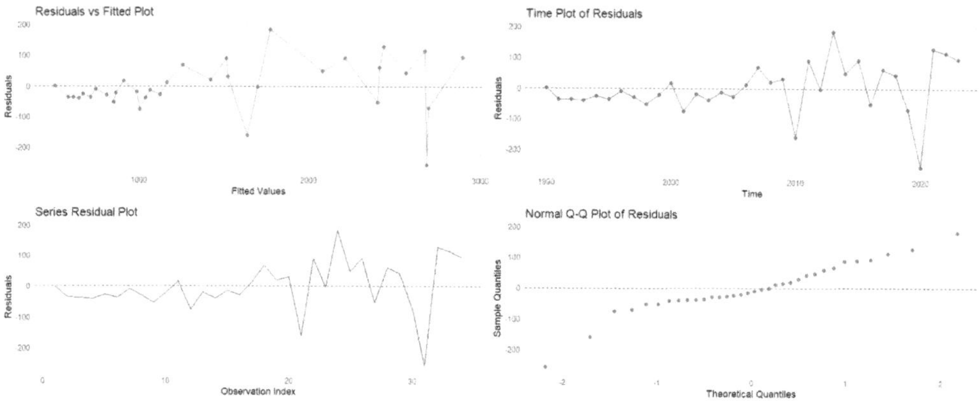

Figure 3. Normality plot.

Fitted Values against Residuals Plot: There were no obvious patterns or trends in the plot of residuals against fitted values, suggesting that the ARIMA model successfully represented the structure of the data. Random distribution close to zero indicated that no significant explanatory information was left unmodeled and that the model effectively captured all detectable information in the data. The absence of residual trends demonstrates that the model was not subjected to structural bias or underfitting.

Residuals Time Series Plot: The residuals moved wildly about zero, with no obvious periodic patterns or systematic shifts over time, according to the time series plot. This residual variation's randomness demonstrated that the model's proper differencing and order selection had adequately removed autocorrelations from the original series.

ACF Plot: The residuals' unpredictability was further confirmed by the ACF plot. The idea that the residuals resembled white noise was supported by the fact that no discernible autocorrelation was seen at any latency. This suggests that there is no residual structure to explain because the ARIMA (0, 2, 4) model accurately captured all pertinent information from historical data.

Normal Q-Q Plot of Residuals: The Q-Q plot showed that the points mainly lined up along the reference line, indicating that the residuals were normally distributed. In real-world applications, small tail variations are to be expected, but they have little effect on the model's dependability. Forecasts and confidence intervals from the model are guaranteed to be statistically valid and understandable due to the residuals' normality.

Collectively, the diagnostic evaluation verified the ARIMA (0, 2, 4) model's robustness and efficiency. The model avoided overfitting and maintained simplicity while producing accurate forecasts by attaining unpredictability and white noise behavior in residuals. The residual analysis also shows that the model accurately depicts the dynamics of India's historical CO_2 emissions.

The chosen ARIMA (0, 2, 4) model, which was determined to be the best match based on its AIC value, has been used to create forecasts for the year 2021–2023. These projections go beyond the documented trends in historical data to offer expected CO_2 emission amounts in 2021–2023. Based on historical data, the ARIMA model accurately predicts probable future patterns in CO_2 emissions.

Table 1 compares the projected CO_2 emissions for 2021–2023 with the actual observed emissions to assess the predictive accuracy of the ARIMA (0, 2, 4) model. In an 80:20 ratio, the data are divided into a training set and a test set. The degree to which the model's predictions match the actual CO_2 emissions is evaluated in this examination. The model appears to be successful in identifying underlying trends in CO_2 emissions, as evidenced by the excellent connection between expected and actual values. On the other hand, notable disparities could indicate areas for possible enhancement or different modelling approaches and expose constraints in the model's predictive power.

Table 1. Comparison between actual and forecasted CO_2 emissions.

Year	Actual	Forecasted	Accuracy
2021	2571.40	2988.94	83.76
2022	2794.83	3066.43	90.28
2023	2994.79	3179.42	93.83

The discrepancies between predicted and actual CO_2 emissions are shown in Figure 4. Some projections show significant differences from actual values, while others closely match them. These discrepancies are explained by the intrinsic difficulties of predicting environmental data, which are impacted by a wide range of uncontrollable variables. This comparative study is essential for verifying the forecasting model's accuracy and boosting trust in its capacity to anticipate future CO_2 emissions.

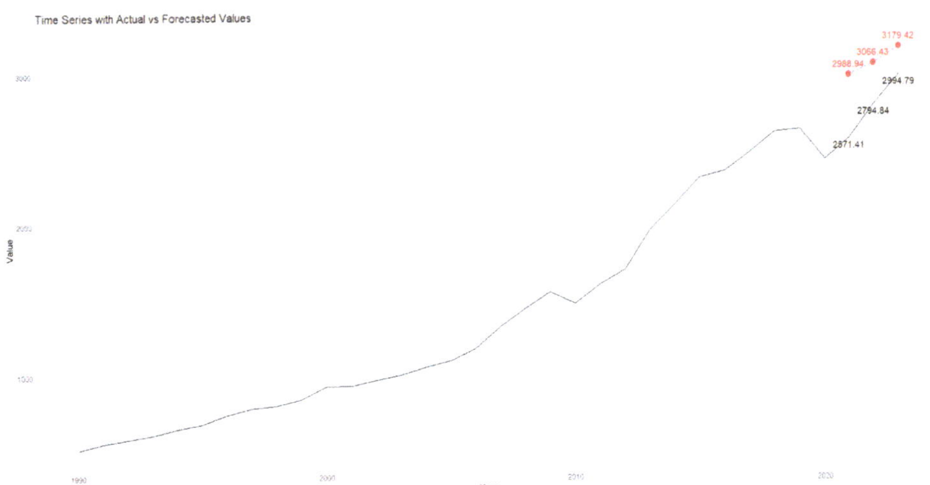

Figure 4. Comparison between actual and forecasted CO_2 emissions.

Using the well-known ARIMA (0, 2, 4) model, forecasts for the years 2024–38 involve estimating CO_2 emissions for each upcoming year. The model utilizes patterns and trends identified in historical data during the analysis to generate these estimates. By extending existing trends, the ARIMA model aims to provide insights into future CO_2 emission levels, helping to inform environmental policy and strategic planning.

From the plotted time series forecast, the following is clearly shown in Figure 5:

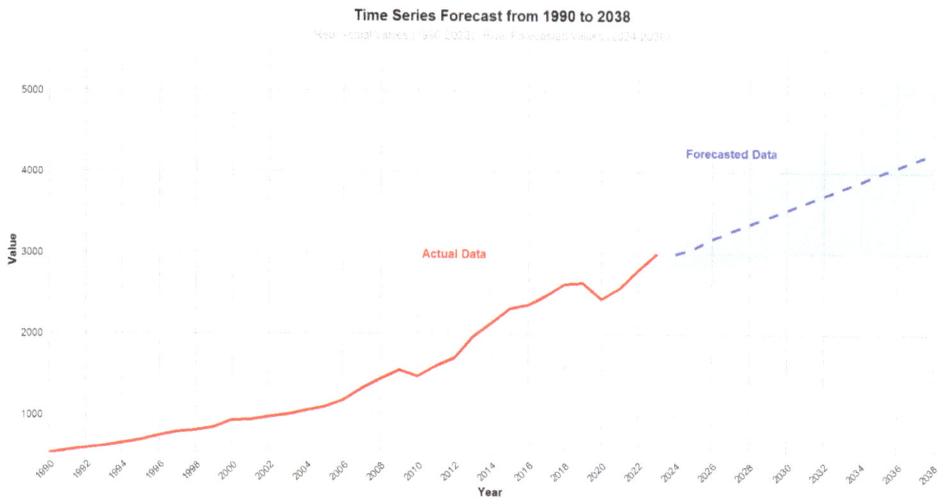

Figure 5. CO_2 emission forecasts for years 2024–2038.

Rising Trend: The predicted values exhibit a distinct rising trend, indicating a steady increase in CO_2 emissions. This pattern is consistent with past data as well as the anticipated expansion of India's GDP and energy usage.

Increasing Rate: As the projected period draws to a close, the rate of increase appears to be quickening, suggesting a possible spike in emissions in the years to follow.

This study provides a comprehensive interpretation of the ARIMA model's output and forecast trends for CO_2 emissions in India. The time series analysis effectively highlights key patterns and future projections. Comparing the observed trends to underlying economic or policy issues is crucial, nevertheless, to deepen the research. For example, times of urbanization, industrial expansion, or rising energy consumption brought on by economic growth may all be associated with higher CO_2 emissions. On the other hand, any reductions or slower rise in emissions that are seen could be related to the adoption of clean energy laws, advances in technology, or global agreements like the Paris Agreement. An analysis that ties these findings to economic or policy developments in India, like the encouragement of renewable energy sources or more strict environmental laws, would give a more insightful and defined understanding of the trends and provide important information about the factors influencing India's trajectory of CO_2 emissions.

3.7. Error Metrics

Kumari and Singh (2023), observed that ARIMA's predicted values are too far from the actual CO_2 emissions. The authors used ARIMA (1, 2, 1) for prediction. It can be seen that the ARIMA model has a MAPE value of 98.969%, which is inaccurate. Therefore, the authors conclude that forecasting for 2020 to 2030 will not be appropriate [38].

In this paper, the model successfully captures the underlying trends with acceptable errors, as evidenced by the Root Mean Square Error (RMSE) value of 306.69, which is a reasonable degree of accuracy for CO_2 emissions forecasting. The model's dependability is demonstrated by the Mean Absolute Error (MAE) value of 291.25, which indicates high predictive performance because the average error is small in relation to the amount of CO_2 emissions. The Mean Absolute Percentage Error (MAPE) value is approximately near 10, which indicates high model accuracy, as shown in Table 2.

Table 2. Error metrics for CO_2 emission.

Error Metrics	ARIMA
RMSE	306.69
MAE	291.25
MAPE	10.71%

4. Discussion

Using an ARIMA (0, 2, 4) model, the time series projection for CO_2 emissions in India shows a worrying rising trend. According to this model's particular specifications, the previous four error terms have an impact on the current emissions value. Furthermore, to obtain stationarity, the necessary condition for efficient time series analysis, the data require second-order differencing.

A mix of industrial and socioeconomic variables are responsible for the reported increase in CO_2 emissions in India. The dependence on fossil fuels like coal and petroleum, which are major contributors to greenhouse gas emissions, has grown dramatically because of rapid industrialization. Population growth and urbanization have increased the requirement for energy for residential, transportation, and electrical purposes. The expansion of India's economy has also led to a rise in energy-intensive industries like manufacturing and construction. Also, because renewable energy technologies like solar and wind have not been widely adopted, the nation's energy mix still significantly favors non-renewable sources. The issue is made worse by inefficient policies, such as those that postpone the implementation of clean energy transitions or the enforcement of pollution restrictions. Another significant industry, agriculture, indirectly contributes through methane emissions, and the increasing number of cars raises CO_2 levels through transportation-related emissions. The persistent rise in CO_2 emissions is caused by these coupled variables, which calls for extensive action measures to slow the trend.

These elements are probably going to have a long-term impact on emissions levels as India develops more. The forecast line's shaded confidence interval illustrates the inherent unpredictability of future projections. Numerous factors, such as shifts in legislation, technical breakthroughs, and economic swings, contribute to this unpredictability. The degree of uncertainty rises as we go farther into the future, according to a broader confidence interval toward the end of the prediction period [39–41].

Several assumptions need to be considered when using time series analysis to predict CO_2 emissions in India over the long run (2024–2038). First, the model assumes that historical trends, patterns, and connections in the data will persist in the future, which might not take into consideration prospective structural shifts in the economy, advances in technology, or changes in policy. One of the main causes of CO_2 emissions is economic activity, which includes increases in energy use, transportation, and industrial production. Future economic developments, such as a move toward more environmentally friendly sectors of the economy or modifications to the way people use energy, could, nevertheless, drastically modify the emissions trajectory.

This forecast is important and has broad significance. Rising sea levels, harsh weather, and biodiversity loss are just a few of the environmental effects of climate change and global warming brought on by rising CO_2 emissions. Furthermore, rising CO_2 emissions have the potential to worsen air pollution, which would be detrimental to people's health and welfare. Implementing strong climate policies and regulations, encouraging the use of renewable energy sources, promoting energy efficiency, and fostering international collaboration are all essential to lessening these effects.

However, it is critical to recognize the ARIMA model's shortcomings. In practical situations, the model's assumptions of stationarity, linearity, and constant variance might not always hold true. Furthermore, emissions trends can be greatly impacted by unforeseen external variables like technological advancements or geopolitical events. To adjust to changing conditions, it is crucial to continuously monitor CO_2 emissions and update the forecast on a frequent basis.

It is essential to concentrate on industries and transportation that contribute the most to CO_2 emissions to improve recommendations. For example, increasing the infrastructure for electric vehicles (EVs), encouraging public transit, or implementing more stringent fuel efficiency regulations could all help cut emissions in the transportation sector. Norway's successful promotion of EVs through tax exemptions and subsidies has resulted in a significant reduction in transportation-related emissions. In the industrial sector, deploying energy-efficient technologies and shifting to greener fuels can cut emissions. Similar tactics could be used in India, as evidenced by Germany's legislation encouraging the use of renewable energy in industrial operations, such as its Renewable Energy Sources Act. Policymakers can prioritize effective actions and make quantifiable progress in reducing emissions by customizing such sector-specific recommendations to India's circumstances.

India needs to take a variety of approaches to addressing the issues brought on by growing CO_2 emissions. Strong climate rules and regulations must be put into place, renewable energy sources like solar, wind, and hydro power must be promoted, energy efficiency must be encouraged, and international collaboration must be fostered. To adjust to shifting conditions, it is also essential to regularly update the forecast and monitor CO_2 emissions continuously. India can strive toward a more sustainable future by being aware of the consequences of the CO_2 emission prediction and taking proactive steps. To lessen the negative consequences of climate change and guarantee a healthy planet for future generations, a coordinated effort involving the government, business community, and citizens is required [4].

While the ARIMA model has proved efficient in forecasting based on the given historical data, there are several drawbacks with this approach that may affect precision in the CO_2 emission forecasts. For example, sudden changes in trends such as government policies increasing stringency in emissions regulations or promotion of renewable energy technologies can easily result in abrupt changes in the trends. ARIMA cannot consider such external changes since it relies on pattern continuity from the past and hence can fail to give forecasts with accuracy. Of course, other challenges will be economic shocks, such as recession. This could temporarily lower the industrial activities and, in turn, reduce emissions, misleading ARIMA to think that a long-term decline has started. Likewise, with sudden economic growth, the emissions could shoot up, resulting in discrepancies in the forecast.

Technological changes, the sudden and widespread diffusion of renewable energy technologies or major successes in carbon capture can lead to nonlinear variations in the emissions trend, for which ARIMA is unprepared. For example, there was a reduction in global emissions due to COVID-19 because of reduced industrial activity; this was an unexpected event which ARIMA would not predict. This uncertainty increases with longer-term forecasts, as external factors such as geopolitical events and resource availability become more influential. Integration of ARIMA with machine learning models or hybrid approaches that incorporate nonlinear interactions and external factors can be carried out to improve reliability for more robust and dynamic predictions.

Limitation and Future Scope

Although a strong tool, the ARIMA model has drawbacks that may affect how accurately CO_2 emissions are predicted. Among these restrictions are the presumptions of linearity, stationarity, and outlier sensitivity. Furthermore, the model can have trouble capturing intricate patterns and nonlinear interactions. While linearity suggests that the correlations between present and historical values are linear, stationarity argues that the time series' statistical characteristics, such as mean and variance, stay constant over time. These assumptions might not hold true in real-world situations because of things like structural breaks, seasonal effects, or outside shocks that affect the data [42].

Unexpected economic events, such as a recession, global pandemic, or rapid adoption of renewable energy, disrupt normal historical patterns in emissions and make ARIMA's forecasts inefficient. Sometimes, during an economic decline, industrial activities decrease, which is usually accompanied by a reduction in energy use, temporarily reducing CO_2 emissions. Only given information about historical patterns, ARIMA might overgeneralize and become overly pessimistic once the economy has recovered. Analogously, the rapid growth of renewable energy or the sudden changes in technology abruptly alter the trend of emissions, which ARIMA models cannot grasp; hence, it also faces limitation for such dynamic and nonlinear scenarios.

The prediction power of the model can be increased by incorporating external elements like economic data, climate legislation, and technological improvements. By combining several models, ensemble approaches can lessen the influence of the biases and uncertainties of individual models. To deal with non-stationarity, techniques like seasonal decomposition or transformations (such as log or power transformations) could be used. Similarly, by identifying nonlinear relationships or dynamic patterns in the data, other models like ARIMA-GARCH or machine learning techniques like LSTMs could enhance ARIMA. Prior knowledge and uncertainty can be included in the forecasting process using Bayesian approaches. Finally, data-driven strategies like machine learning can assist in locating intricate links and patterns in the data, producing projections that are more accurate [43–45].

Different policy measures are suggested as effective in containing CO_2 emissions. Taxes related to high-ecological-impact activities should be expensive, carbon taxes should be levied, and cap-and-trade systems, together with carbon offsets and environmental technology standards, will steer pro-sustenance behaviors. Initiatives on educating the public about sources of pollution and the resultant environmental impacts are much needed. The facilitation of free public transit and promotion of electric vehicles can reduce national fuel consumption, resulting in lower carbon footprints. There are opportunities to diversify into low-carbon energy technologies, such as carbon-free hydrogen and low-carbon biofuels. Gradually, the dependency on coal will be reduced, evolving a leading position in renewable energy. Furthermore, industrial emissions abatement by a voluntary approach must be pursued [45].

The forecast of CO_2 emission in this study is univariate and does not consider exogenous factors that might emanate from population growth, economic development and/or technological change, along with shifts to renewable energy sources, and even government policies that are likely to evolve in the future. Such a future study could also develop more comprehensive, multifactorial models for emission forecasting, which will be useful for policymakers by giving them a fuller picture.

5. Conclusions

A disturbing increasing trend is revealed by the study of the time series forecast for CO_2 emissions in India, which was produced using an ARIMA (0, 2, 4) model. The main forces behind this development are industrialization, rising energy use, and economic expansion. Although the model offers insightful information, it also draws attention to the inherent uncertainties in future projections. The trajectory of emissions can be greatly impacted by variables such as changes in legislation, technical breakthroughs, and economic swings. This prediction has wide-ranging effects. Increasing CO_2 emissions cause climate change and global warming, which have several negative effects on the ecosystem. Furthermore, rising CO_2 emissions have the potential to worsen air pollution, which would be detrimental to people's health and welfare. Implementing strong climate policies and regulations, encouraging the use of renewable energy sources, promoting energy efficiency, and fostering international collaboration are all essential to lessening these effects.

The analysis emphasizes the significance of using data-driven policymaking to address CO_2 emissions. Important quantitative findings, including anticipated emission levels by 2038, indicate the pressing need for efficient actions. Prioritizing emissions reduction in high-impact industries and transportation, encouraging the use of renewable energy sources, and implementing stronger environmental laws are examples of concrete actions. Additionally, as variables like policy changes, economic shifts, and technology breakthroughs evolve, projections stay accurate and pertinent thanks to adaptive forecasting, which updates models with fresh data on a regular basis. This ongoing monitoring strategy strengthens India's commitment to addressing climate change by allowing policymakers to assess progress toward emission objectives and make appropriate adjustments.

Increased sea levels, biodiversity loss, and extreme weather events that endanger ecosystems and livelihoods are just a few of the dire effects of unchecked CO_2 emissions, which worsen climate change. The financial consequences, such as skyrocketing catastrophe recovery costs and medical bills linked to pollution, further emphasize how urgently sustainable emission control measures are needed to protect the environment and the general welfare.

Nonetheless, it is critical to recognize the ARIMA model's limitations. In practical situations, the model's assumptions of stationarity, linearity, and constant variance might not always hold true. Furthermore, emissions trends can be greatly impacted by unforeseen external variables like technological advancements or geopolitical events. To adjust to changing conditions, it is crucial to continuously monitor CO_2 emissions and update the forecast on a frequent basis. India can strive toward a more sustainable future by being aware of the consequences of the CO_2 emission prediction and taking proactive steps. To lessen the negative consequences of climate change and guarantee a healthy planet for future generations, a coordinated effort involving the government, business community, and citizens is required.

Author Contributions: Methodology, A.A.D.; software, H.P.M. and T.W.A.; validation, T.W.A.; investigation, A.A.D. and H.P.M.; writing—review and editing, A.A.D., T.W.A., M.Z.R. and H.P.M.; visualization, T.W.A. All authors have read and agreed to the published version of the manuscript.

Funding: This research was funded by Researchers Supporting Project number (RSPD2024R1038), King Saud University, Riyadh, Saudi Arabia.

Data Availability Statement: The original contributions presented in this study are included in the article. Further inquiries can be directed to the corresponding author.

Acknowledgments: The authors extend their sincere appreciation to the Researchers Supporting Project number (RSPD2024R1038), King Saud University, Riyadh, Saudi Arabia.

Conflicts of Interest: The authors declare no conflict of interest.

References

1. Dar, A.A.; Jain, A.; Malhotra, M.; Farooqi, A.R.; Albalawi, O.; Khan, M.S.; Hiba. Time Series Analysis with ARIMA for Historical Stock Data and Future Projections. *Soft Comput.* **2025**. [CrossRef]
2. Kazemzadeh, E.; Lotfalipour, M.R.; Shirazi, M.; Sargolzaie, A. Heterogeneous effects of energy consumption structure on ecological footprint. *Environ. Sci. Pollut. Res.* **2023**, *30*, 55884–55904. [CrossRef] [PubMed]
3. Dao, N.B.; Dogan, B.; Ghosh, S.; Kazemzadeh, E.; Radulescu, M. Toward sustainable ecology: How do environmental technologies, green financial policies, energy uncertainties, and natural resources rents matter? *Clean Technol. Environ. Policy* **2024**, 1–19. [CrossRef]
4. Silva, N.; Fuinhas, J.A.; Koengkan, M.; Kazemzadeh, E. What are the causal conditions that lead to high or low environmental performance? A worldwide assessment. *Environ. Impact Assess. Rev.* **2024**, *104*, 107342. [CrossRef]
5. Awasthi, A.; Ranjan, R. A hybrid ARIMA-GARCH approach for forecasting CO_2 emissions in India. *Environ. Sci. Pollut. Res.* **2021**, *28*, 11658–11669.
6. Bhatia, R.; Gupta, R. Forecasting CO_2 emissions using ARIMA and machine learning approaches: A case study of India. *Sustain. Prod. Consum.* **2022**, *32*, 647–659. [CrossRef]
7. Chaudhry, S.; Mohsin, M. Time series forecasting of CO_2 emissions in India: An ARIMA model approach. *Environ. Monit. Assess.* **2023**, *195*, 1–14.
8. Raihan, A.; Begum, R.A.; Nizam, M.; Said, M.; Pereira, J.J. Dynamic impacts of energy use, agricultural land expansion, and deforestation on CO_2 emissions in Malaysia. *Environ. Ecol. Stat.* **2022**, *29*, 477–507. [CrossRef]
9. Raihan, A.; Tuspekova, A. Nexus between emission reduction factors and anthropogenic carbon emissions in India. *Anthr. Sci.* **2022**, *1*, 295–310. [CrossRef]

10. Raihan, A.; Tuspekova, A. The nexus between economic growth, renewable energy use, agricultural land expansion, and carbon emissions: New insights from Peru. *Energy Nexus* **2022**, *6*, 100067. [CrossRef]
11. Jayasinghe, M.; Selvanathan, E.A. Energy consumption, tourism, economic growth and CO_2 emissions nexus in India. *J. Asia Pac. Econ.* **2021**, *26*, 361–380. [CrossRef]
12. Das, S.; Sahu, K. Analyzing CO_2 emissions in India using ARIMA modeling: Implications for policy. *Energy Rep.* **2021**, *7*, 123–135. [CrossRef]
13. Ahmed, N.K.; Atiya, A.F.; Gayar, N.E.; El-Shishiny, H. An empirical comparison of machine learning models for time series forecasting. *Econom. Rev.* **2010**, *29*, 594–621. [CrossRef]
14. Iftikhar, H.; Khan, M.; Żywiołek, J.; Khan, M.; López-Gonzales, J.L. Modeling and Forecasting Carbon Dioxide Emission in Pakistan Using a Hybrid Combination of Regression and Time Series Models. *Heliyon* **2024**, *10*, e33148. [CrossRef]
15. Tanania, V.; Shukla, S.; Singh, S. Time series data analysis and prediction of CO_2 emissions. In Proceedings of the 2020 10th International Conference on Cloud Computing, Data Science & Engineering (Confluence), Noida, India, 29–31 January 2020; pp. 665–669.
16. Merchante, L.F.; Clar, D.; Carnicero, A.; Lopez-Valdes, F.J.; Jimenez-Octavio, J.R. Real-time CO_2 emissions estimation in Spain and application to the COVID-19 pandemic. *J. Clean. Prod.* **2021**, *296*, 126425. [CrossRef]
17. Wen, L.; Yuan, X. Forecasting CO_2 emissions in Chinas commercial department, through BP neural network based on random forest and PSO. *Sci. Total Environ.* **2020**, *718*, 137194. [CrossRef]
18. Alam, T.; AlArjani, A. A Comparative Study of CO_2 Emission Forecasting in the Gulf Countries Using Autoregressive Integrated Moving Average, Artificial Neural Network, and Holt-Winters Exponential Smoothing Models. *Adv. Meteorol.* **2021**, *2021*, 8322590. [CrossRef]
19. Linardatos, P.; Papastefanopoulos, V.; Panagiotakopoulos, T.; Kotsiantis, S. CO_2 concentration forecasting in smart cities using a hybrid ARIMA–TFT model on multivariate time series IoT data. *Sci. Rep.* **2023**, *13*, 17266. [CrossRef]
20. Dutta, A.; Chatterjee, S. Forecasting carbon dioxide emissions in India: A comparative analysis of ARIMA and SARIMA models. *Appl. Energy* **2022**, *301*, 117447. [CrossRef]
21. Gupta, K.R.; Kumar, A. Time series analysis of CO_2 emissions in India: An ARIMA model approach. *J. Environ. Manag.* **2021**, *285*, 112203. [CrossRef]
22. Jain, P.; Sharma, S. Forecasting CO_2 emissions in India: A time series analysis using ARIMA and neural networks. *J. Clean. Prod.* **2023**, *276*, 123143. [CrossRef]
23. Joshi, P.; Singh, R. Forecasting carbon emissions in India using ARIMA: A time series analysis. *Sustain. Cities Soc.* **2022**, *78*, 103528. [CrossRef]
24. Kumar, S.; Yadav, A. Time series forecasting of CO_2 emissions in India using ARIMA and seasonal decomposition. *Energy Policy* **2022**, *162*, 112772.
25. Kumar, V.; Prakash, A. A comparative study of ARIMA and exponential smoothing methods for CO_2 emissions forecasting in India. *Environ. Sci. Pollut. Res.* **2021**, *28*, 9503–9515.
26. Malakar, A.; Bera, A. ARIMA forecasting of CO_2 emissions in India: An integrated approach with climatic factors. *Atmos. Environ.* **2023**, *276*, 119013.
27. Mittal, S.; Kumar, S. CO_2 emissions forecasting in India: An ARIMA model analysis. *Int. J. Energy Econ. Policy* **2021**, *11*, 569–578.
28. Mukherjee, A.; Bhanja, S. Modeling and forecasting CO_2 emissions in India: An ARIMA approach. *Renew. Sustain. Energy Rev.* **2022**, *158*, 112036. [CrossRef]
29. Prakash, A.; Sharma, R. Using ARIMA for time series forecasting of CO_2 emissions in India. *J. Environ. Manag.* **2022**, *310*, 114778. [CrossRef]
30. Albeladi, K.; Zafar, B.; Mueen, A. Time Series Forecasting using LSTM and ARIMA. *Int. J. Adv. Comput. Sci. Appl.* **2023**, *14*, 313–320. [CrossRef]
31. Ranjan, R.; Awasthi, A. Time series modeling of CO_2 emissions in India: A comprehensive approach using ARIMA. *J. Clean. Prod.* **2021**, *278*, 123944. [CrossRef]
32. Sharma, A.; Singh, S. Forecasting CO_2 emissions using ARIMA model: Evidence from India. *Environ. Sci. Pollut. Res.* **2022**, *29*, 9932–9941.
33. Sharma, R.; Gupta, K.R. ARIMA model for forecasting carbon emissions in India: An empirical analysis. *Energy Rep.* **2023**, *9*, 99–107. [CrossRef]
34. Shukla, A.; Gupta, S. ARIMA modeling for CO_2 emissions in India: A time series approach. *J. Clean. Prod.* **2021**, *327*, 129450.
35. Singh, R.; Kumar, A. Time series analysis and forecasting of CO_2 emissions in India: An ARIMA approach. *Int. J. Clim. Change Strateg. Manag.* **2021**, *13*, 267–281.
36. Singh, S.P.; Jain, A. An ARIMA-based approach to forecast CO_2 emissions in India. *Environ. Sci. Pollut. Res.* **2023**, *30*, 17516–17528.
37. Sudhakar, M.; Babu, G. Time series forecasting of CO_2 emissions in India using ARIMA and Holt-Winters methods. *Energy Sources Part B Econ. Plan. Policy* **2022**, *17*, 32–42.
38. Kumari, S.; Singh, S.K. Machine learning-based time series models for effective CO_2 emission prediction in India. *Environ. Sci. Pollut. Res.* **2023**, *30*, 116601–116616. [CrossRef]
39. Tiwari, R.; Kumar, V. Forecasting CO_2 emissions in India: A time series approach using ARIMA. *Environ. Sci. Pollut. Res.* **2021**, *28*, 4420–4431.

40. Tripathi, R.; Kumar, R. ARIMA-based forecasting of CO_2 emissions in India: A sectoral analysis. *Energy Rep.* **2022**, *8*, 175–182. [CrossRef]
41. Verma, M.; Bansal, R. A comparative analysis of ARIMA and exponential smoothing methods for forecasting CO_2 emissions in India. *J. Clean. Prod.* **2023**, *362*, 132556.
42. Wang, Q.; Zhang, L. ARIMA forecasting of CO_2 emissions in India: Integrating climatic factors. *Energy Econ.* **2022**, *103*, 105557.
43. Yadav, A.; Gupta, R. Time series forecasting of CO_2 emissions in India using ARIMA model: Insights from data analysis. *Environ. Sci. Pollut. Res.* **2021**, *28*, 18953–18965.
44. Yadav, P.; Ghosh, S. CO_2 emissions forecasting using ARIMA model: Evidence from India. *Sustain. Cities Soc.* **2022**, *84*, 103044.
45. Zhang, Y.; Zhao, Y. Forecasting CO_2 emissions in India: An ARIMA approach. *Environ. Sci. Pollut. Res.* **2021**, *28*, 39872–39884.

processes

MDPI

Article

Modeling the Benefits of Electric Cooking in Ecuador: A Long-Term Perspective

Veronica Guayanlema [1], Javier Martínez-Gómez [2,3,*], Javier Fontalvo [1] and Vicente Sebastian Espinoza [1]

1 Instituto de Investigación Geológico y Energético, Quito 170518, Ecuador; veronica.guayanlema@gmail.com (V.G.); javier.fontalvo@geoenergia.gob.ec (J.F.); sebastian.espinoza@geoenergia.gob.ec (V.S.E.)
2 Departamento de Teoría de la Señal y Comunicación, (Área de Ingeniería Mecánica) Escuela Politécnica, Universidad de Alcalá, 28805 Alcalá de Henares, Madrid, Spain
3 Facultad de Arquitectura e Ingenierías, Universidad Internacional SEK, Albert Einstein s/n and 5th, Quito 170302, Ecuador
* Correspondence: javier.martinezgomez@uah.es

Abstract: The study quantifies the benefits of expanding electric cooking in the residential sector in replacement of liquefied petroleum gas (LPG), including economic savings and the avoided emissions resulting from this transition, viewed through the perspective of a long-range optimal energy system model developed for the Ecuadorian energy system under the LEAP (Long-range Energy Alternative Planning) framework. In Ecuador, electricity generation is predominantly based on hydropower obtained from run-of-the-river schemes. The model results indicate that a sectorial-level policy to promote electric cooking reduces the use of LPG per annum, which consequently leads to reductions in greenhouse gas emissions. Additionally, the electric cooking scenario also complements the Ecuadorian vision of reducing deforestation and reaching carbon neutrality. Furthermore, the subsidies to LPG will be reduced, improving energy sovereignty. Finally, the paper discusses the effects and implications of this policy implementation over the nationally determined contributions (NDC).

Keywords: electric cooking; emissions; perspective; subsidies

1. Introduction

Electricity is an essential requirement for sustainable development. Around the world, 1.5 billion people lack access to electricity, of which more than 600 million are in sub-Saharan Africa and the other 300 million in India alone. Also, three billion people rely on wood, coal, or animal droppings for cooking and heating. Access to electricity and clean-cooking equipment is necessary to eradicate poverty and build shared prosperity [1]. Therefore, the General Assembly of the United Nations declared in 2012 that the decade between 2014 and 2024 would be the Decade of Sustainable Energy for All. While the extensive and inefficient use of fossil fuels, which increases greenhouse gas emissions and contributes to Global Warming and Climate Change, puts all of us at risk, it does so to a more harmful and greater extent to the poor [2].

The World Health Organization (WHO) estimates that 1.5 million premature deaths a year are directly associated with the use of solid fuels for cooking activities [3]. Children, elder people, and women are the most affected, given that they spend more time near the stove [4]. To overcome this situation, several programs have been carried out to improve cooking techniques worldwide. In the case of China, different initiatives have been taken to increase the number of new efficient stoves since the 1980s [5]. In rural areas of India, studies have found that around 80% of total energy consumption is for cooking [6], with several programs carried out to improve efficiency and mitigate the harmful effects on people from biomass stoves. Moreover, solar cooktops and induction cooktops have been developed and tested [6,7]. In the case of Burkina Faso, studies have been promoted to understand household energy preferences for cooking in the city of Ouagadougou [8]. Once those preferences were outlined, efforts were addressed to influence the population to adopt clean-cooking alternatives. In Ethiopia, studies have been carried out to introduce solar cookers for the development of clean cooking [9].

Despite the aforementioned efforts, governments in the majority of developing countries are still lagging in devoting stronger efforts to introduce electricity as a fuel option for cooking, which is also related to sustainable development. Electric stoves, and specifically induction cooktops, have advantages over biomass, natural gas, and LPG stoves in aspects such as safety, cooking time, cleaning after use and energy efficiency, as reported in [10]. One example where this transition is encouraged occurs in India, where a pilot project was conducted to introduce 4000 induction stoves in a rural area with encouraging results [7].

In Ecuador, around 86.56% of the demanded LPG is imported, creating a dependence on foreign fuel availability and a significant outflow of domestic funds that affects the trade balance of the country [11,12]. LPG 15-kg cylinders for domestic use are subsidized by the state, an assistance that roughly costs US $716 million per year to the national coffers [13]. LPG cylinders are sold at an official price of USD $1.60, whereas in the neighboring countries of Peru and Colombia, prices for the same quantity of fuel can be up to thirteen times higher (US $17 and US $23, respectively). Hence, around 20% of the dispatched LPG is usually lost in contraband or non-domestic use [14].

As one of the policies established to face this problem, the National Plan of Efficient Cooking (NECP) induction stove adopters were subsidized by the Ecuadorian government by not charging the initial 80 kWh consumed on their electric bills, which is thought to be enough for an average family [14]. This government help has been in implementation until 2023. But even with this benefit, citizens in Ecuador have shown a marked reluctance to migrate to induction stoves. Until 2018, 595,343 families were enrolled in the NECP Program [15] and benefited from the discount rate. In addition, in areas where induction stoves have been donated by the state, as is the case of border regions included in the Plan Frontera, residents have remained using LPG stoves due to the low cost of LPG fuel and the fact that some zones have reliability issues in electricity provision [16].

The Ecuadorian government launched a program for clean cooking with induction stoves named the National Plan of Efficient Cooking (NECP). The NECP was linked to the transformation of the national electricity mix, which aims to reach energy sovereignty and access to clean energy for Ecuador [10]. Furthermore, the initiative is part of the compromises required by the country to reduce its emissions of greenhouse gases in its first nationally determined contributions (NDCs) [17].

The aim of the ECP program was to replace LPG cookers with induction cookers. In addition, it sought to achieve three large-scale changes:

1. Replace imported fossil fuels with electricity from hydro sources in the residential sector.
2. Phase out the LPG subsidies.
3. Propel the local industry through technological development.

Initially, the goal of the program was to provide 3,500,000 appliances in 2017 through an investment of about $2.4 billion. To further this measure, a tax increase of almost 200% on LPG stoves was set in order to encourage the sale of induction cookers. The strategy counted on the completion of large hydro projects in 2017, such as the Coca Codo Sinclair Power Plant [18]. This goal was not reached because of the delays in the construction of several hydro projects and the lack of sales of induction cooktops. At present, it is expected that 2,000,000 induction cookers will have been introduced by 2023 [19].

The Ministry of Energy and Non-Renewable Resources (now the Ministry of Energy and Mines) has also organized several demonstrative workshops to instruct on the use of induction stoves and highlight the benefits obtained when using this technology. However, the LPG subsidy represents a great barrier to shifting to other cooking fuels in Ecuador since society is not aware of its real price and has gotten used to using cheap LPG with its consequent inefficient and improper usage. This is a very important issue for the current national energy policy because it affects the entire country, especially the low-income population.

So far, Ecuador presents a unique case of a governmental policy aimed at promoting the widespread implementation of induction cooking at a national level. Hence, the analysis of the characteristics of the NECP program, how it worked during its implementation years, and the causes behind its phase-out provide novel insights given its particular context. Furthermore, the germinal implementation of the program in the country provides data that is unavailable in other geographical regions. Thus, the medium- and long-term perspectives modeled in this paper will prove highly beneficial for energy researchers worldwide who are interested in the dynamics of future mass adoption of induction cooking within a reference framework of energy transition and resource scarcity.

The aim of this research is to analyze the implementation of the NECP program, in the context of the long-term goals defined for both energy consumption and greenhouse gas emissions. This study quantifies the environmental benefits of expanding electric cooking in Ecuador, as opposed to the current dominant practice of burning LPG.

The present paper consists of six sections. Background information on Ecuador and its energy system is presented in Section 2. Section 3 describes the methodology (modeling in LEAP and end-use demand projections) and the base case scenario description. Section 4 explains the alternative electric cooking scenario. Section 5 discusses the model result of promoting electric cooking and its implications for energy system development with a focus on reduction in fuelwood consumption and reduction in the emission of pollutants. The concluding remarks are presented in Section 6.

2. Background of the Ecuadorian Energy System

Ecuador started the transition towards a more renewable-based power grid in 2009, with the first issuance of the National Plan for Good Living (FNPGL) [19]. By 2009, nameplate capacity in Ecuador was led by conventional thermal generation, with 53.99% of the share, followed by hydroelectric energy with 46.69% and nonconventional renewable energy (represented by solar photovoltaic, wind, geothermal and biomass processing) with 0.02% of the total [20]. The goals for the electric sector in the FNPGL were to guarantee energy sovereignty, offer a trustworthy and quality service, and achieve a cleaner electricity system. This set of policies drove the growth of renewable energy sources in the electricity generation mix.

Following the development of several large generation projects, mainly hydroelectric, the amount of renewable energy in the electric power mix increased (Table 1). The share of renewable energy reached 76.11%, followed by internal combustion engines with 16.26%, gas turbines with 4.53%, and steam turbines with 3.09% [18].

Table 1. Gross generation based on [18].

Energy Source	Central Type	Gross Generation	
		GWh	%
Renewable	Hydro	24,635.16	74.63
	Biomass	348.08	1.05
	Wind	60.6	0.18
	Biogas	41.59	0.13
	Solar PV	38.5	0.12
Renewable Total		**25,123.93**	**76.11**
Non-Renewable	Combustion Engine	5366.38	16.26
	Gas turbine	1496.46	4.53
	Stream Turbine	1021.54	3.09
Non-Renewable Total		**7884.37**	**23.89**
Total		**33,008.30**	**100**

From Table 1, it can be noticed that 76.13% of total electricity production was renewable, while 23.87% was non-renewable. While renewable energy sources turned to produce the majority of the electricity in Ecuador, in recent years, it is important to note that most of this renewable production comes from hydroelectric power plants, which makes the integration of renewables less challenging. However, the reliance on hydroelectric power for electricity production must also be assessed, especially since the country experiences a dry season between October and February that reduces river flow and subsequently lowers the output of these plants. Figure 1 shows the electric energy consumption and electric losses by sector during 2022 in GWh. The three largest sectors of electricity consumption were industry with 11,802 GWh, and a share of 35.6%, households with 7833 GWh and a 23.6% share, and commercial venues and public services including public lighting which used 7387 GWh, which corresponds to 22.3%. The remaining share grouped transport, other sectors, and transmission and distribution losses.

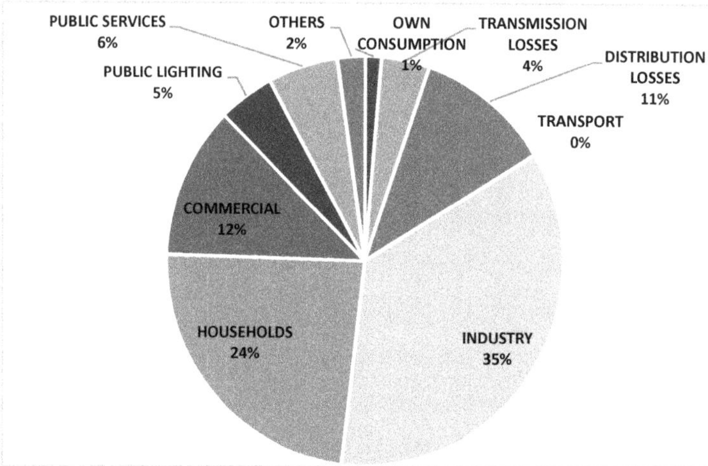

Figure 1. Electric energy consumption and electric losses by sector based on [20].

Data on the Ecuadorian residential sector shows that its energy consumption has grown along with the increase of both new constructions and population, as shown in Figure 3. Hence, one hypothesis to be considered is that energy subsidies in the country have been tightly linked to energy consumption. Particularly, subsidies for oil products have been part of the national energy policy since 1972, mainly as a compensation system during periods of high international oil prices. This decision was first aimed at meeting the requirements of military transport and then extended to benefit both public and private road transport. Afterwards, the subsidy policy was expanded in order to support disadvantaged social groups, specifically through the application of a subsidy in the price of liquefied petroleum gas (LPG), designed to compensate for imported LPG prices that were higher than those of the same fuel that was locally produced [21].

The production of petroleum products in Ecuador is constrained by the refining capacity in the country. The Ecuadorian economy is highly reliant on crude oil exports, limiting the supply for secondary energy conversion. In 2023, imports of oil derivatives reached 61,437 kboe, out of a total demand of 121,556 kboe [12]. Figure 2 shows the structure of the country's energy consumption, with transport as the main energy user, followed by industry and households.

Subsidies to oil products (mainly LPG, gasoline, and diesel) contributed to a constant increase in demand, which has been covered by a higher share of imports throughout the years. Due to capacity constraints and refining profile, refining infrastructure in the country has become insufficient to supply the growing domestic demand. In the case of LPG, the share of domestic production in the total supply went from 40% in 2000 to approximately

13% in 2022, as shown in Figure 4. In addition, diesel imports with respect to the total supply in the country increased its share from 17% to 69% in the same period [13].

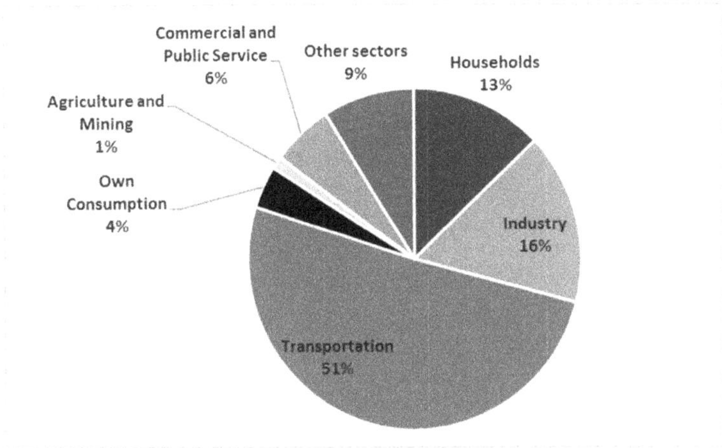

Figure 2. Structure of the country's energy consumption [18].

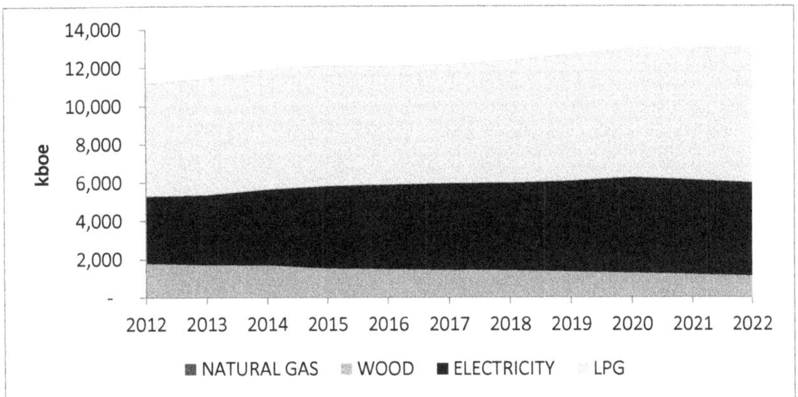

Figure 3. Evolution of the residential energy demand in Ecuador, by source.

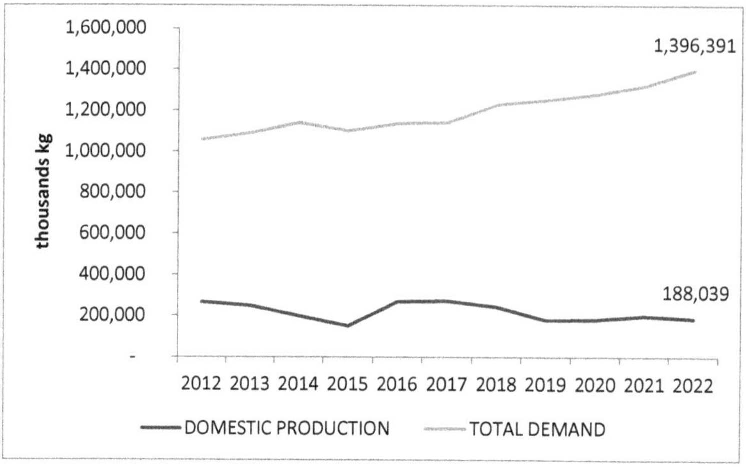

Figure 4. GLP demand and domestic production in Ecuador.

3. Methodology

The Long-range Energy Alternative Planning System (LEAP) model was used in order to estimate the energy demand in the country and the corresponding environmental impact. LEAP is a bottom-up, demand-driven model developed by the Stockholm Environment Institute (SEI) in Boston, Massachusetts that assigns energy flows to the energy supply technologies in a geographical region [22]. In this sense, LEAP is not a model of a particular energy system, but rather a tool that can be used to create models according to the available information [23]. The LEAP model block diagram is shown in Figure 5.

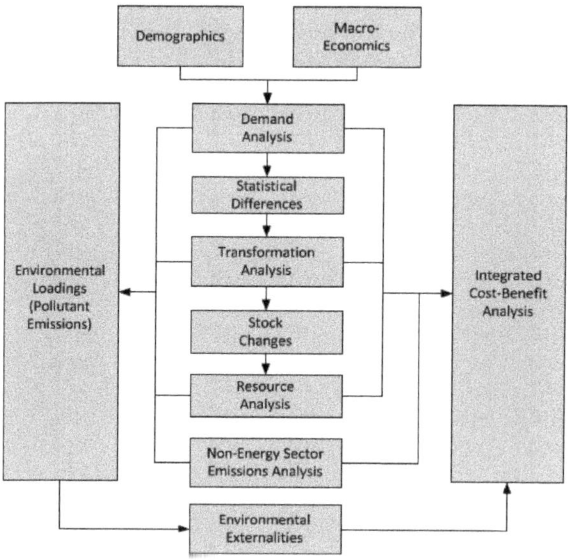

Figure 5. LEAP model block diagram [24].

In the case of Ecuador, data included in the National Development Plan and the National Efficient Energy Plan (PLANEE in Spanish) have been considered to lay out the model of household energy demand. The PLANEE was published by the Ministry of Energy in 2017, with the aim of achieving a more efficient use of energy through the execution of public programs and projects. Other targets of the plan were to reduce the import of oil products and contribute to the mitigation of climate change in the country. In the end, the PLANEE seeks to promote a culture of energy efficiency among the population, supported by a solid legal and institutional framework. The PLANEE is structured in six chapters considering energy supply, energy demand, legal considerations, institutional aspects and informational elements as follows: (1) Legal, Institutional framework and Information Access; (2) Households, Commercial and Public sectors; (3) Industry; (4) Transport, (5) Own Use, and (6) Galapagos. Each chapter has a general objective, specific objectives, goals, and action lines [15]. Some strategies included in the PLANEE look to improve the energy intensity of households, with PEC being the most influential and the one with the broadest scope.

3.1. General Energy Assumptions

One of the main variables in energy demand models is the gross domestic product (GDP) of a determined geographical region. Historical GDP data taken from the World Bank [25] for the period 1961–2007, and from the Central Bank of Ecuador [26] for the 2008–2015 period was used in the present study. The National Development Plan (NDP) has as one of the main goals for the country to reach a GDP per capita growth of 5.0% by 2025. Given the inertia of the Historical GDP scenario, it has been considered that these growth rates might be plausible after 2023. Thus, both scenarios will have the same GDP up to the aforementioned year. The CAGR of GDP for the period 2020–2030 was estimated to reach 4.4% using NDP information [11]. Figure 6 shows the GDP from Ecuador provided in the referenced documents.

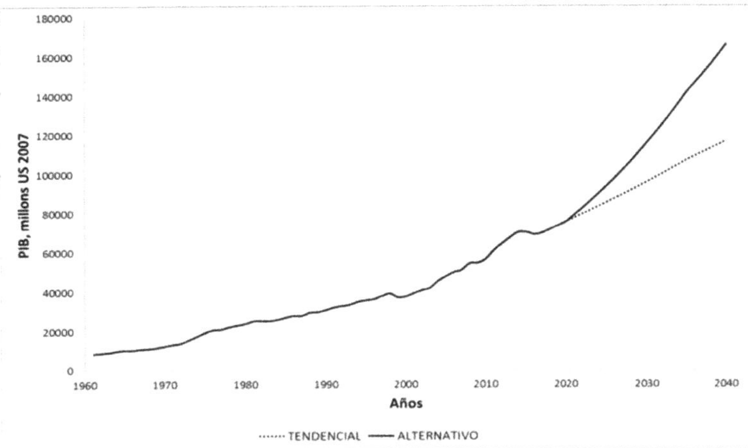

Figure 6. Ecuadorian GDP projections used in the modeling.

Population is another variable of interest. Data from the National Institute of Statistics and Census of Ecuador (INEC) was included in the model of the national energy system (Figure 7). Based on INEC information, the country's population was projected to grow with an average annual rate of 1.33% for the period 2007–2050. These estimations included the birth rate and population projections for each Ecuadorian province.

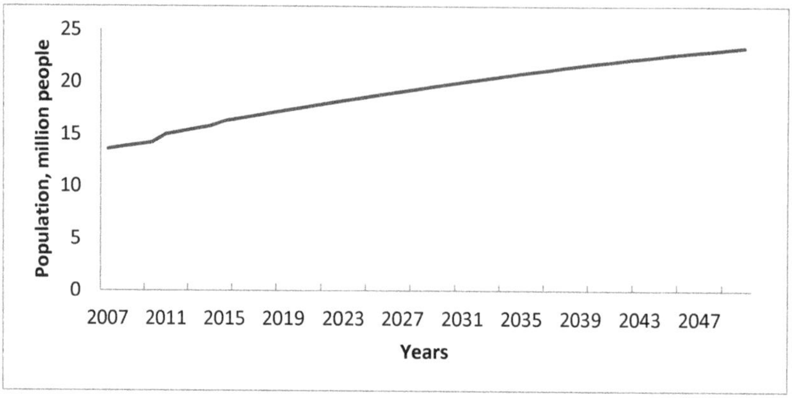

Figure 7. Population in Ecuador 2007–2050 [27].

For population projections, the year 2010 is defined by the INEC as the base year for the calculations. The projection assesses population by age and sex with data taken from the 2010 Census of Population and Households. Based on the population numbers of 1990 and 2001 censuses, along with demographic dynamics from the 1990–2010 period, the historical data is corrected and birth–death and migration data are added to the population in the base year (Figure 7) [27].

The number of households was forecasted using the 1990, 2001 and 2010 censuses. According to the 2001 and 2010 censuses, the total number of households in the country was around 2.9 million, and 3.8 million, respectively. The used annual growth rate of 1.53% has been set by data included in the "National Energy Agenda" for the period 2008–2050 [28], specifically population growth and number of family members in a household.

3.2. *Reference Energy System*

The developed energy system includes the sectorial demand structure, transformation processes, and energy sources, and is implemented through the LEAP model tools. In order to analyze the effects of energy policies and plans, the LEAP model comprises the energy chains of the whole country; the structure of the household sector demand inside the model is depicted in Figure 8. In the case of cooking activities, the final energy use of the sector was determined by aggregating the consumption of each technology, considering the percentage of households that use water heaters, GLP stoves, refrigerators, and other equipment—information included in the PLANEE. The total energy demand of households was obtained from the Ecuadorian Energy Balance [12]. In the case of induction stoves, it should be noted that electricity in the country is generated mainly from hydro sources, with thermal sources in second place. The share of electricity production by source is presented in Section 2.

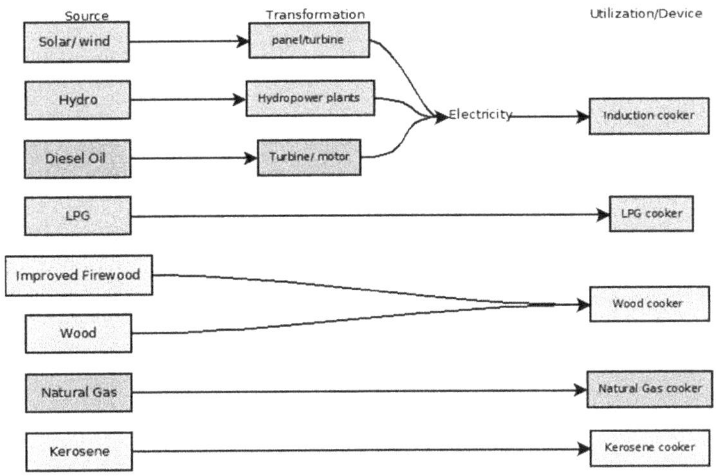

Figure 8. Structure energy demand in cooking.

3.3. Base Scenario Description

The base scenario (BS) takes into account the trends like the business as usual (BAU) scenario. The demand and supply of energy behave under the conditions of the year 2022, in which no type of action is carried out. Energy demand in the sectors keeps a growth tendency like the data ranging from 2007 to 2022. The fossil fuel necessary to meet the demand is imported in the case of secondary sources, as the investment in new refinery capacity is not a policy priority in the model. In this scenario, electricity is mainly generated by thermal plants, with an estimate of 56% on average [12]. The BS scenario is necessary to compare other scenarios that include assumptions related to the NECP program.

In Ecuador, LPG is the first energy source for cooking, with around 90% of this fossil fuel used in households. In terms of performance, induction stoves have an estimated efficiency of 84%, while an LGP efficient stove shows an average of 40% efficiency [29]. The BS scenario assumes each household is represented by one stove. In this case, the total energy demand was calculated using activity levels for each stove. The useful energy intensity is the variable that leads to consumption, along with growth in GDP per capita. In the cooking subsector, sales of induction stoves keep the implementation trend of the first two years (around 5.2% share of households using induction stoves).

3.4. Efficiency Scenario

The efficiency scenario considers four alternatives to introduce induction cooktops in the country. The four hypotheses are depicted by scenario in Figure 9. These figures illustrate the percentage of each technology simulated in the model.

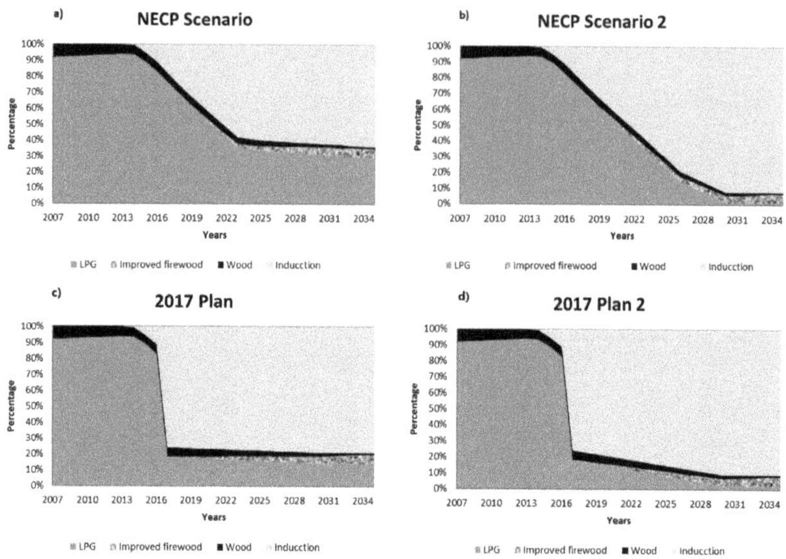

Figure 9. Considerations in the cooking sub-sector in (**a**) NECP Scenario, (**b**) NECP Scenario 2, (**c**) 2017 Plan 1, (**d**) 2017 Plan 2.

The share of households that use induction technology was estimated using the number of sales of cooktops and the households that cook. This relation is a percentage taken from all the sales of induction cooktops, in which this is the main cooking technology. Nonetheless, the forecast includes the development goals from 2020, and yearly data is calculated using a linear model (1).

$$y = mx + b \tag{1}$$

The first scenario (Figure 9a) includes the NECP goal of 2 million stoves sold by 2022. From this year onwards, the sales increase according to the growth in the number of households.

The hypothesis for the second scenario is continuing with the NECP to 2030 and replacing 92% of LPG stoves in households (Figure 9b, NECP 2). This option will require more investment in clean energy supply. The projections were obtained using a linear model with historical data. The second, third and fourth scenarios consider the opportunity to leverage financing in a conditional scenario of the NDC.

The third scenario consists of reaching the first aim of NECP, depicted in Figure 9c (2017 PLAN 1). This alternative includes reaching the goal of 3 million cooktops sold by 2017. Initially, the NECP pretended to replace 3 million cooktops by 2017; however, this goal was modified due to economic problems combined with a poor campaign to encourage the sales of induction stoves and is awaiting to be resumed. However, it is important to consider this alternative to analyze the opportunity to reduce energy consumption in the residential sector. Finally, the last alternative maintains the growth of the third scenario (2017 PLAN 2) as shown in Figure 9d.

The LEAP model calculated energy demand by relating the share of households that use a specific technology with energy efficiency and useful energy intensity projections. GHG emissions were estimated through IPCC emission factors that are included in the modeling software used (version 2020.1.0.98).

4. Results

Figure 10 shows the energy consumption of households for each scenario. Figure 10a depicts the consumption in millions of barrels of oil equivalent in the national energy demand, and it is observed that, in the BS scenario, demand growth keeps going up, reaching around 16 million BOE by 2035. In the case of the "NECP" and "NECP 2" scenarios, the trend remains similar until 2023, with the consumption from 2013 up to that point reaching between 12 and 13 million barrels, respectively. From then on, in the case of the "NECP", consumption increases from 12 million to almost 13.5 million BOE, while in the "NECP 2" scenario, demand decreases to around 11 million barrels in 2030 before growing back with a similar trend as the "PLAN 2". In the case of the "PLAN" and "PLAN 2" scenarios, there is a decrease in household energy demand from 12.3 million to 10.5 million BOE and 10 million as of 2019, respectively. Additionally, electricity demand in the "PLAN" and in "PLAN 2" scenarios reach almost 13 million and 12 million BOE respectively. Figure 10b depicts the modeled projection of electricity demand as a whole by the residential sector.

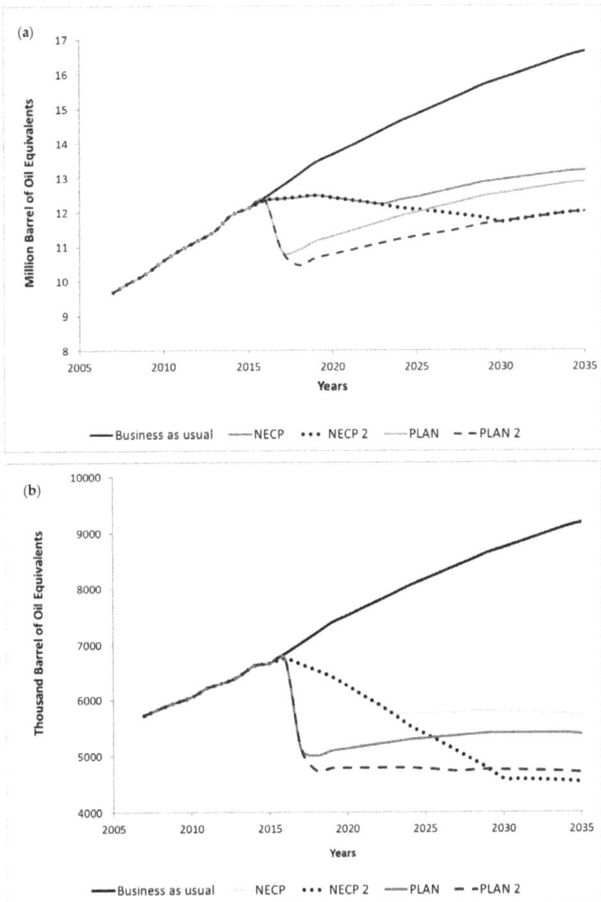

Figure 10. (**a**) Total energy demand of households for each scenario in million barrels of oil equivalent (**b**) Total electricity demand of households for each scenario in thousand barrels of oil equivalent.

Figure 11 shows the electricity demand only for cooking in thousands of barrels of oil equivalent. In the case of the BS scenario, electricity consumption remains constant below 0.2 million barrels of oil equivalent from 2016 to 2035. In the case of the "NECP" and "NECP 2" scenarios, the trend remains similar until 2023, with electricity consumption increasing to almost 2 million barrels of oil equivalent by 2023. Thereafter, in the case of the "NECP", electricity demand reaches almost 2.5 million BOE in 2035, while in the case of "NECP 2", cooking electricity consumption increases up to 3.2 million by 2030 and then grows back with a similar trend to the "PLAN 2". In the case of the 2017 "PLAN" and "PLAN 2" scenarios, there is growth from 0.2 million barrels of oil equivalent in 2016 to 2 million and 2.5 million barrels of oil equivalent respectively. Afterwards, demand in 2035 reaches 2.3 million barrels of oil equivalent in the "PLAN" and nearly 3.2 million barrels of oil equivalent in the "PLAN 2".

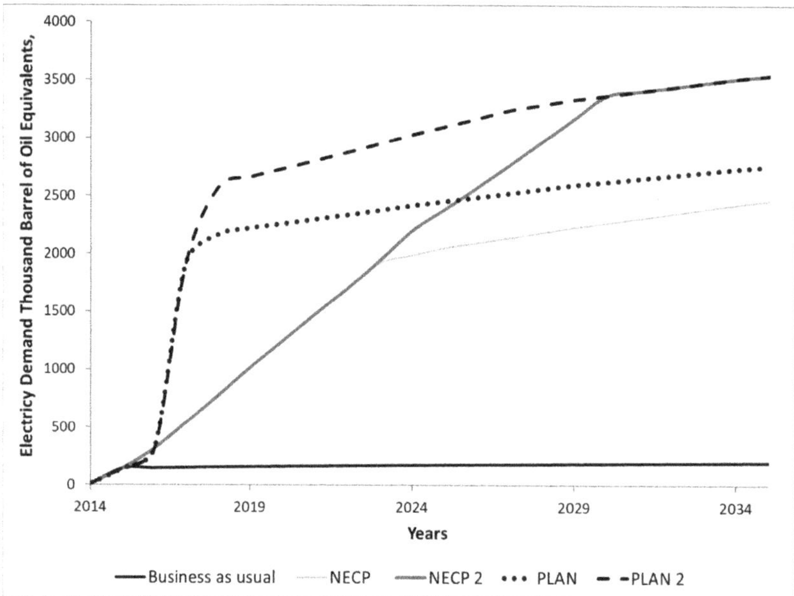

Figure 11. Electricity demand in cooking scenarios.

Figure 12 shows the GHG emissions for the residential sector in each scenario. It is observed that emissions increase from 2000 kt CO_2e in 2007 to 3000 kt CO_2e by 2035 in the BS scenario. This result is related to a high consumption of GLP.

In the case of the "NECP" and "NECP 2", the trend remains similar until 2023, with GHG emissions increasing from 2000 kt CO_2e in 2007 to almost 2600 kt CO_2e in 2017. From this year onwards, in the case of the "NECP", it drops to 1600 kt CO_2e in 2023 and remains relatively constant until 2035. Meanwhile, in the "NECP 2" scenario, emissions decrease to 600 kt CO_2e in 2030 and remain constant until 2035.

In the case of the 2017 and "PLAN 2" scenarios, they follow a growth trend for the years 2005 and 2017 similar to the BS scenario. In the "PLAN", emissions reduce from 2600 kt CO_2e in 2017 to below 1200.00 kt CO_2e in 2035, while in the "PLAN 2", it goes below 450 kt CO_2e.

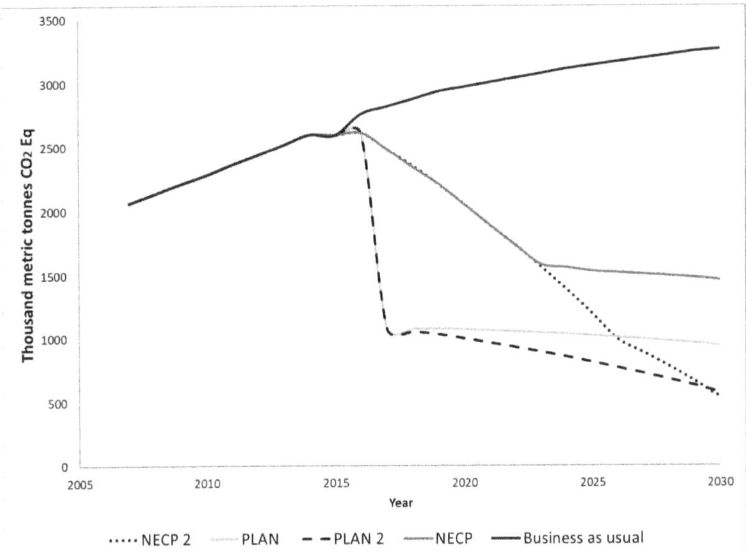

Figure 12. GHG emissions in cooking sceneries.

5. Discussion

This research sheds light on the environmental benefits of induction cookers and clean-cooking programs, particularly in regions where hydroelectric power is the primary source of electricity. Notably, this study represents one of the first large-scale initiatives in Latin America to promote the adoption of induction cookers.

According to the results obtained from the modeling and simulation of the Ecuadorian residential sector, scenarios that include the implementation of clean-cooking programs show a reduction in the energy consumption for households that by 2035 could reach 28%, compared against the consumption of the base case with no efficient cooking policies. This reduction in total energy consumption by households is reached despite an increase in electricity demand that in the most ambitious case is 18 times higher than the base case, showing that the electrification of a sector can result in total energy savings even if electricity use sharply rises.

In terms of environmental impact, greenhouse gas emissions also fall with the implementation of induction cooking, as in 2035 emissions of 3300 kt CO_2e in the BS scenario surpass the 1500 kt in the "NECP" scenario, 900 kt in the "PLAN" scenario and 450 kt in both the "NECP 2" and "PLAN 2" scenarios. Hence, this reduction in the emission of greenhouse gases offers an opportunity for future national governments to offer 1.8 million tCO_2 on the international carbon market during the studied period, whose revenues could reinforce the "NECP" implementation and its sustainability. These actions not only mitigate greenhouse gas emissions but also alleviate indoor air pollution and enhance health conditions.

6. Conclusions and Policy Implications

This research helps make the case for the use of induction cookers for clean-cooking programs when the source of electricity is produced mainly by hydroelectric power plants. According to the results obtained by modeling the experience of the promotion of induction cooking in Ecuador, this research exposes that electricity can be of importance in fueling a clean-cooking environment. To this end, the LEAP tool was used to quantify the environmental benefits of the expansion of induction cooking in Ecuador, replacing in part the current dominant practice of LPG burning. Four scenarios were proposed under different hypotheses in order to compare induction-cooking policies to a base scenario where no induction cooking was implemented.

Ecuador's pursuit of clean-cooking technology between 2016 and 2019 through the introduction of induction cookstoves was driven largely by new hydropower facilities and looked to replace the often-abused subsidy for LPG. The substitution of this widely imported fossil fuel could lead to a reduction in public spending, greater energy sovereignty, and improvements in the health and safety of families (mainly women and children) and represents a feasible option to end the current subsidy of the GLP. This program shares the economic impact of an important technological migration with the population of the country, without generating the financial difficulties that could result in an abrupt end of the subsidy to LPG. It should also be taken into account that the feasibility of the massive introduction of induction heaters depends to a large extent on the necessary corrections that must be made to the relative price structures of LPG and electric power. Moreover, the impact of these policies in terms of reducing greenhouse gas emissions is an alternative to contribute to the country's decarbonization goals as presented in the 2070 climate commitment scenario.

While the modeling developed in this paper shows that a mass introduction of induction stoves would bring a net reduction in energy usage and greenhouse emissions, it also highlights that a considerable growth in electricity consumption is to be expected. Hence, technological constraints imposed by such a transition would appear in the form of growing requirements in electricity production capacity (which should be renewable to keep the estimated environmental gains), transmission, and distribution.

Furthermore, as the stoves constitute inductive loads, a comprehensive plan to guarantee the electric service quality should be developed. Finally, the NECP program implementation lost momentum with fading interest among the Ecuadorian population, accustomed to the long-term use of LPG stoves. While an interesting case study due to its novelty, induction cooking has proven to be difficult to escalate above a certain threshold of the population, and further studies will be focused on the conciliation between the electrification of cooking and the social and technical requirements of a successful transition.

Author Contributions: Conceptualization, V.G., J.M.-G. and V.S.E.; Methodology, V.G. and J.F.; Software, V.G. and J.F.; Validation, V.G., J.M.-G., J.F. and V.S.E.; Formal analysis, V.G., J.M.-G., J.F. and V.S.E.; Investigation, V.G., J.M.-G. and J.F.; Resources, J.F.; Writing—original draft, V.G., J.M.-G. and J.F.; Writing—review & editing, V.S.E.; Visualization, V.G.; Supervision, V.S.E.; Project administration, J.M.-G. and V.S.E.; Funding acquisition, J.M.-G. and V.S.E. All authors have read and agreed to the published version of the manuscript.

Funding: This research takes part in project P121819, Parque de Energías Renovables, founded by Universidad International SEK.

Data Availability Statement: Data are contained within the article.

Conflicts of Interest: The authors declare no conflict of interest.

References

1. Martínez-Gomez, J.; Guerron, G.; Riofrio, A.J. Analysis of the" Plan Fronteras" for Clean Cooking in Ecuador. *Int. J. Energy Econ. Policy* **2017**, *7*, 135–146. Available online: http://www.econjournals.com/index.php/ijeep/article/view/3574 (accessed on 19 October 2023).
2. Smith, K.R.; Sagar, A. Making the clean available: Escaping India's Chulha Trap. *Energy Policy* **2014**, *75*, 410–414. [CrossRef]
3. Smith, K.R. In praise of power. *Science* **2014**, *345*, 603. [CrossRef] [PubMed]
4. Goldemberg, J.; Martinez-Gomez, J.; Sagar, A.; Smith, K.R. Household air pollution, health, and climate change: Cleaning the air. *Environ. Res. Lett.* **2018**, *13*, 030201. Available online: http://iopscience.iop.org/article/10.1088/1748-9326/aaa49d/meta (accessed on 19 October 2023). [CrossRef]
5. Sinton, J.E.; Smith, K.R.; Peabody, J.W.; Yaping, L.; Xiliang, Z.; Edwards, R.; Quan, G. An assessment of programs to promote improved household stoves in China. *Energy Sustain. Dev.* **2004**, *8*, 33–52. [CrossRef]
6. Purohit, P.; Kumar, A.; Rana, S.; Kandpal, T. Using renewable energy technologies for domestic cooking in India: A methodology for potential estimation. *Renew. Energy* **2002**, *26*, 235–246. [CrossRef]

7. Banerjee, M.; Prasad, R.; Rehman, I.H.; Gill, B. Induction stoves as an option for clean cooking in rural India. *Energy Policy* **2016**, *88*, 159–167. [CrossRef]
8. Ouedraogo, B. Household energy preferences for cooking in urban Ouagadougou, Burkina Faso. *Energy Policy* **2006**, *34*, 3787–3795. [CrossRef]
9. Kebede, K.Y.; Mitsufuji, T.; Yemiru, B.S. Diffusion of solar cookers in Africa: Status and prospects. *Int. J. Energy Technol. Policy* **2014**, *10*, 200–220. [CrossRef]
10. Kastillo, J.P.; Martínez-Gómez, J.; Villacis, S.P.; Riofrio, A.J. Thermal Natural Convection Analysis of Olive Oil in Different Cookware Materials for Induction Stoves. *Int. J. Food Eng.* **2017**, *13*, 3. Available online: https://www.degruyter.com/view/j/ijfe.ahead-of-print/ijfe-2016-0065/www.degruyter.com/view/j/ijfe.ahead-of-print/ijfe-2016-0065/ijfe-2016-0065.xml (accessed on 19 October 2023). [CrossRef]
11. Espinoza, V.S.; Guayanlema, V.; Martínez-Gómez, J. Energy Efficiency Plan Benefits in Ecuador: Long-range Energy Alternative Planning Model. *Int. J. Energy Econ. Policy* **2018**, *8*, 52–64. Available online: http://www.econjournals.com/index.php/ijeep/article/view/6503/3784 (accessed on 22 March 2024).
12. Ministerio de Energía y Minas, Balance Energético Nacional 2023, Quito-Ecuador. Available online: https://www.recursosyenergia.gob.ec/5900-2/ (accessed on 12 August 2024).
13. Espinoza, S.; Guayanlema, V. Balance y proyecciones del sistema de subsidios energéticos en Ecuador. In *Análisis*; Friedrich-Ebert-Stiftung: Hannover, Germany, 2017; pp. 1–28. Available online: http://library.fes.de/pdf-files/bueros/quito/13648.pdf (accessed on 12 August 2024).
14. Martínez-Gómez, J.; Ibarra, D.; Villacis, S.; Cuji, P.; Cruz, P. Analysis of LPG, electric and induction cookers during cooking typical Ecuadorian dishes into the national efficient cooking program. *Food Policy* **2016**, *59*, 88–102. [CrossRef]
15. MEER. *Plan Nacional de Eficiencia Energética 2016–2035*; MEER: Quito, Ecuador, 2017.
16. Martínez-Gómez, J. Analysis of physicochemical and microbiological measurements of food prepared using different stoves. *Carpathian J. Food Sci. Technol.* **2017**, *9*, 68–79.
17. Ecuador. *Primera Contribución Determinada a Nivel Nacional del Ecuador Para el Acuerdo de París*; Republic of Ecuador: Quito, Ecuador, 2019.
18. CONELEC. Plan Maestro de Electrificación 2013–2022. Quito. 2014. Available online: https://www.ariae.org/servicio-documental/plan-de-electrificacion-2013-2022 (accessed on 12 August 2024).
19. MEER. *Informe de Ventas Del "Programa de Eficiencia Energética Para Cocción Por Inducción y Calentamiento de Agua con Electricidad en Sustitución del Gas Licuado de Petróleo en el Sector Residencial-PEC"*; MEER: Quito, Ecuador, 2018.
20. ARCONEL-(2018) Estadísticas del Sector Eléctrico. Balance Multianual de Energía. Available online: http://www.regulacionelectrica.gob.ec/estadistica-del-sector-electrico/balance-multianual-de-energia/ (accessed on 24 April 2018).
21. Andrade, S. El Precio Social del Gas Licuado de Petróleo en el Ecuador. Crisis de Gobernanza. Master's Thesis, Facultad Latinoamericana de Ciencias Sociales SEDE Ecuador, Diego de Almagro, Quito, 2011.
22. Sbroiavacca, N. Shale Oil y Shale Gas en Argentina. Estado de Situacion y Prospectiva. 2011. Available online: https://www.encyclopedie-energie.org/wp-content/uploads/2018/09/art020_di-sbroiavacca-nicolas_shale-oil-shale-gas-argentina.pdf (accessed on 12 August 2024).
23. Heaps, C.G. *Long-Range Energy Alternative Planning (LEAP) System*; Energy Community: Somerville, MA, USA, 2013.
24. SEI. *Fundación Bariloche*; User Manual LEAP: Boston, MA, USA, 2004.
25. World Bank. GDP (Constant LCU) Ecuador. 2016. Available online: http://data.worldbank.org/indicator/NY.GDP.MKTP.KN?end=2015&locations=EC&start=1960&view=chart (accessed on 21 March 2017).
26. Banco Central del Ecuador. Producto Interno Bruto-PIB. Available online: https://contenido.bce.fin.ec/documentos/PublicacionesNotas/Catalogo/CuentasNacionales/PIBProduccionConstante2020p.xlsx (accessed on 12 August 2024).
27. INEC. Censo de Población y Vivienda 2007. Quito. 2010. Available online: https://www.ecuadorencifras.gob.ec/censos/ (accessed on 12 August 2024).
28. MICSE. Agenda Nacional de Energía 2016–2040. Quito. 2016. Available online: https://www.ariae.org/servicio-documental/agenda-nacional-de-energia-2016-2040 (accessed on 12 August 2024).
29. Villacís, S.; Martínez, J.; Riofrío, A.; Carrión, D.; Orozco, M.; Vaca, D. Energy Efficiency Analysis of Different Materials for Cookware Commonly Used in Induction Cookers. *Energy Procedia* **2015**, *75*, 925–930. [CrossRef]

processes

Article

Life Cycle Assessment of CO_2-Based and Conventional Methanol Production Pathways in Thailand

Adeel Rafiq [1,2], Ahsan Farooq [1,2] and Shabbir. H. Gheewala [1,2,*]

[1] The Joint Graduate School of Energy and Environment, King Mongkut's University of Technology Thonburi, Bangkok 10140, Thailand; adeel.rafi@kmutt.ac.th (A.R.); ahsan.faro@mail.kmutt.ac.th (A.F.)

[2] Center of Excellence on Energy Technology and Environment (CEE), Ministry of Higher Education, Science, Research and Innovation, Bangkok 10400, Thailand

[*] Correspondence: shabbir.ghe@kmutt.ac.th or shabbirg@hotmail.com; Tel.: +66-24708309

Abstract: Methanol production through carbon capture and utilization technologies offers promising alternatives to traditional natural-gas-based methods, potentially mitigating climate change impacts and improving resource efficiency. This study evaluates four methanol production pathways: CO_2 hydrogenation, tri-reforming of methane, electrochemical CO_2 reduction, and co-electrolysis of CO_2 and water. The analysis covers 19 scenarios, combining three electricity mixes (100% Thai grid mix, 50% Thai grid mix and 50% renewable energy, and 100% renewable energy) with two hydrogen production technologies (alkaline water electrolysis and grey hydrogen). Environmental life cycle assessment results showed that most pathways perform well when using the 100% renewable energy with co-electrolysis (CE-100%) showing the most substantial reductions across all impact categories as compared conventional methanol production. Electrochemical reduction demonstrated the poorest environmental performance for all scenarios. In Thailand, implementing the CE-100% pathway could potentially yield 12.4 million tonnes of methanol annually from the cement industry's CO_2 emissions, with an estimated value of approximately USD 5.4 billion, while reducing emissions from the industrial processes and product use (IPPU) sector by 75%. The findings provide valuable insights for policymakers, industry stakeholders, and researchers, supporting Thailand's transition towards sustainable methanol production and broader climate goals.

Keywords: methanol; carbon capture; electrolysis; hydrogenation; life cycle assessment

1. Introduction

Fossil fuels, while currently dominant, are finite resources, and their diminishing availability has led to increased prices and economic instability in the global energy sector [1–3]. This scarcity not only threatens energy security but also worsens environmental concerns due to higher emissions from harder-to-reach fossil fuel reserves [4]. Compounding this issue, global energy consumption continues to rise, driven by population growth, industrialization, and improving living standards in developing countries. According to the latest OPEC World Oil Outlook, oil demand is expected to rise from approximately 97 million barrels per day (mb/d) in 2021 to around 110 mb/d by 2045 [5]. In light of these trends, the urgent need for renewable energy alternatives has become vital to address both environmental concerns and the growing energy demand [6,7].

Methanol, a versatile chemical compound, emerges as a potential solution to these energy security challenges. When produced through green pathways, methanol can serve as a carbon-neutral energy carrier and storage medium, offering a means to balance intermittent energy sources and provide a stable energy supply [8–11]. The global methanol industry is rapidly expanding, with around 90 plants producing 148 million tonnes annually as of 2019. Industry projections suggest that production capacity may increase significantly by 2030, potentially reaching 311 million tonnes per year [12]. Traditionally, methanol is produced from natural gas through a reforming process, which heavily relies on fossil fuels and contributes to greenhouse gas emissions [13,14].

As the world tackles the challenges of climate change and the need for sustainable development, alternative methods for producing methanol have gained significant attention [15]. Carbon capture and utilization (CCU) emerges as a promising approach to address these concerns [16–18]. By capturing carbon dioxide (CO_2) from industrial sources and utilizing it as a feedstock for methanol production, CCU not only reduces the dependence on fossil fuels but also offers a potential means of mitigating greenhouse gas emissions [19–21]. Carbon capture technologies, such as post-combustion capture, are integral to these pathways, enabling the extraction of CO_2 from flue gases emitted by power plants and industrial processes, which is then transformed into valuable products through CCU processes [22–25].

In the literature, several CO_2-based methanol production pathways have been explored, including CO_2 hydrogenation, tri-reforming of methane, electrochemical reduction of CO_2, and co-electrolysis. CO_2 hydrogenation involves the catalytic conversion of CO_2 and hydrogen (H_2) to methanol. This process typically employs copper-based catalysts and operates at elevated temperatures and pressures [26]. Tri-reforming of methane combines steam reforming, dry reforming, and partial oxidation of methane in the presence of CO_2 and oxygen, producing a syngas mixture suitable for methanol synthesis [12]. Electrochemical reduction of CO_2, on the other hand, utilizes renewable electricity to convert CO_2 directly to methanol in an electrochemical cell, offering the potential for a more sustainable and flexible production route [27].

Solid-oxide electrolysis (SOE) at high temperatures offers a promising solution for renewable energy storage and CO_2 conversion. This process co-electrolyzes water and CO_2 directly into syngas, achieving the desired H_2 to carbon monoxide ratio for downstream applications [28]. In comparison to low-temperature electrolysis technologies, SOE offers enhanced electrical efficiency and the capability for heat integration with subsequent synthesis processes, significantly improving overall system performance [29–31]. Originally proposed by the National Aeronautics and Space Administration (NASA) in the 1960s for oxygen supply and spacecraft propulsion, SOE was overlooked for many years due to inexpensive fossil fuels [32]. Nevertheless, increasing concerns regarding climate change and fossil fuel shortages have brought SOE back into focus in recent years [32].

However, to comprehensively evaluate the environmental implications of CO_2-based methanol production, conducting a life cycle assessment (LCA) that considers the entire production process, from raw material extraction to the final product, is essential [33]. LCA allows for the quantification of environmental impacts across various categories, such as global warming potential, acidification, and resource depletion, enabling a holistic comparison of different production pathways.

As a rapidly developing country with a growing chemical industry, Thailand has significant potential for implementing CO_2-based methanol production technologies [34]. To support the transition towards sustainable methanol production in Thailand, a thorough understanding of the environmental impacts associated with different production pathways is crucial. To date, life cycle assessment studies for methanol production have predominantly focused on comparing a single CO_2-based route with conventional methods, such as CO_2 hydrogenation versus steam methane reforming [13,35,36]. Recently, Rosental et al. (2020) conducted a life cycle assessment of CO_2-based production of methanol, olefins, and aromatics using CO_2 captured from industrial point sources and hydrogen derived from wind-powered electrolysis. The cradle-to-gate analysis demonstrated an 88–97% reduction in greenhouse gas (GHG) emissions compared to fossil-based production routes when wind power was employed. However, the study also noted increases in other environmental impacts, such as eutrophication and ozone depletion [35]. Similar results were reported by Hoppe et al. (2017), highlighting the potential of CO_2-based production to mitigate GHG emissions while highlighting trade-offs in resource use [33]. In addition, Win et al. (2023) explored methanol and formic acid production through CO_2 hydrogenation at a power plant. Their study found that CO_2-based methanol had higher environmental impacts in most categories compared to conventional production, while CO_2-based formic acid showed lower impacts in several categories. These studies collectively indicate that while CO_2-based processes offer substantial GHG reductions, they also present trade-offs in terms of resource use and other environmental impacts [13].

However, the rapid evolution of diverse and efficient pathways necessitates a more comprehensive sustainability evaluation. Although some studies have examined multiple methanol production routes, they have largely emphasized CO_2 utilization potential rather than CO_2 reduction potential [12,27]. This narrow focus leaves a critical gap in understanding the overall environmental impacts of emerging methanol production technologies. This study addresses these limitations by conducting Thailand's first comparative LCA of four distinct CO_2-based methanol production technologies alongside the conventional route. By simultaneously evaluating multiple pathways, including scenarios utilizing wind energy as a renewable power source, this research offers a more nuanced and holistic perspective on the environmental implications of emerging methanol production methods, extending its relevance beyond Thailand. Furthermore, this study transcends the typical assessment of CO_2 reduction potential by identifying the most environmentally advantageous scenario and quantifying its CO_2 emissions reduction capability relative to traditional methanol production methods. This multifaceted approach yields invaluable insights into the comparative environmental performance of various methanol production technologies, thereby facilitating evidence-based decision making in the pursuit of more sustainable production practices. Ultimately, this comprehensive analysis contributes significantly to the body of literature surrounding sustainable methanol production, offering a robust foundation for future research and industrial applications. The objectives of this study are threefold. First, it aims to evaluate and compare the environmental impacts of four CO_2-based methanol production technologies, CO_2 hydrogenation, tri-reforming of methane, electrochemical reduction of CO_2, and solid-oxide electrolysis, with the conventional natural-gas-based methanol production method. Second, it seeks to assess the influence of different electricity sources, including renewable and Thailand's grid mix, on the environmental performance of each CO_2-based methanol production pathway. Lastly, the study aims to examine the environmental impacts of different hydrogen production methods, specifically alkaline water electrolysis and conventional methods, within the CO_2 hydrogenation and tri-reforming pathways, to determine their contribution to the overall environmental footprint of methanol production.

By quantitatively assessing these production pathways, the study identifies the most environmentally advantageous route for methanol synthesis. The results provide critical insights into the feasibility and potential environmental impacts of implementing CO_2-based methanol production technologies in Thailand. This comprehensive analysis supports Thailand's efforts in transitioning towards a low-carbon economy and achieving its climate goals.

2. Materials and Methods

This study employed a cradle-to-gate LCA approach following ISO 14040 and 14044 standards to ensure a rigorous and comprehensive evaluation of the environmental impacts associated with alternative and conventional methanol production [37,38].

2.1. Goal and Scope Definition

The study assesses the environmental sustainability of various methanol production routes in Thailand, focusing on CO_2-based and conventional pathways. The functional unit for this assessment is defined as one kilotonne (kt) of methanol produced. The system boundaries include all relevant stages, viz., raw material extraction, CO_2 capture, hydrogen production, and methanol synthesis.

For a comprehensive assessment, 19 scenarios (see Table 1) were developed by combining different methanol production routes, electricity mixes, and hydrogen production methods. The criteria for selecting these scenarios included: (1) inclusion of all major CO_2-based methanol production routes currently under research and development; (2) representation of varying degrees of renewable energy integration in electricity supply; (3) consideration of different hydrogen production methods for relevant pathways; and (4) inclusion of the conventional production method as a baseline for comparison. The routes assessed are CO_2 hydrogenation, tri-reforming of methane, direct electrochemical reduction of CO_2, co-electrolysis of CO_2 and H_2O in a solid-oxide electrolyzer, and the conventional natural gas reforming method. The study evaluates each production route under three distinct electricity generation scenarios: 100% Thai grid mix, a hybrid mix comprising 50% Thai grid and 50% additional renewable energy, and 100% renewable energy. For the hybrid scenario, it is assumed that half of the energy is generated by the plant itself from renewable sources, while the other half is purchased from the Thai grid mix. Additionally, the study explores hydrogenation and tri-reforming processes using two hydrogen production methods: alkaline water electrolysis and hydrogen (reformer), to identify the most environmentally sustainable methanol production route under different energy scenarios. A detailed description of all scenarios follows.

Table 1. Scenarios for methanol production pathways.

Scenarios	100% Renewable	50% Renewable + 50% Grid	100% Grid	Conventional Method
Hydrogenation with alkaline hydrogen	Scenario 1 (Hyd-Alk-100%)	Scenario 2 (Hyd-Alk-50%)	Scenario 3 (Hyd-Alk-Grid)	
Hydrogenation with grey hydrogen	Scenario 4 (Hyd-Grey-100%)	Scenario 5 (Hyd-Grey-50%)	Scenario 6 (Hyd-Grey-Grid)	
Tri-reforming with alkaline hydrogen	Scenario 7 (Tri-Alk-100%)	Scenario 8 (Tri-Alk-50%)	Scenario 9 (Tri-Alk-Grid)	Scenario 19 (Conv-SMR)
Tri-reforming with grey hydrogen	Scenario 10 (Tri-Grey-100%)	Scenario 11 (Tri-Grey-50%)	Scenario 12 (Tri-Grey-Grid)	
Co-electrolysis	Scenario 13 (CE-100%)	Scenario 14 (CE-50%)	Scenario 15 (CE-Grid)	
Electrochemical reduction	Scenario 16 (ER-100%)	Scenario 17 (ER-50%)	Scenario 18 (ER-Grid)	

Note: Alkaline hydrogen represents hydrogen generated through water electrolysis using an alkaline electrolyte, whereas grey hydrogen represents hydrogen produced from natural gas via steam methane reforming. Hyd-Alk and Hyd-Grey refer to CO_2 hydrogenation using alkaline and grey hydrogen, respectively. Tri-Alk and Tri-Grey denote tri-reforming processes utilizing alkaline and grey hydrogen. CE stands for co-electrolysis, which generates both methanol and electricity, while ER indicates electrochemical reduction, producing methanol and hydrogen. Conv-SMR represents conventional steam methane reforming of natural gas to produce methanol, a process widely used in industry. The figure also categorizes energy sources as 100% renewable (entirely sourced from renewable energy), 50% (half renewable, half grid), and fully grid-based (relying entirely on the existing power grid).

2.1.1. Route 1: Methanol Production from CO_2 Hydrogenation

This route covers the first six scenarios. Methanol production through CO_2 hydrogenation is a multistep process that begins with the capture of CO_2 from industrial sources, such as flue gas from cement plants. For hydrogen production, this case utilizes alkaline water electrolysis (AE), which was chosen for its technological maturity and lower capital cost compared to other electrolysis technologies [39]. The process employs a liquid alkaline electrolyte, commonly a 20–40 wt.% solution of potassium hydroxide, which can be highly corrosive and may lead to increased maintenance expenses [40].

AE systems are suitable for large-scale applications and can achieve stack efficiencies of up to 67% based on the lower heating value of the produced hydrogen [41]. These systems typically operate at temperatures ranging from 60 °C to 80 °C with electric current densities of approximately 0.2–0.4 A/cm^2 [42]. The overall energy consumption of the system, including both stack and peripheral components, falls within the range of 4.4−6.6 kWh per cubic meter of hydrogen produced. The resulting hydrogen gas purity exceeds 99.5%, and the operating pressure is typically maintained below 30 bar [42–44].

In the AE process, hydrogen is generated at the cathode, while oxygen is produced at the anode. The hydroxide ions (OH^-) serve as the charge carriers, transporting charge from the cathode to the anode through the diaphragm separating the two electrodes. The hydrogen produced in the process is combined with captured carbon dioxide, and the resulting mixture undergoes compression and preheating to temperatures reaching 280 °C [45] (Figure 1). This heated and pressurized blend of CO_2 and H_2 is then introduced into a fixed-bed flow reactor containing a $CuO/ZnO/Al_2O_3$ catalyst, which is comparable to the catalyst employed in the well-established methanol synthesis process using syngas [13,19,35,45]. Within the reactor, the exothermic hydrogenation of CO_2 to methanol occurs according to the following reaction (Equation (1)):

$$CO_2 + 3H_2 \rightarrow CH_3OH + H_2O \quad (1)$$

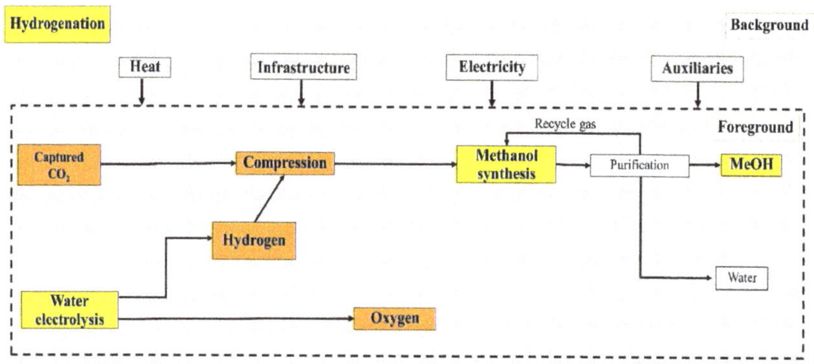

Figure 1. System boundary diagram illustrating the hydrogenation process.

This reaction attains equilibrium alongside various side reactions, yielding methanol, water, and byproducts such as carbon monoxide and methane. The reactor effluent, which comprises methanol, water, unreacted gases, and byproducts, undergoes cooling to separate the liquid methanol and water from the gaseous components [45]. Subsequently, the methanol/water mixture is subjected to distillation to obtain pure methanol product, while the water can be recycled back to the electrolyzer to enhance overall efficiency.

2.1.2. Route 2: Tri-Reforming of Methane with CO_2 Utilization

Syngas production can be achieved through various methodologies; tri-reforming of methane (TRM) was selected for this study due to its superior thermodynamic efficiency and operational flexibility. TRM integrates steam reforming, partial oxidation, and CO_2 reforming in a single process, utilizing natural gas, steam, carbon dioxide, and oxygen as feedstocks (Figure 2). A key innovation in this implementation is the utilization of oxygen generated from water electrolysis, eliminating the need for an energy-intensive air separation unit. The TRM process offers precise control over the H_2/CO ratio in the produced syngas by modulating the steam and CO_2 input ratios, enabling optimization for downstream methanol synthesis [46,47]. Compared to dry reforming, which also involves CO_2 conversion, the presence of steam in TRM significantly reduces coke formation [48,49].

The primary reactor in the TRM system resembles an auto-thermal reformer and comprises three distinct reaction zones [50]. The initial zone is the burner, where the incoming feed combines within a turbulent diffusion flame. Next is the combustion zone, where a partial oxidation reaction (POX) takes place. Finally, there is the catalytic zone, where both steam reforming and dry reforming reactions occur. In this study, a commercial Ni/Al_2O_3

catalyst is utilized in the main reactor to facilitate the specified reactions [51–53]. To extend the operational lifespan of this type of reactor, the incoming natural gas must undergo feed preconditioning. This procedure usually involves desulfurization and prereforming processes. The unreacted gases and gaseous byproducts are also separated, recompressed, and recycled back to the reactor inlet to maximize CO_2 conversion and minimize the carbon footprint of the process. The utilization of recycling streams for both water and unreacted gases contributes to improving the efficiency and sustainability of the process, rendering it an attractive option for the production of this crucial chemical feedstock and fuel. In addition to alkaline hydrogen, grey hydrogen, which is produced from natural gas through a process called steam methane reforming, is also considered. This method is currently common and relatively inexpensive compared to AE technology.

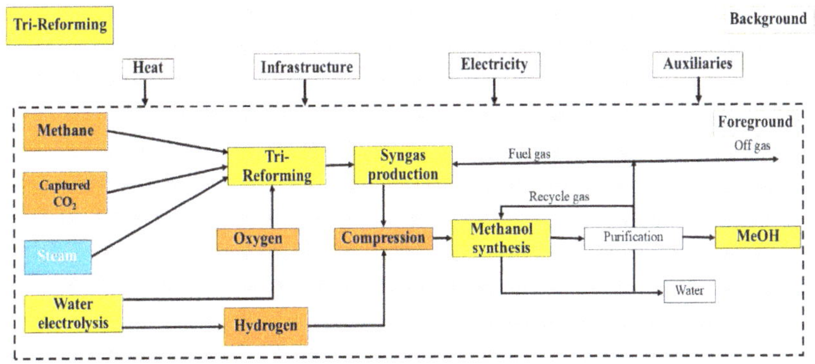

Figure 2. System boundary diagram depicting the tri-reforming process.

During the desulfurization phase, hydrogen reacts with sulfur-containing compounds present in the feed, forming hydrogen sulfide, which is then removed from the primary process. This step prevents catalyst degradation in the main reactor, allowing for extended operation without the need for catalyst substitution [54]. In the prereforming stage, hydrocarbons with two or more carbon atoms are removed via reactions with steam, thus avoiding soot creation and accumulation within the main reactor. The prereformer utilizes the same nickel on alumina catalyst as the primary TRM reactor. The sources of hydrogen and methanol synthesis and purification steps are similar to the CO_2 hydrogenation pathway [55]. Heat integration is employed throughout the process to maximize energy efficiency [12]. This route includes scenarios 7 through 12: three scenarios with hydrogen generated from water electrolysis and three scenarios with grey hydrogen.

2.1.3. Route 3: Co-Electrolysis of CO_2 to Methanol

The methanol production process via co-electrolysis of CO_2 and H_2O in a solid-oxide electrolyzer (SOE) involves several key steps. Initially, CO_2 and H_2O are co-electrolyzed at 750 °C and 1.31 bar in an SOE, producing a syngas mixture of H_2, CO, and O_2 (Figure 3). The H_2-rich syngas is then compressed to 54 bars, cooled to 40 °C, and further compressed to 56 bars. In the methanol synthesis reactor, operating at 230 °C and 56 bar, the syngas is converted to methanol using a $Cu/ZnO/Al_2O_3$ catalyst [56–58]. The reaction mixture is subsequently cooled and separated, with the liquid phase containing crude methanol (91 wt%) depressurized to 4 bar and purified to 99.5 wt% in two distillation columns. This system effectively integrates heat from the exothermic methanol synthesis process, significantly enhancing overall efficiency. This route covers scenarios 13 through 15. The key chemical reactions involved in the process are as below (Equations (2)–(6)) [59–63]:

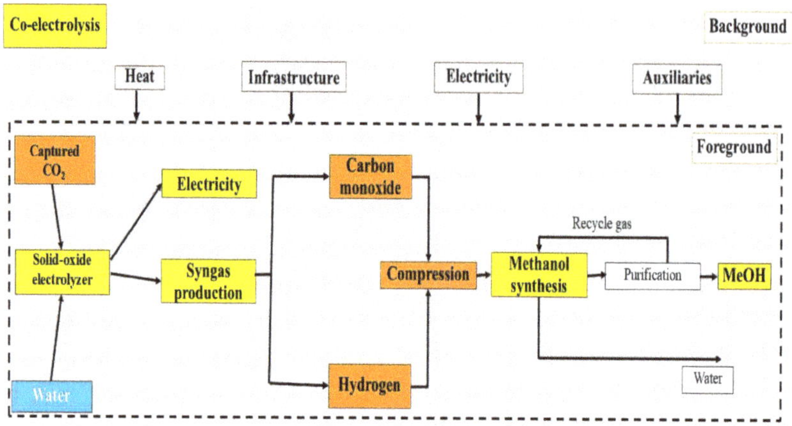

Figure 3. System boundary diagram illustrating the co-electrolysis process.

Electrolysis reactions:

$$H_2O + 2e^{2-} \rightarrow H_2 + O^{2-} \quad (2)$$

$$CO_2 + 2e^{2-} \rightarrow CO + O^{2-} \quad (3)$$

$$2O^{2-} \rightarrow O_2 + 4e^- \quad (4)$$

Methanol synthesis reactions:

$$CO + 2H_2 \leftrightarrow CH_3OH \quad (5)$$

$$CO_2 + 3H_2 \leftrightarrow CH_3OH + H_2O \quad (6)$$

2.1.4. Route 4: Direct Electrochemical Reduction of CO_2 to Methanol

The primary output is a methanol–water mixture, necessitating subsequent separation and purification to achieve commercial-grade methanol, commonly through energy-intensive distillation processes. A significant challenge in ER technology is the low methanol concentration in the cathode effluent, which necessitates extensive purification. Increasing the methanol yield at the electrolyzer output is critical for minimizing distillation energy requirements and improving the overall efficiency and economic viability of ER for industrial-scale methanol synthesis [27,64].

Concurrently, the competing hydrogen evolution reaction at the cathode produces hydrogen as a byproduct, where protons and electrons combine to form hydrogen gas (see Equations (7)–(10)). This side reaction can affect methanol yield and must be carefully managed through catalyst design and reaction conditions optimization (Figure 4). This route spans scenarios 16 through 18.

Cathodic reaction:

$$CO_2 + 6H^+ + 6e^- \leftrightarrow CH_3OH + H_2O \quad (7)$$

$$2H^+ + 2e^- \leftrightarrow H_2 \quad (8)$$

Anodic reaction:

$$3H_2O \leftrightarrow 1.5O_2 + 6H^+ + 6e^- \quad (9)$$

Overall cell reaction:

$$CO_2 + 2H_2O \leftrightarrow CH_3OH + 1.5O_2 \quad (10)$$

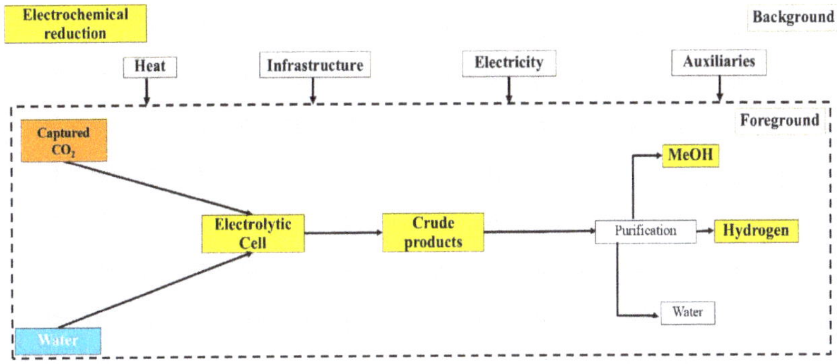

Figure 4. System boundary diagram showing the electrochemical reduction process.

2.1.5. Route 5: Conventional Method

This route begins with syngas production, where methane from natural gas undergoes a reforming process with steam and/or oxygen at elevated temperatures (800–1000 °C) and pressures (20–30 bar) in the presence of a nickel catalyst (Figure 5). This process yields syngas, a mixture of H_2 and CO. The syngas is subsequently directed to the methanol synthesis phase, where it undergoes compression and is subjected to a reaction in the presence of a copper–zinc oxide catalyst.

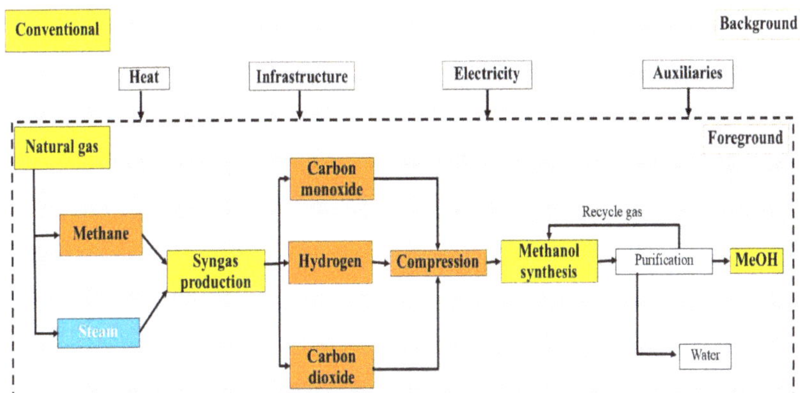

Figure 5. System boundary diagram depicting natural gas reforming for methanol production.

This process takes place at temperatures spanning from 200 °C to 300 °C and pressures ranging between 50 and 100 bar, ultimately yielding methanol as the product. In cases where the syngas has an excess of hydrogen, carbon dioxide can be introduced to adjust the composition, ensuring optimal conditions for methanol synthesis [65–67]. This conventional route addresses scenario 19. The reaction equation for methanol production via natural gas reforming is as follows (see Equations (11) and (12)):

$$\text{1st step: } CH_4 + H_2O \rightarrow CO + 3H_2 \quad (11)$$

$$\text{2nd step: } CO + 2H_2 \rightarrow CH_3OH \quad (12)$$

2.2. Source of Inputs and Allocation Method

Carbon capture is a critical step in the utilization of CO_2 as a feedstock for producing valuable chemicals. The selection of an appropriate carbon capture system depends on factors such as the CO_2 source, the composition of the flue gas, and the desired CO_2 purity and pressure [68–70]. The study focused on a cement plant as one of the primary sources of CO_2 emissions. This sector was chosen because it is a significant contributor to Thailand's CO_2 emissions and is considered difficult to decarbonize due to its high carbon intensity [71].

An amine-based postcombustion capture system was considered for the cement plant. Amine-based capture is a mature and widely used technology that involves the absorption of CO_2 from the flue gas using an aqueous amine solution, typically monoethanolamine [72–74].

The process consists of two main steps: absorption and regeneration. In the absorption step, the flue gas is in contact with the amine solution in an absorption column, where CO_2 reacts with the amine to form a carbonate salt. The CO_2-rich amine solution is then sent to a regeneration column, where it is heated to release the captured CO_2 and regenerate the amine solution [75,76]. The regenerated amine solution is cooled and recycled back to the absorption column, while the concentrated CO_2 stream is compressed and purified for utilization or storage [77]. The CO_2 point source such as cement plant was considered a separate system, with the environmental burdens associated with CO_2 production and main products allocated to the cement plant. Two pathways for hydrogen production were considered: alkaline water electrolysis and grey hydrogen from natural gas reforming, as described in Section 2.1.1.

Figure 6 delineates various process pathways for CO_2-based methanol (MeOH) production, presenting 18 distinct scenarios with their respective reactants and avoided products. The left side specifies the reactants for each pathway. For example, the Hyd-Alk pathway utilizes hydrogen from alkaline water electrolysis and CO_2, while the Hyd-Grey pathway employs grey hydrogen from natural gas reforming and CO_2. Tri-reforming pathways incorporate methane, water vapor, and oxygen, combined with either alkaline (Tri-Alk) or grey hydrogen (Tri-Grey). Co-electrolysis pathways use water to simultaneously produce methanol and electricity, whereas electrochemical reduction pathways generate methanol and hydrogen from water. The right side of the figure elucidates the avoided products and associated CO_2 emissions reductions for each scenario. Brown boxes signify the avoidance of conventional methanol production and its CO_2 emissions, applicable to scenarios such as MeOH (Hyd-Alk-100%), MeOH (Hyd-Alk-50%), and MeOH (Hyd-Alk-Grid). Purple boxes represent the combined avoidance of conventional methanol production and coal-based electricity, pertinent to co-electrolysis scenarios like MeOH (CE-100%) + electricity. Grey boxes indicate the simultaneous avoidance of conventional methanol production and fossil-based hydrogen, observed in electrochemical reduction scenarios such as MeOH (ER-100%) + hydrogen. In each pathway, the product and co-product replace their conventional counterparts. Environmental credits are assigned to each pathway based on the amount and type of conventional products replaced.

Carbon capture credits were allocated across all scenarios to account for diverted CO_2 emissions. Scenarios 1 through 12 were assigned displacement credits for substituting conventional methanol production (Figure 6).

For scenarios 13 through 15, which involve co-electrolysis, credits were allocated for both methanol production and the displacement of coal-based electricity, reflecting the dual outputs of this process. Scenarios 16 through 18 received credits for methanol production and the substitution of grey hydrogen, acknowledging the environmental benefits of replacing carbon-intensive hydrogen production methods. In the life cycle assessment of Tri-Alk scenarios, a greenhouse gas reduction credit was incorporated, equivalent to the carbon intensity of electricity consumption by an air separation unit (ASU) for producing an equivalent oxygen quantity (245 kWh/tonne oxygen) [78]. This credit accounts for the avoided emissions from conventional oxygen production. Conversely, in Tri-Grey scenarios, this ASU-related credit was omitted to accurately reflect the carbon intensity of the process. These differentiated credit allocations enable a more nuanced and comprehensive evaluation of the environmental impacts across the various methanol production pathways.

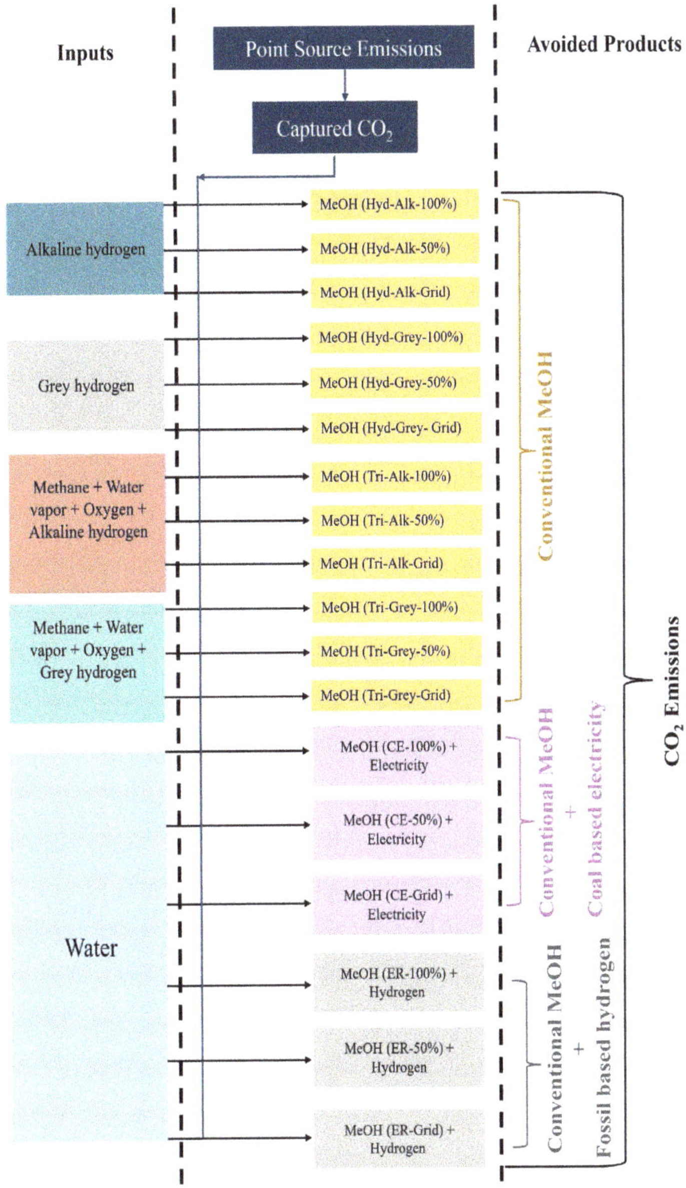

Figure 6. Production pathways for CO_2-based methanol synthesis.

2.3. *Life Cycle Inventory (LCI) and Impact Categories*

The LCI phase involves collecting data for all inputs and outputs associated with the methanol production pathways. Foreground inventory data were sourced from various studies in the literature. For hydrogenation (Table S1), data came from [36,79–82]. Table S1 also provides the inventory data for tri-reforming, sourced from [12,55,81,82].

Co-electrolysis data (Table S1) were derived from [83], while electrochemical reduction data came from [27,64]. The ecoinvent 3 database was predominantly used for background inventory data, including data for the conventional methanol production process. The collected data were analyzed using SimaPro 9.5 LCA software and the ReCiPe 2016 life cycle impact assessment method to calculate environmental impacts at both midpoint and endpoint levels [84]. The analysis primarily focused on two midpoint impact categories: global warming (human health) and global warming (terrestrial ecosystems). These categories were selected due to their relevance in evaluating the performance of CCU technologies, which aim to capture and utilize CO_2, thereby potentially reducing global warming impacts. To provide a more comprehensive assessment, an endpoint analysis was also conducted, evaluating three impact categories: human health (quantified in disability-adjusted life years, DALY), ecosystem quality (expressed in Potentially Disappeared Fraction of species.m^2.year (species.yr)), and resource scarcity (measured in USD2013). This comprehensive approach allowed for the assessment of environmental impacts at both specific (midpoint) and broader (endpoint) levels, providing a nuanced understanding of the product's environmental performance across different impact dimensions. The analysis was conducted using the hierarchist perspective, a consensus model frequently used in scientific studies and considered the default model in ReCiPe 2016.

2.4. Net Reduction in Life Cycle CO_2 Emissions

To estimate the net benefit of avoided CO_2 emissions for the best-case scenario, the analysis includes CO_2 emissions such as those arising from fossil fuel combustion, land transformation, as well as those avoided from the cement production process. The methodology employs a comparative assessment of the total life cycle CO_2 emissions between conventional methanol production and an alternative CO_2-based methanol production pathway as illustrated in Figure 7. This method quantifies the net climate impact, providing a clear measure of the reduction in overall CO_2 emissions attributable to the alternative production method.

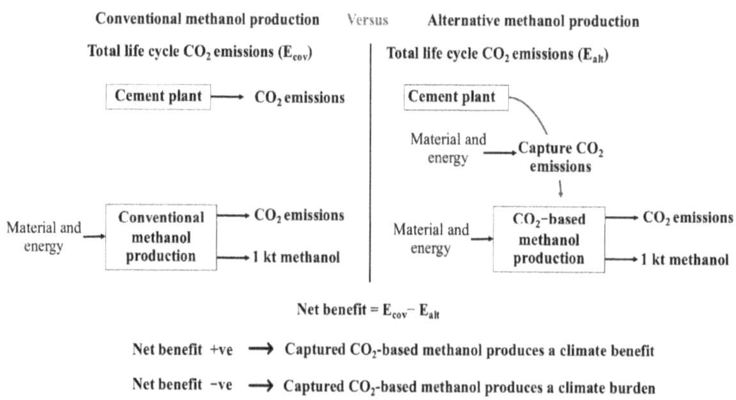

Figure 7. Comparison of total life cycle CO_2 emissions between conventional and CO_2-based methanol production.

The study has a few limitations worth mentioning. First, it relies on data from various literature sources rather than primary data collection, which may introduce some uncertainties. However, it must also be noted that primary data collection would also have limitation of scope and representativeness. Second, the analysis is based on current technologies and does not account for potential technological improvements which may be considered in future research.

3. Results

3.1. Midpoint Impact Categories

The analysis of global warming impacts on human health (measured in DALY) across different methanol production technologies reveals significant insights into the effectiveness of various approaches and the impact of different electricity mixes and hydrogen sources (Figure 8). For the co-electrolysis process, CE-100% is the best-performing scenario, showing a substantial reduction in global warming impacts (−2 DALY), translating to a 420% decrease compared to conventional steam methane reforming (Conv-SMR). In contrast, CE-50% and CE-Grid show increased impacts (7.00×10^{-1} DALY, 11% and 3.50 DALY, 443%, respectively), emphasizing the negative effects of relying on grid electricity predominantly sourced from fossil fuels.

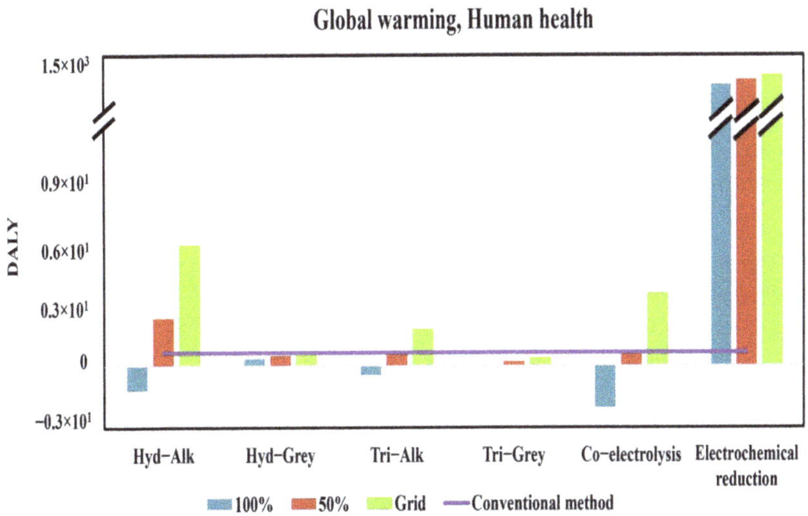

Figure 8. Global warming, human health (DALY) impacts across different methanol production technologies, considering electricity mixes and hydrogen sources.

In the CE-100% scenario, the major contributors to the reduction in global warming impacts on human health are carbon capture (−1.21 DALY) and the displacement of conventional methanol (-6.38×10^{-1} DALY). Additionally, electricity replacement from coal significantly contributes to the reduction (-3.64×10^{-1} DALY). Minor contributions to the impacts include electricity use (1.64×10^{-1} DALY), water use (2.55×10^{-3} DALY), copper oxide (4.89×10^{-4} DALY), and zinc oxide (5.76×10^{-5} DALY). These factors collectively contribute to the overall impact reduction in the CE-100% scenario. High performance in co-electrolysis of CO_2 and H_2O in SOEs is achieved through optimal electrode materials, efficient heat integration, and maintaining appropriate operating temperatures and pressures [83].

In the hydrogenation process, Hyd-Alk-100% significantly reduces global warming impacts (−1.20 DALY, −290%) compared to the conventional steam methane reforming method. However, as the renewable mix decreases, impacts for alkaline hydrogen rise sharply, with Hyd-Alk-50% and Hyd-Alk-Grid showing increased impacts (2.30 DALY, 259% and 5.80 DALY, 810%) compared to Conv-SMR. On the other hand, grey hydrogen scenarios, such as Hyd-Grey-100% (3.00×10^{-1} DALY, −46%), Hyd-Grey-50% (5.00×10^{-1} DALY, −25%), and Hyd-Grey-Grid (6.00×10^{-1} DALY, −4%), demonstrate lower impacts compared to their alkaline counterparts in similar conditions, and they are also better than the conventional method. This indicates that while both alkaline and grey hydrogen can reduce global warming impacts compared to Conv-SMR, grey hydrogen scenarios show relatively lower impacts when partial or full grid reliance is involved.

In the Hyd-Alk-100% scenario, global warming impacts on human health are significantly mitigated by carbon capture (−1.19 DALY) and the displacement of conventional methanol (-6.38×10^{-1} DALY). Major contributors to global warming include hydrogen production via alkaline water electrolysis (2.08×10^{-1} DALY) and methanol from CO_2-based sources (8.35×10^{-2} DALY). Minor impacts come from materials such as copper oxide (1.42×10^{-1} DALY) and heat from steam (1.71×10^{-1} DALY). The factors contributing to optimal performance in hydrogenation include high thermal efficiency, efficient integration of heat pumps and access to low-carbon or renewable electricity sources, and a substantial CO_2 utilization rate compared to other pathways [12,13,19].

For tri-reforming, Tri-Alk-100% shows a significant reduction in impacts (-4.00×10^{-1} DALY, −168%) compared to both Conv-SMR and Tri-Grey-100% (9.20×10^{-3} DALY, −99%). However, as the renewable energy mix decreases, the impacts for Tri-Alk-50% and Tri-Alk-Grid increase (7.00×10^{-1} DALY, 2% and 1.70 DALY, 173%) compared to Conv-SMR. In contrast, Tri-Grey-50% and Tri-Grey-Grid have impacts of 1.74×10^{-1} DALY (−83%) and 3.39×10^{-1} DALY (−63%), respectively, showing better performance than the conventional method.

In the Tri-Alk-100% scenario, global warming impacts on human health are primarily alleviated by carbon capture (-3.30×10^{-1} DALY) and the displacement of conventional methanol (-6.38×10^{-1} DALY). Significant contributors include methanol production from CO_2-based sources (1.80×10^{-1} DALY) and natural gas (2.83×10^{-1} DALY). Minor impacts come from hydrogen production via alkaline water electrolysis (5.94×10^{-2} DALY) and other materials like copper oxide and zinc oxide. The tri-reforming process performed well because it effectively integrates steam and CO_2 reforming, improving syngas quality and overall thermal efficiency [12].

The electrochemical reduction process demonstrates the poorest performance, with ER-Grid having the highest impact (1.04×10^{3} DALY). ER-100% and ER-50% also show severe impacts (1.01×10^{3} DALY and 1.03×10^{3} DALY, respectively), indicating the inefficiency and high environmental damage when using grid electricity heavily sourced from fossil fuels. ER is less efficient due to its significantly higher energy consumption (50.5 kWh/kg) compared to alternatives. It produces a very low methanol concentration (0.05% wt.), requiring extensive purification and high steam usage. The process lacks heat integration and has a lower technological readiness level, indicating it needs substantial development to become competitive [27].

An evaluation of global warming impacts on terrestrial ecosystems (measured in species.yr) across methanol production technologies unveils considerable differences attributable to varying electricity mixes and hydrogen sources (Figure 9). Within this context, CE-100% emerges as particularly effective, achieving a substantial impact reduction (-6.16×10^{-3} species.yr). This represents a 420% decrease compared to Conv-SMR. Conversely, CE-50% and CE-Grid exhibit increased impacts (2.14×10^{-3} species.yr, 11% and 1.05×10^{-2} species.yr, 443%), indicating the negative influence of grid electricity predominantly sourced from fossil fuels. Similarly, in the hydrogenation process, Hyd-Alk-100% shows a substantial reduction in global warming impacts (-3.66×10^{-3}species.yr, −290%) compared to Conv-SMR.

The major contributors to the reduction in global warming impacts for the CE-100% scenario include carbon capture, which accounts for a significant reduction of -3.65×10^{-3} species.yr. The shift from conventional methanol contributes -1.93×10^{-3} species.yr, and electricity replacement from coal adds another -1.10×10^{-3} species.yr. Smaller contributions come from electricity use (4.96×10^{-4} species.yr), copper oxide (1.48×10^{-6} species.yr), zinc oxide (1.74×10^{-7} species.yr), and water use (7.69×10^{-6} species.yr). These factors collectively result in a total net reduction of -6.16×10^{-3} species.yr for global warming impacts in terrestrial ecosystems.

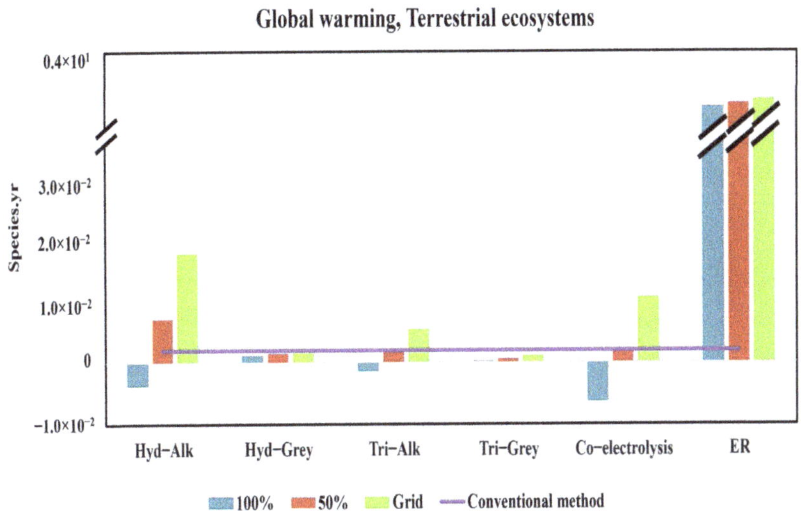

Figure 9. Global warming, terrestrial ecosystem impacts (species.yr) for different methanol production technologies, accounting for various electricity mixes and hydrogen sources.

For the Hyd-Alk-100% scenario, the main drivers for the reduction in global warming impacts (species.yr) are carbon capture (-3.60×10^{-3} species.yr) and the shift from conventional methanol (-1.93×10^{-3}). Hydrogen production via alkaline water electrolysis (6.27×10^{-4} species.yr), copper oxide (4.29×10^{-4} species.yr), and heat from steam (5.15×10^{-4} species.yr) contribute positively but are outweighed by the larger reductions. Smaller contributions include methanol (CO_2-based) (2.52×10^{-4} species.yr), zinc oxide (1.90×10^{-5} species.yr), aluminum oxide (1.47×10^{-5} species.yr), and electricity (1.18×10^{-5} species.yr). These combined elements result in a total net reduction of -3.66×10^{-3} species.yr.

For tri-reforming, Tri-Alk-100% shows a significant reduction in impacts (-1.32×10^{-3} species.yr, −168%) compared to both Conv-SMR and Tri-Grey-100% (2.60×10^{-5} species.yr, −100%). However, as the renewable energy mix decreases, impacts rise for Tri-Alk-50% and Tri-Alk-Grid (1.97×10^{-3} species.yr, 2% and 5.26×10^{-3} species.yr, 173%) compared to Conv-SMR. In contrast, Tri-Grey−50% and Tri-Grey-Grid show reductions in impacts (5.24×10^{-4} species.yr, −73% and 1.02×10^{-3} species.yr, −47%), demonstrating better performance than both their alkaline counterparts and the conventional method. In the Tri-Alk-100% scenario, the most significant contributors to reducing global warming impacts (species.yr) are carbon capture (-9.94×10^{-4} species.yr) and the transition from conventional methanol (-1.93×10^{-3} species.yr). The use of CO_2-based methanol (5.43×10^{-4}) and natural gas (8.56×10^{-4}) adds to the impact but is offset by the larger reductions. Other contributions come from hydrogen produced via alkaline water electrolysis (1.79×10^{-4}) and electricity (1.96×10^{-5}). Smaller contributions are from materials such as copper oxide (6.70×10^{-8}), zinc oxide (5.27×10^{-9}), aluminum oxide (9.24×10^{-8}), and nickel (1.43×10^{-7}) (Table S2).

3.2. Endpoint Impact Categories

Figure 10 presents the human health impacts, measured in DALY, for various methanol production technologies. The data illustrate significant differences in health implications across production methods, influenced by the choice of electricity mix and hydrogen source. In different scenarios, CE-100% and Tri-Alk-100% exhibit the most substantial reductions in DALY, demonstrating decreases of −2.25 DALY (−322%) and -3.91×10^{-1} DALY (−138%), respectively. These findings emphasize the critical role of utilizing 100% renewable energy sources to optimize environmental performance. Conversely, the electrochemical reduction scenarios consistently show the highest DALY impacts, regardless of the energy mix. The ER-Grid scenario, in particular, demonstrates a concerning increase to 1.27×10^3 DALY compared to the conventional method. The ER-100% scenario shows an increase to 1.22×10^3 DALY, and the ER-50% scenario indicates a rise to 1.25×10^3 DALY. These results highlight the inefficiency and high environmental cost associated with this technology, indicating that ER is currently the least viable option for sustainable methanol production, primarily due to its high electricity consumption and low technological maturity. The impact of hydrogen technologies varies significantly with the energy source. Hydrogenation with alkaline hydrogen performs well with 100% renewable energy, showing a reduction of 7.17×10^{-1} DALY (−29%). However, impacts increase drastically with grid reliance, reaching 1.14×10^1 DALY. In comparison, hydrogenation with grey hydrogen scenarios, although less effective with renewable energy (2.54 DALY, 150%) as compared to the alkaline pathway, demonstrate more resilience to grid reliance, with impacts ranging from 2.75 DALY (over 170% higher than the conventional method) to 2.95 DALY (191%). Tri-reforming with grey hydrogen also consistently reduces DALY impacts across all energy mixes, with reductions of 1.30×10^{-1} DALY (−87%) for 100% renewable energy, 3.24×10^{-1} DALY (−68%) for 50% renewable, and 5.18×10^{-1} DALY (−49%) for grid reliance, outperforming Tri-Alk when partial or full grid reliance is involved. This suggests that while renewable pathways are ideal, grey hydrogen pathways offer lower impacts when grid electricity is used.

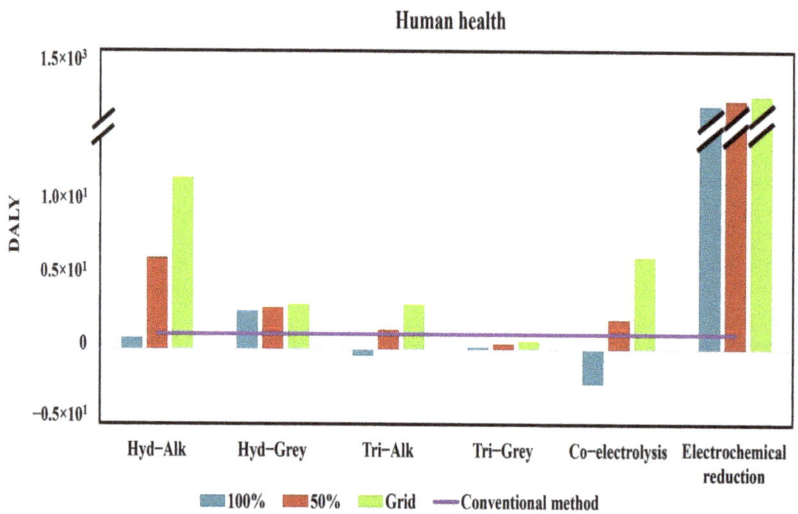

Figure 10. Comparison of human health impacts (DALYs) among diverse methanol production pathways, incorporating variations in power sources and hydrogen generation methods.

Figure 11 displays the ecosystem impacts, quantified in species.yr, for different methanol production technologies. The results demonstrate notable variations in environmental consequences, reflecting the effects of diverse electricity sources and hydrogen production methods. Among the scenarios, CE-100% and Tri-Alk-100% exhibit the most substantial reductions in species.yr, with CE-100% showing a significant decrease of -6.49×10^{-3} species.yr (−357%) and Tri-Alk-100% demonstrating a reduction of -1.30×10^{-3} species.yr (−151%), emphasizing the effectiveness of utilizing 100% renewable energy sources to minimize ecosystem impacts. Conversely, electrochemical reduction scenarios consistently show the highest impacts on ecosystems, with the ER-Grid scenario displaying a dramatic increase to 3.47 species.yr, markedly higher than the conventional method, indicating a stark increase in ecosystem impacts due to high electricity consumption and low efficiency associated with ER technology. ER-100% and ER-50% scenarios also exhibit similarly high impacts, with values of 3.35 and 3.41 species.yr, respectively. Hydrogenation technologies present varying impacts depending on the hydrogen source and energy mix. Hydrogenation with alkaline hydrogen performs well with 100% renewable energy, showing a reduction of 4.41×10^{-4} species.yr (−82%). However, ecosystem impacts increase significantly with grid reliance, reaching 2.71×10^{-2} species.yr (975%).

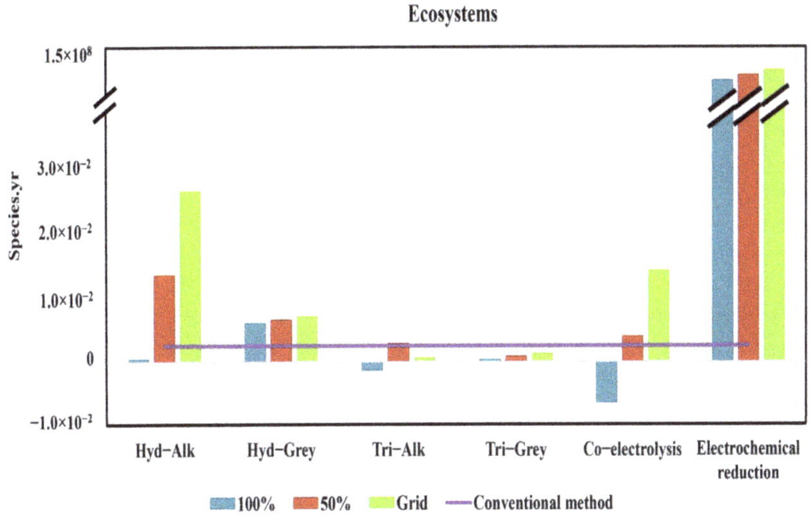

Figure 11. Assessment of ecosystem impacts (species.yr) for multiple methanol synthesis routes, factoring in different electricity origins and hydrogen production techniques.

In comparison, hydrogenation with grey hydrogen scenarios are more resilient to grid reliance, with impacts ranging from 6.09×10^{-3} species.yr (141%) to 7.12×10^{-3} species.yr (182%). Tri-reforming with grey hydrogen consistently reduces ecosystem impacts across all energy mixes, with reductions of 3.09×10^{-4} species.yr (−87%) for 100% renewable energy, 7.92×10^{-4} species.yr (−68%) for 50% renewable, and 1.27×10^{-3} species.yr (−49%) for grid reliance. These results indicate that grey hydrogen pathways are more robust and have lower impacts under grid electricity conditions compared to their alkaline counterparts.

Figure 12 shows the resource scarcity impacts, expressed in USD2013, associated with various methanol production technologies. The analysis reveals the economic implications of resource depletion, highlighting the influence of electricity mix and hydrogen source choices on long-term resource availability. Co-electrolysis and hydrogenation with alkaline hydrogen continue to demonstrate substantial reductions in resource costs. CE-100% achieves a notable decrease of -2.43×10^5 USD2013, while Hyd-Alk-100% shows a reduction of -1.97×10^5 USD2013 compared to the conventional method. These findings underscore the effectiveness of utilizing 100% renewable energy in minimizing resource scarcity impacts. Conversely, electrochemical reduction scenarios remain the most resource intensive, with the ER-Grid scenario seeing a dramatic increase to 1.37×10^8 USD2013, significantly surpassing the conventional method. ER-100% and ER-50% scenarios also show considerable impacts, with values of 1.35×10^8 USD2013 and 1.36×10^8 USD2013, respectively.

Comparing alkaline and grey hydrogen technologies reveals further insights. For hydrogenation, Hyd-Alk is highly effective with 100% renewable energy, however, its impact rises with grid reliance, peaking at 3.58×10^5 USD2013. Conversely, hydrogenation with grey hydrogen consistently shows reductions across all energy mixes: -8.93×10^4 USD2013 for 100% renewable energy, -7.86×10^4 USD2013 for 50% renewable, and -6.79×10^4 USD2013 for grid reliance, indicating superior performance under less favorable energy conditions as compared to alkaline counterparts. Similarly, tri-reforming technologies display varied performance. Tri-reforming with alkaline hydrogen reduces impacts by -7×10^4 USD2013 using 100% renewable energy but shows increased impacts under grid reliance, rising to 1×10^5 USD2013.

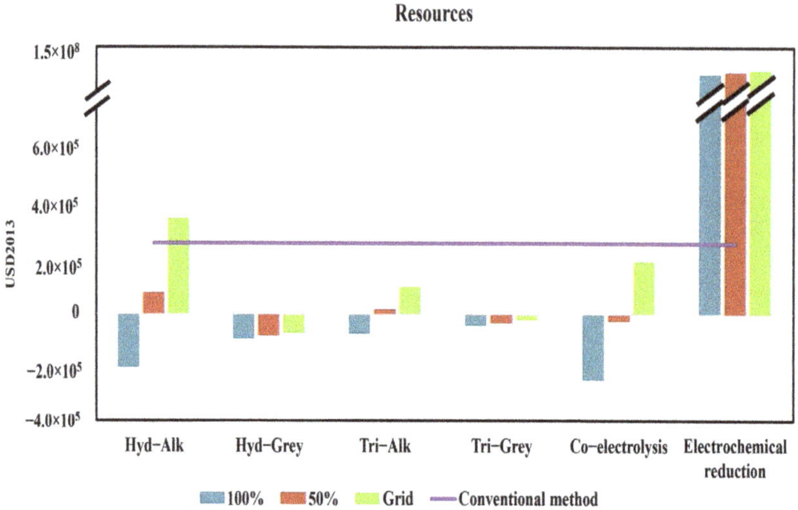

Figure 12. Evaluation of resource scarcity impacts (USD2013) across varied methanol manufacturing processes, considering the influence of distinct power grids and hydrogen feedstock options.

In comparison, tri-reforming with grey hydrogen achieves reductions in resource scarcity across all energy mixes, with -3.92×10^4 USD2013 for 100% renewable, -2.91×10^4 USD2013 for 50% renewable, and -1.91×10^4 USD2013 for grid reliance. These results indicate that grey hydrogen pathways generally offer a more balanced performance with lower resource scarcity impacts, particularly when grid electricity is used, compared to their alkaline counterparts.

In the literature, various studies have reported that CO_2 hydrogenation for methanol production outperforms conventional methods in terms of reducing greenhouse gas emissions [35,36,85]. This process has garnered significant attention as a promising strategy for mitigating climate change. However, a closer examination of endpoint impacts unveils a more complex pattern in its environmental performance. This divergence between midpoint and endpoint results necessitates a comprehensive analysis of various methanol production methods, including tri-reforming, co-electrolysis, and electrochemical reduction of CO_2. Table 2 presents a comparative analysis of the advantages and disadvantages for the four alternative methanol production routes considered.

The results demonstrate significant differences in environmental performance between hydrogenation and tri-reforming processes for methanol synthesis, primarily attributed to their respective energy and resource requirements. Hydrogenation, particularly when utilizing alkaline electrolysis (Hyd-Alk scenarios), exhibits high sensitivity to the electricity mix composition. While Hyd-Alk-100% shows substantial reductions in global warming potential at the midpoint level, its performance at the endpoint level is less favorable compared to Tri-Alk-100%. This discrepancy is attributed to the higher energy demands of the hydrogenation process. CO_2 hydrogenation necessitates approximately 0.21 kWh of electricity per kg of methanol produced [80]. Additionally, green hydrogen production via alkaline water electrolysis is energy-intensive, consuming 52 kWh per kg of hydrogen [81].

Table 2. Advantages and disadvantages of the four alternative methanol production routes.

Technologies	Advantages	Disadvantages
Hydrogenation	• Established process • High product purity • High CO_2 reduction potential with renewable H_2 • Moderate energy requirements • Environmental impacts depend on hydrogen source (better with renewable hydrogen)	• Requires separate H_2 source
Tri-reforming	• High product purity • Moderate CO_2 reduction potential with renewable H_2 • Moderate energy requirements • Syngas ratio can be adjusted • Moderate environmental impacts (due to natural gas usage)	• Emerging process • Uses natural gas (fossil fuel) • Requires separate H_2 source
Co-electrolysis	• High product purity • High CO_2 reduction potential with renewable energy • Moderate energy requirements • Does not require separate H_2 source • Low environmental impacts with 100% renewable energy	• Emerging technology
Electrochemical reduction	• In situ H_2 generation • Potential for high CO_2 reduction	• Low product purity • Extensive purification needed • Emerging technology • Higher environmental impacts compared to other methods

The elevated energy requirements of hydrogenation corroborate the findings of Win et al. (2023), who observed that CO_2-based methanol production resulted in higher environmental impacts across multiple categories due to its energy-intensive nature [13]. This elucidates why hydrogenation, despite its potential benefits in CO_2 utilization, can result in higher impacts for certain environmental indicators, particularly when considering endpoint categories. Conversely, tri-reforming exhibits superior overall environmental performance across various energy scenarios. Tri-Alk-100% demonstrates significant reductions in global warming potential for both human health and ecosystem endpoints. The enhanced performance of tri-reforming can be attributed to its lower resource and energy requirements. Tri-reforming integrates dry reforming of methane with steam reforming and partial oxidation, resulting in more efficient resource utilization. Notably, tri-reforming requires lower CO_2 and hydrogen inputs compared to hydrogenation, translating to reduced overall energy demands [12].

The CE-100% scenario emerges as the optimally performing option among all scenarios examined. This superior performance is attributed to multiple factors. Primarily, the co-electrolysis process requires lower energy input compared to the hydrogenation method (Table S1). Additionally, it eliminates the necessity for external hydrogen production, a significant energy-intensive step in alternative processes. While CO_2 utilization in co-electrolysis is comparable to the hydrogenation process, a key advantage is the generation of electricity as a co-product [13,83]. These factors synergistically contribute to the enhanced environmental performance of the CE-100% scenario.

The current performance of electrochemical reduction of CO_2 to methanol is suboptimal when compared to alternative methods due to several critical limitations. These include low methanol concentration (0.05% wt compared to the desired 40–67% wt), high energy demand (50.5 kWh/kg MeOH versus 0.21 kWh/kg in hydrogenation method), substantial steam consumption for purification (>10,000 MJ/kg versus 1.44 MJ/kg in hydrogenation method), and low technological readiness level [13,27,80]. Enhancing ER efficiency necessitates advancements in electrocatalyst design for improved selectivity and efficiency, reactor configuration and reaction condition optimization, integration of renewable energy and process heat, and Faradaic efficiency improvements coupled with reduced cell potential. Achieving a minimum methanol concentration of 40% wt at the reactor output is crucial for ER to become a competitive low-carbon solution for methanol production [27]. However, substantial research and development efforts are imperative to address these challenges and elevate ER to industrial viability.

3.3. *Application of Best-Case Scenario*

Thailand's energy landscape is evolving rapidly, as outlined in the 2024 Power Development Plan [86]. While natural gas and coal dominated the electricity generation mix in 2020, their roles are set to diminish substantially by 2037. The plan projects a decrease in natural gas usage from 57% to 41% and coal from around 20% to 7%, while renewable energy is expected to surge to 51% of the total energy mix [86]. Carbon capture and utilization technologies for coal and natural gas plants may offer limited long-term benefits, as the energy sector increasingly shifts toward renewable sources. However, CCU remains crucial for hard-to-decarbonize industries such as cement production.

The cement industry in Thailand is a significant contributor to the country's greenhouse gas emissions, with approximately more than 18 million tonnes of CO_2 emitted in 2019 [71]. This substantial carbon footprint presents both a challenge and an opportunity for innovative solutions in Thailand's transition towards sustainability. The concept of CCU technology, specifically co-electrolysis, to convert these emissions into methanol offers an attractive pathway for emissions reduction and value creation.

While traditional methanol production relies heavily on fossil fuels and generates substantial life cycle CO_2 emissions, the CE-100% approach utilizes CO_2 from cement plants as a feedstock. This innovative method can avoid approximately 1.96 kt of CO_2 per kt of methanol produced. This approach addresses three critical environmental concerns simultaneously: it mitigates emissions from cement production and avoids the associated emissions from conventional methanol production and coal-based electricity generation. By repurposing industrial CO_2 emissions and decreasing reliance on fossil fuels, the CE-100% technology represents a significant step towards more sustainable industrial practices and circular economy principles.

Ideally, the full conversion of cement industry emissions could yield an impressive 12.4 million tonnes of methanol annually, while simultaneously reducing CO_2 emissions by approximately 24.3 million tonnes per year, equivalent to a 75% reduction of emissions from the industrial processes and product use (IPPU) sector. This methanol production capacity, however, far exceeds Thailand's projected domestic demand of 800,000 tonnes by 2034 [87], potentially positioning the country as a significant methanol exporter. Methanol's versatility as a fuel additive, chemical feedstock, and industrial solvent further enhances its appeal as a CCU product. The economic implications of such a transformation are also substantial with potential annual revenue reaching USD 5.4 billion (assuming a conservative methanol price of USD 440 per tonne) [88]; this approach could significantly impact Thailand's trade balance, shifting the country from a methanol importer to a major exporter. However, the practical implementation of this ambitious plan faces considerable challenges. It would require extensive infrastructure development, including large-scale CO_2 capture facilities at cement plants, electrolysis plants for hydrogen production, methanol synthesis plants, and associated transportation and storage systems. The necessary technology investments are substantial, containing not only CCU-specific equipment but also significant upgrades to the power grid to accommodate the increased renewable energy capacity required to power the CE process.

Market dynamics present another crucial consideration. While Thailand's domestic methanol demand is growing, absorbing the potential excess production would necessitate developing new international markets and competing with established methanol producers. This may also involve creating new domestic applications for methanol to increase local consumption. Regulatory support would play a vital role in making this vision a reality. Implementing carbon pricing mechanisms, providing subsidies or tax incentives for CCU and methanol production, mandating methanol blending in fuels, and streamlining permitting processes for new facilities would all be essential steps in creating a favorable environment for this transition. Despite these challenges, the potential benefits are compelling. This approach could lead to a significant reduction in industrial CO_2 emissions, create a new sustainable industry and export market, support Thailand's transition to a low-carbon economy, generate jobs in CCU and related sectors, and reduce dependence on fossil fuel imports.

4. Conclusions

Carbon capture and utilization (CCU) for methanol production offers a promising solution to address greenhouse gas emissions and reduce fossil fuel dependence in the chemical sector. This research employed a comprehensive life cycle assessment to evaluate the environmental impacts of CO_2-based and conventional methanol production methods in Thailand, focusing on CCU technologies. Among the 19 evaluated pathways, co-electrolysis of CO_2 and water using solid-oxide electrolyzers demonstrated the most substantial environmental benefits, particularly under a 100% renewable energy scenario (CE-100%). This process showed the highest reductions across all impact categories compared to conventional steam methane reforming (Conv-SMR), attributed to its high electrical efficiency and effective heat integration.

CO_2 hydrogenation using hydrogen from alkaline water electrolysis (Hyd-Alk-100%) also showed significant benefits when powered by renewable energy, though its advantages diminish with grid electricity use. Tri-reforming of methane, especially Tri-Alk-100%, presented a balanced and robust performance across various scenarios. In contrast, electrochemical reduction of CO_2 consistently exhibited the worst environmental performance due to high energy consumption and poor methanol yield. Implementing CE-100% methanol production using CO_2 emissions from Thailand's cement industry could potentially produce 12.4 million tonnes of methanol annually, valued at approximately USD 5.4 billion, while reducing CO_2 emissions by 24.3 million tonnes per year. This represents a 75% reduction in emissions from the industrial processes and product use (IPPU) sector. This study provides crucial insights for policymakers, industry stakeholders, and researchers, aiding Thailand's transition to greener methanol production and advancing its climate change goals. Carbon capture and utilization for the production of bulk chemicals is an emerging field with significant potential. To make informed decisions, future research should incorporate economic assessments to identify cost-effective solutions for sustainable methanol production.

Supplementary Materials: The following supporting information can be downloaded at: https://www.mdpi.com/article/10.3390/pr12091868/s1.

Author Contributions: A.R. contributed to Conceptualization, Writing the Original Draft, Methodology, Data Acquisition and Formal Analysis, and Writing—Review and Editing. A.F. contributed to Conceptualization, Methodology, and Writing—Review and Editing. S.H.G. contributed to Conceptualization, Methodology, Writing—Review and Editing, and Supervision. All authors have read and agreed to the published version of the manuscript.

Funding: This research received no external funding.

Data Availability Statement: Data are contained within the article and Supplementary Materials.

Acknowledgments: The authors would like to acknowledge King Mongkut's University of Technology Thonburi (KMUTT), the Joint Graduate School of Energy and Environment (JGSEE), and the Center of Excellence on Energy Technology and Environment.

Conflicts of Interest: The authors declare no conflicts of interest.

References

1. Kreps, B.H. The rising costs of fossil-fuel extraction: An energy crisis that will not go away. *Am. J. Econ. Sociol.* **2020**, *79*, 695–717. [CrossRef]
2. Day, C.; Day, G. Climate change, fossil fuel prices and depletion: The rationale for a falling export tax. *Econ. Modell.* **2017**, *63*, 153–160. [CrossRef]
3. Wang, J.; Azam, W. Natural resource scarcity, fossil fuel energy consumption, and total greenhouse gas emissions in top emitting countries. *Geosci. Front.* **2024**, *15*, 101757. [CrossRef]
4. Debating Science. The Effects of Arctic Offshore Drilling on Marine Ecosystems and Wildlife. 2016. Available online: https://websites.umass.edu/natsci397a-eross/the-effects-of-arctic-offshore-drilling-on-marine-ecosystems-and-wildlife/ (accessed on 2 August 2024).
5. Organization of the Petroleum Exporting Countries. World Oil Outlook 2045. 2023. Available online: https://www.opec.org/opec_web/en/press_room/7042.htm (accessed on 24 August 2024).
6. Sayed, E.T.; Olabi, A.G.; Alami, A.H.; Radwan, A.; Mdallal, A.; Rezk, A.; Abdelkareem, M.A. Renewable energy and energy storage systems. *Energies* **2023**, *16*, 1415. [CrossRef]
7. Chia, S.R.; Nomanbhay, S.; Ong, M.Y.; Chew, K.W.; Show, P.L. Renewable diesel as fossil fuel substitution in Malaysia: A review. *Fuel* **2022**, *314*, 123137. [CrossRef]
8. Bertau, M.; Offermanns, H.; Plass, L.; Schmidt, F.; Wernicke, H.-J. *Methanol: The Basic Chemical and Energy Feedstock of the Future*; Springer: Berlin/Heidelberg, Germany, 2014; Volume 1.
9. Dalena, F.; Senatore, A.; Marino, A.; Gordano, A.; Basile, M.; Basile, A. *Chapter 1-Methanol Production and Applications: An Overview*; Methanol; Elsevier: Amsterdam, The Netherlands, 2018; pp. 3–28.
10. Simon Araya, S.; Liso, V.; Cui, X.; Li, N.; Zhu, J.; Sahlin, S.L.; Jensen, S.H.; Nielsen, M.P.; Kær, S.K. A review of the methanol economy: The fuel cell route. *Energies* **2020**, *13*, 596. [CrossRef]
11. Olah, G.A.; Goeppert, A.; Prakash, G.S. *Beyond Oil and Gas: The Methanol Economy*; John Wiley & Sons: Hoboken, NJ, USA, 2011.
12. Khojasteh-Salkuyeh, Y.; Ashrafi, O.; Mostafavi, E.; Navarri, P. CO_2 utilization for methanol production; Part I: Process design and life cycle GHG assessment of different pathways. *J. CO_2 Util.* **2021**, *50*, 101608. [CrossRef]
13. Win, S.Y.; Opaprakasit, P.; Papong, S. Environmental and economic assessment of carbon capture and utilization at coal-fired power plant in Thailand. *J. Clean. Prod.* **2023**, *414*, 137595. [CrossRef]
14. Tabibian, S.S.; Sharifzadeh, M. Statistical and analytical investigation of methanol applications, production technologies, value-chain and economy with a special focus on renewable methanol. *Renew. Sust. Energy Rev.* **2023**, *179*, 113281.
15. Artz, J.; Müller, T.E.; Thenert, K.; Kleinekorte, J.; Meys, R.; Sternberg, A.; Bardow, A.; Leitner, W. Sustainable conversion of carbon dioxide: An integrated review of catalysis and life cycle assessment. *Chem. Rev.* **2018**, *118*, 434–504.
16. Solomon, M.B.; Rabha, S.S.; Fimbres-Weihs, G.; Goyal, H.; Taghikhah, F.R.; Varghese, J.J.; Wenger, S.R.; Liang, W.; Kearns, E.R.; Huang, J. Decarbonization in Australia and India: Bilateral opportunities and challenges for the net zero transformation. *ACS Eng. Au* **2024**, *4*, 295–311. [CrossRef]
17. Langie, K.M.G.; Tak, K.; Kim, C.; Lee, H.W.; Park, K.; Kim, D.; Jung, W.; Lee, C.W.; Oh, H.-S.; Lee, D.K. Toward economical application of carbon capture and utilization technology with near-zero carbon emission. *Nat. Commun.* **2022**, *13*, 7482. [CrossRef]
18. Song, B.; Liang, Y.; Zhou, Y.; Zhang, L.; Li, H.; Zhu, N.-X.; Tang, B.Z.; Zhao, D.; Liu, B. CO_2-Based Stable Porous Metal–Organic Frameworks for CO_2 Utilization. *J. Am. Chem. Soc.* **2024**, *146*, 14835–14843. [CrossRef]
19. Pérez-Fortes, M.; Schöneberger, J.C.; Boulamanti, A.; Tzimas, E. Methanol synthesis using captured CO_2 as raw material: Techno-economic and environmental assessment. *Appl. Energy* **2016**, *161*, 718–732. [CrossRef]
20. Rajabloo, T.; Valee, J.; Marenne, Y.; Coppens, L.; De Ceuninck, W. Carbon capture and utilization for industrial applications. *Energy Rep.* **2023**, *9*, 111–116. [CrossRef]
21. Moore, T.; Oyarzun, D.I.; Li, W.; Lin, T.Y.; Goldman, M.; Wong, A.A.; Jaffer, S.A.; Sarkar, A.; Baker, S.E.; Duoss, E.B. Electrolyzer energy dominates separation costs in state-of-the-art CO_2 electrolyzers: Implications for single-pass CO_2 utilization. *Joule* **2023**, *7*, 782–796. [CrossRef]
22. Idem, R.; Supap, T.; Shi, H.; Gelowitz, D.; Ball, M.; Campbell, C.; Tontiwachwuthikul, P. Practical experience in post-combustion CO2 capture using reactive solvents in large pilot and demonstration plants. *Int. J. Greenh. Gas Control* **2015**, *40*, 6–25. [CrossRef]
23. Rezaei, S.; Liu, A.; Hovington, P. Emerging technologies in post-combustion carbon dioxide capture & removal. *Catal. Today* **2023**, *423*, 114286.
24. Theo, W.L.; Lim, J.S.; Hashim, H.; Mustaffa, A.A.; Ho, W.S. Review of pre-combustion capture and ionic liquid in carbon capture and storage. *Appl. Energy* **2016**, *183*, 1633–1663.
25. Theofanidis, S.A.; Antzaras, A.N.; Lemonidou, A.A. CO2 as a building block: From capture to utilization. *Curr. Opin. Chem. Eng.* **2023**, *39*, 100902. [CrossRef]
26. Dang, S.; Yang, H.; Gao, P.; Wang, H.; Li, X.; Wei, W.; Sun, Y. A review of research progress on heterogeneous catalysts for methanol synthesis from carbon dioxide hydrogenation. *Catal. Today.* **2019**, *330*, 61–75. [CrossRef]
27. Rumayor, M.; Dominguez-Ramos, A.; Irabien, A. Innovative alternatives to methanol manufacture: Carbon footprint assessment. *J. Clean. Prod.* **2019**, *225*, 426–434. [CrossRef]
28. Dueñas, D.M.A.; Riedel, M.; Riegraf, M.; Costa, R.; Friedrich, K.A. High temperature co-electrolysis for power-to-X. *Chem. Ing. Tech.* **2020**, *92*, 45–52. [CrossRef]

29. Ni, M.; Leung, M.K.; Leung, D.Y. Technological development of hydrogen production by solid oxide electrolyzer cell (SOEC). *Int. J. Hydrogen Energy* **2008**, *33*, 2337–2354. [CrossRef]
30. Leonzio, G. Design and feasibility analysis of a Power-to-Gas plant in Germany. *J. Clean. Prod.* **2017**, *162*, 609–623.
31. Sterner, M.; Stadler, I. *Energiespeicher-Bedarf, Technologien, Integration*; Springer: Berlin/Heidelberg, Germany, 2014.
32. Zhang, X.; Song, Y.; Wang, G.; Bao, X. Co-electrolysis of CO_2 and H_2O in high-temperature solid oxide electrolysis cells: Recent advance in cathodes. *J. Energy Chem.* **2017**, *26*, 839–853. [CrossRef]
33. Hoppe, W.; Thonemann, N.; Bringezu, S. Life cycle assessment of carbon dioxide–based production of methane and methanol and derived polymers. *J. Ind. Ecol.* **2018**, *22*, 327–340. [CrossRef]
34. Nimmanterdwong, P.; Piumsomboon, P.; Chalermsinsuwan, B. Emergy investigation of carbon dioxide utilization processes for methanol synthesis. *J. Environ. Chem. Eng.* **2022**, *10*, 108063. [CrossRef]
35. Rosental, M.; Fröhlich, T.; Liebich, A. Life cycle assessment of carbon capture and utilization for the production of large volume organic chemicals. *Front. Clim.* **2020**, *2*, 586199. [CrossRef]
36. Michailos, S.; Sanderson, P.; Villa Zaragoza, A.; McCord, S.; Armstrong, K.; Styring, P.; Mason, F.; Stokes, G.; Williams, E.; Zimmermann, A. *Methanol Worked Examples for the TEA and LCA Guidelines for CO_2 Utilization*; University of Michigan Library: Ann Arbor, MI, USA, 2018. [CrossRef]
37. *ISO 14040*; Environmental Management—Life Cycle Assessment: Principles and Framework. International Organization for Standardization: Geneva, Switzerland, 2006.
38. *ISO 14044*; Environmental Management—Life Cycle Assessment: Requirements and Guidelines. International Organization for Standardization: Geneva, Switzerland, 2006.
39. Kreuter, W.; Hofmann, H. Electrolysis: The important energy transformer in a world of sustainable energy. *Int. J. Hydrogen Energy* **1998**, *23*, 661–666. [CrossRef]
40. Mergel, J.; Carmo, M.; Fritz, D. Status on technologies for hydrogen production by water electrolysis. In *Transition to Renewable Energy Systems*; Wiley-VCH Verlag GmbH & Co. KGaA: Weinheim, Germany, 2013; pp. 423–450.
41. Schiebahn, S.; Grube, T.; Robinius, M.; Tietze, V.; Kumar, B.; Stolten, D. Power to gas: Technological overview, systems analysis and economic assessment for a case study in Germany. *Int. J. Hydrogen Energy* **2015**, *40*, 4285–4294. [CrossRef]
42. Schmidt, O.; Gambhir, A.; Staffell, I.; Hawkes, A.; Nelson, J.; Few, S. Future cost and performance of water electrolysis: An expert elicitation study. *Int. J. Hydrogen Energy* **2017**, *42*, 30470–30492. [CrossRef]
43. Zeng, K.; Zhang, D. Recent progress in alkaline water electrolysis for hydrogen production and applications. *Prog. Energy Combust Sci.* **2010**, *36*, 307–326. [CrossRef]
44. Carmo, M.; Fritz, D.L.; Mergel, J.; Stolten, D. A comprehensive review on PEM water electrolysis. *Int. J. Hydrogen Energy.* **2013**, *38*, 4901–4934.
45. Sarp, S.; Hernandez, S.G.; Chen, C.; Sheehan, S.W. Alcohol production from carbon dioxide: Methanol as a fuel and chemical feedstock. *Joule* **2021**, *5*, 59–76.
46. Song, C.; Pan, W. Tri-reforming of methane: A novel concept for catalytic production of industrially useful synthesis gas with desired H_2/CO ratios. *Catal. Today* **2004**, *98*, 463–484. [CrossRef]
47. Cho, W.; Song, T.; Mitsos, A.; McKinnon, J.T.; Ko, G.H.; Tolsma, J.E.; Denholm, D.; Park, T. Optimal design and operation of a natural gas tri-reforming reactor for DME synthesis. *Catal. Today* **2009**, *139*, 261–267. [CrossRef]
48. Goeppert, A.; Czaun, M.; Jones, J.-P.; Prakash, G.S.; Olah, G.A. Recycling of carbon dioxide to methanol and derived products–closing the loop. *Chem. Soc. Rev.* **2014**, *43*, 7995–8048.
49. Arora, S.; Prasad, R. An overview on dry reforming of methane: Strategies to reduce carbonaceous deactivation of catalysts. *RSC Adv.* **2016**, *6*, 108668–108688. [CrossRef]
50. Speight, J.G. *Gasification of Unconventional Feedstocks*; Gulf Professional Publishing: Waltham, MA, USA, 2014.
51. Kumar, R.; Kumar, K.; Pant, K.; Choudary, N. Tuning the metal-support interaction of methane tri-reforming catalysts for industrial flue gas utilization. *Int. J. Hydrogen Energy* **2020**, *45*, 1911–1929. [CrossRef]
52. Challiwala, M.S.; Ghouri, M.M.; Sengupta, D.; El-Halwagi, M.M.; Elbashir, N.O. A process integration approach to the optimization of CO_2 utilization via tri-reforming of methane. *Comput. Aided Chem. Eng.* **2017**, *40*, 1993–1998.
53. Rezaei, E.; Catalan, L.J. Evaluation of CO_2 utilization for methanol production via tri-reforming of methane. *J. CO_2 Util.* **2020**, *42*, 101272. [CrossRef]
54. Usman, M.; Daud, W.W. Recent advances in the methanol synthesis via methane reforming processes. *RSC Adv.* **2015**, *5*, 21945–21972. [CrossRef]
55. Shi, C.; Labbaf, B.; Mostafavi, E.; Mahinpey, N. Methanol production from water electrolysis and tri-reforming: Process design and technical-economic analysis. *J. CO2 Util.* **2020**, *38*, 241–251. [CrossRef]
56. Wang, L.; Rao, M.; Diethelm, S.; Lin, T.-E.; Zhang, H.; Hagen, A.; Maréchal, F. Power-to-methane via co-electrolysis of H_2O and CO_2: The effects of pressurized operation and internal methanation. *Appl. Energy* **2019**, *250*, 1432–1445. [CrossRef]
57. Albarelli, J.Q.; Onorati, S.; Caliandro, P.; Peduzzi, E.; Meireles, M.A.A.; Marechal, F.; Ensinas, A.V. Multi-objective optimization of a sugarcane biorefinery for integrated ethanol and methanol production. *Energy* **2017**, *138*, 1281–1290. [CrossRef]
58. Phillips, V.; Kinoshita, C.; Neill, D.; Takahashi, P. Thermochemical production of methanol from biomass in Hawaii. *Appl. Energy* **1990**, *35*, 167–175. [CrossRef]

59. Becker, W.; Braun, R.; Penev, M.; Melaina, M. Production of Fischer–Tropsch liquid fuels from high temperature solid oxide co-electrolysis units. *Energy* **2012**, *47*, 99–115. [CrossRef]
60. Wang, L.; Pérez-Fortes, M.; Madi, H.; Diethelm, S.; Maréchal, F. Optimal design of solid-oxide electrolyzer based power-to-methane systems: A comprehensive comparison between steam electrolysis and co-electrolysis. *Appl. Energy* **2018**, *211*, 1060–1079.
61. Ertl, G.; Knözinger, H.; Schüth, F.; Weitkamp, J. *Handbook of Heterogeneous Catalysis*; Wiley-VCH Verlag GmbH & Co. KGaA: Weinheim, Germany, 2008.
62. Tijm, P.; Waller, F.; Brown, D. Methanol technology developments for the new millennium. *Appl. Catal.* **2001**, *221*, 275–282. [CrossRef]
63. Wilkinson, S.; Van De Water, L.; Miller, B.; Simmons, M.; Stitt, E.H.; Watson, M. Understanding the generation of methanol synthesis and water gas shift activity over copper-based catalysts–A spatially resolved experimental kinetic study using steady and non-steady state operation under $CO/CO_2/H_2$ feeds. *J. Catal.* **2016**, *337*, 208–220. [CrossRef]
64. Albo, J.; Irabien, A. Cu_2O-loaded gas diffusion electrodes for the continuous electrochemical reduction of CO_2 to methanol. *J. Catal.* **2016**, *343*, 232–239. [CrossRef]
65. Deka, T.J.; Osman, A.I.; Baruah, D.C.; Rooney, D.W. Methanol fuel production, utilization, and techno-economy: A review. *Environ. Chem. Lett.* **2022**, *20*, 3525–3554. [CrossRef]
66. Pirola, C.; Bozzano, G.; Manenti, F. *Chapter 3-Fossil or Renewable Sources for Methanol Production? Methanol*; Elsevier: Amsterdam, The Netherlands, 2018; pp. 53–93.
67. González-Garay, A.; Frei, M.S.; Al-Qahtani, A.; Mondelli, C.; Guillén-Gosálbez, G.; Pérez-Ramírez, J. Plant-to-planet analysis of CO_2-based methanol processes. *Energy Environ. Sci.* **2019**, *12*, 3425–3436. [CrossRef]
68. Maina, J.W.; Pringle, J.M.; Razal, J.M.; Nunes, S.; Vega, L.; Gallucci, F.; Dumée, L.F. Strategies for integrated capture and conversion of CO_2 from dilute flue gases and the atmosphere. *ChemSusChem* **2021**, *14*, 1805–1820. [CrossRef]
69. Zhao, H.; Luo, X.; Zhang, H.; Sun, N.; Wei, W.; Sun, Y. Carbon-based adsorbents for post-combustion capture: A review. *Greenhouse Gas. Sci. Technol.* **2018**, *8*, 11–36. [CrossRef]
70. Wu, C.; Huang, Q.; Xu, Z.; Sipra, A.T.; Gao, N.; de Souza Vandenberghe, L.P.; Vieira, S.; Soccol, C.R.; Zhao, R.; Deng, S. A comprehensive review of carbon capture science and technologies. *Carbon Capture Sci. Technol.* **2024**, *11*, 100178.
71. BUR4. The UNFCCC Secretariat (UN Climate Change). Thailand. 2022. Available online: https://unfccc.int/sites/default/files/resource/Thailand_BUR4_final_28122022.pdf (accessed on 15 June 2024).
72. Rochelle, G.T. Amine scrubbing for CO_2 capture. *Science* **2009**, *325*, 1652–1654. [CrossRef]
73. Abu-Zahra, M.R.; Schneiders, L.H.; Niederer, J.P.; Feron, P.H.; Versteeg, G.F. CO_2 capture from power plants: Part I. A parametric study of the technical performance based on monoethanolamine. *Int. J. Greenh. Gas Control* **2007**, *1*, 37–46. [CrossRef]
74. Liang, Z.H.; Rongwong, W.; Liu, H.; Fu, K.; Gao, H.; Cao, F.; Zhang, R.; Sema, T.; Henni, A.; Sumon, K. Recent progress and new developments in post-combustion carbon-capture technology with amine based solvents. *Int. J. Greenh. Gas Control* **2015**, *40*, 26–54.
75. Wu, X.; Yu, Y.; Qin, Z.; Zhang, Z. The advances of post-combustion CO2 capture with chemical solvents: Review and guidelines. *Energy Proc.* **2014**, *63*, 1339–1346. [CrossRef]
76. Dave, N.; Do, T.; Palfreyman, D.; Feron, P. Impact of liquid absorption process development on the costs of post-combustion capture in Australian coal-fired power stations. *Chem. Eng. Res. Des.* **2011**, *89*, 1625–1638. [CrossRef]
77. Dumée, L.; Scholes, C.; Stevens, G.; Kentish, S. Purification of aqueous amine solvents used in post combustion CO_2 capture: A review. *Int. J. Greenh. Gas Control* **2012**, *10*, 443–455. [CrossRef]
78. Beysel, G. Enhanced cryogenic air separation—a proven process applied to oxyfuel. In Proceedings of the 1st Oxyfuel Combustion Conference, Cottbus, Germany, 8–11 September 2009.
79. Patel, G.H.; Havukainen, J.; Horttanainen, M.; Soukka, R.; Tuomaala, M. Climate change performance of hydrogen production based on life cycle assessment. *Green Chem.* **2024**, *26*, 992–1006. [CrossRef]
80. Rodríguez-Vallejo, D.F.; Valente, A.; Guillén-Gosálbez, G.; Chachuat, B. Economic and life-cycle assessment of OME 3–5 as transport fuel: A comparison of production pathways. *Sustain. Energy Fuels* **2021**, *5*, 2504–2516. [CrossRef]
81. Gerloff, N. Comparative Life-Cycle-Assessment analysis of three major water electrolysis technologies while applying various energy scenarios for a greener hydrogen production. *J. Energy Storage* **2021**, *43*, 102759. [CrossRef]
82. Koornneef, J.; van Keulen, T.; Faaij, A.; Turkenburg, W. Life cycle assessment of a pulverized coal power plant with post-combustion capture, transport and storage of CO_2. *Int. J. Greenh. Gas Control* **2008**, *2*, 448–467. [CrossRef]
83. Zhang, H.; Desideri, U. Techno-economic optimization of power-to-methanol with co-electrolysis of CO_2 and H_2O in solid-oxide electrolyzers. *Energy* **2020**, *199*, 117498. [CrossRef]
84. Huijbregts, M.A.; Steinmann, Z.J.; Elshout, P.M.; Stam, G.; Verones, F.; Vieira, M.; Zijp, M.; Hollander, A.; Van Zelm, R. ReCiPe 2016: A Harmonized Life Cycle Impact Assessment Method at Midpoint and Endpoint Level, Report I: Characterization. Available online: https://www.rivm.nl/bibliotheek/rapporten/2016-0104.pdf (accessed on 21 July 2024).
85. Ravikumar, D.; Keoleian, G.A.; Miller, S.A.; Sick, V. Assessing the relative climate impact of carbon utilization for concrete, chemical, and mineral production. *Environ. Sci. Technol.* **2021**, *55*, 12019–12031. [CrossRef]

86. Reccessary. Thailand's New Power Plan Aims for 51% Renewable Energy by 2037. 2024. Available online: https://www.reccessary.com/en/news/th-regulation/thailand-new-power-development-plan-raises-renewable-energy-targets (accessed on 8 August 2024).
87. Chemanalyst. Decode the Future of Methanol. 2024. Available online: https://www.chemanalyst.com/industry-report/thailand-methanol-market-200 (accessed on 7 August 2024).
88. Schorn, F.; Breuer, J.L.; Samsun, R.C.; Schnorbus, T.; Heuser, B.; Peters, R.; Stolten, D. Methanol as a renewable energy carrier: An assessment of production and transportation costs for selected global locations. *Adv. Appl. Energy* **2021**, *3*, 100050. [CrossRef]

processes

MDPI

Review

re-ISSUES—Renewable Energy-Linked Interoperable Smart and Sustainable Urban Environmental Systems

Raúl Pastor [1,*], Antonio Lecuona [2] and Anabel Fraga [3]

[1] Comisión de Energía, Colegio Oficial de Ingenieros de Madrid (COIIM), 28002 Madrid, Spain
[2] Instituto de Desarrollo Tecnológico y Promoción de la Innovación Pedro Juan de Lastanosa, Universidad Carlos III de Madrid, 28911 Leganés, Spain; lecuona@ing.uc3m.es
[3] Computer Science and Engineering, Universidad Carlos III de Madrid, 28911 Leganés, Spain; afraga@inf.uc3m.es
* Correspondence: raulpastorgarcia@coiim.es

Abstract: Smart cities will be smart if they improve their citizens' quality of life; to do so, it is essential to listen to citizens and collaborate with service and technological companies. For that, digitalization seems essential. Environmental management systems are complex and expensive. If their lifecycle costs are reduced, these systems would be more sustainable. This can be achieved through citizen collaboration (CS), the use of low-cost Internet of Things (IoT) devices, and collaboration with local renewable energy businesses. All this leads to a real interoperability challenge. Systems engineering offers a valid framework for managing information and knowledge for environmental systems. It offers a range of guides for processes that can improve the quality of the related information and the reusability of knowledge throughout the lifecycles of these systems. After quantifying the opportunity and the cost for a motivational case of atmospheric neighborhood odor impact and introducing trends and opportunities in energy management, the authors propose a model for renewable energy-linked interoperable smart and sustainable urban environmental systems (re-ISSUES). The model's ontology is used to discover research trends and potential for improvements to the model itself, enabling semantic interoperability and knowledge reuse.

Keywords: smart city; citizen science; IoT; integrated cost; environmental systems; energy management; systems engineering; re-ISSUES

1. Introduction

Listening to citizens and collaborating with service and technological companies as an integral approach is essential for Smart Cities.

First, waste and water treatment facilities may cause undesirable odors because of underestimated uncertainties in using models and data during the design stage. Central heating using biomass boilers may be closer than expected by design to homes and other essential services (health, education, etc.). This is the bad side, but there is a good side because these industrial activities contribute to urban decarbonization by reducing fossil resource consumption. Presumably, they operate within the neighborhood, whose evolution is under continuous adjustments because of the growth of residential areas [1]. It is mandatory to make them compatible, with the help of further public or private activities and developments. At the same time, renewable energies like photovoltaic or solar thermal can be implemented in the city.

As another example, the so-called local or proximity economy after COVID-19 increased so-called "ghost kitchens", industrial kitchens without an in situ restaurant but for remote consumption, emitting smoke and noise within urban districts [2] and also increasing urban traffic. Usually, they are installed in the basement of a community residence building. In certain extreme cases, not necessarily unlikely ones, homeowners may suffer not only because they must fight to defend their rights to a home not invaded by air pollution, but also because of the loss of property value. A consequence of not attending to their claims can be that the homeowners move, causing re-urbanization, which means urbanizing unnecessarily in other nearby districts. This could be considered a type of public management inefficiency.

The above-mentioned motivating examples are just some problems that can happen, even concurrently. As a general comment, to avoid unfair persecution of industries or human suffering, it is mandatory to identify odor sources with professional measurement and verification services, from the first instant that people complain. For this, digitalization seems essential because it enables data traceability, a requisite for conducting any reliable verification activity.

2. Materials and Methods

The authors hypothesize that promoting interoperability between renewable energy generation and environmental management systems in cities will contribute to the sustainability of both types of systems, at least in the economic and social aspects of their operation. The main purpose of this work is to demonstrate the hypothesis and motivate further research efforts reusing the methods described in this work.

The workflow, summarized in Table 1, starts by introducing a context, a suitable engineering framework, and a conceptual model of an environmental measurement and verification system (EMVS), presented in Section 3. After that, a motivational example for promoting interoperability between solar energy generation systems and environmental management systems is defined, first stating the environmental problem, followed by proposing the solution's economics in Section 4, and then adding realistic cost optimization and opportunities for the environmental measurement and verification system (EMVS) economics in Section 5.

Table 1. Workflow Summary.

Content	Section	Section Title
Context, engineering framework, and conceptual model of an environmental measurement and verification system (EMVS).	Section 3	Environmental Management and the Smart City.
Motivational example of an EMVS and economics.	Section 4	Motivational Example: An Urban Odor Measurement and Verification.
EMVS economic optimization.	Section 5	Cost Optimization and New Opportunities.
High-level description of the re-ISSUES model.	Section 6	Purpose of the re-ISSUES Model.
Key questions formulation (research questions).	Section 7	Research Questions.
Populated EMVS ontology extended for bibliography search in ScienceDirect, and natural language processing (NLP).	Section 8	Ontology For Using the re-ISSSUES Model.
Textual alerts definition by extending the EMVS ontology.	Section 9	Textual Alerts Definition.
Expert analysis of interest in the literature for each question.	Section 10	Results And Discussion.

In Section 6, a high-level model of an EMVS is defined in terms of an ontology. In Section 7, some key questions to promote interoperability between renewable energy generation and environmental management systems are introduced. The questions are intended to be answered with a set of scientific literature from the ScienceDirect database and its search portal by using the ontology. Then, the ontology is extended in Section 8 to detect research trends through specific questions using the natural language processing (NLP) method as described in Section 9. Acting as experts, in Section 10, the authors verify the degree of interest for each work in each question to find trends, highlighting the main results analyzed to find potential improvements for the model.

3. Environmental Management and the Smart City

This section introduces the context of environmental management in the Smart City, a possible engineering framework for managing complexity, and concludes with an approach for an environmental measurement and verification system (EMVS).

3.1. Environmental Management Context

Environmental management systems are complex and expensive. They involve different devices, simulation models, methods, processes, and regulations. If the lifecycle cost, which is the cost of procuring, operating, and maintaining them, is reduced, these systems can be more sustainable in the economic dimension. However, the social dimension also requires attention. The authors believe that citizen collaboration (CS), the use of low-cost Internet of Things (IoT) devices, and collaboration with local renewable energy businesses can contribute to their sustainability.

One type of environmental management system is an environmental measurement and verification system (EMVS) for air pollution. Although odors can be non-polluting, they can damage private life inside dwellings. EMVS is also considered as a subsystem in larger environmental systems. At the same time, innovative EMVSs can be used for odor measurement.

Although there are several methodologies to assess an odor's impact, such as dynamic olfactometry (UNE-EN 13725:2022) [3] coupled with the use of dispersion modelling or field inspections (UNE-EN 16841:2017) [4], none of these techniques directly involve the citizens who suffer the nuisance. The recent Spanish standard UNE 77270:2023 on citizen science (CS) [5] is the first standard of its type relating to this topic. This standard describes a method for building collaborative odor maps to assess odor annoyance through real-time communication by participating citizens. This is a term used to define the participatory process through which citizens actively engage in a project, either with their intellectual effort, associated knowledge, or with tools and resources. A CS project should meet the minimum requirements based on the 10 principles of CS established by the European Citizen Science Association (ECSA) [6]. The UNE 77270:2023 standard is already being implemented using mobile apps, and citizen associations have the right to be listened to, implying a political risk for those who do not attend and react to reasonable claims [7]. Nevertheless, the citizens' co-creation processes have some intricacies because of diverse cultural, societal, and regulatory backgrounds, as the scaling-up co-creation project described previously [8] demonstrates.

The good news is that by using models for forecasting effects, an EMVS can provide alerts for industrial activities to stop operating and start assessing corrective measures to avoid shutdowns when the opportunity cost (or the cost of the risk) is too high. According to experts, low-cost sensors and IoT, in general, can contribute too [9,10].

Urban IoT devices to measure odors should be installed on buildings' roofs, eventually near solar energy systems and or micro wind energy systems. The reason is presented as follows. These systems need to be maintained regularly and, for this reason, it is expected that renewable energy service providers benefit from the action of measuring odors in the smart city, and vice versa. Communications like 5G and other key enabling technologies should be considered early for this purpose [11]. Combining data from service providers, apps in mobiles (the most used IoT devices), sensors, and calibrated models can be a real interoperability challenge. Herewith, interoperability is the ability of systems or systems' parts to exchange and use the exchanged information.

3.2. Systems Engineering as a Value-Added Framework for Environmental Management Systems and Environmental Safety

The systems engineering discipline offers a valid framework for managing information and knowledge. It offers a range of guides for processes that can improve the quality of related information and knowledge reuse throughout the lifecycle of these systems.

On the road towards sustainability, we, as technology specialists, can find support in using best practices, such as those stated in the International Council on Systems Engineering's Guidelines (INCOSE) [12]. These guidelines define a set of processes for engineering successful systems.

Interoperability management processes are key for the different stakeholders, systems, subsystems, and parts, providing an understanding of the information exchange necessary at any stage of the system's lifecycle.

Currently, the European Administration is progressing in promoting interoperability management, regulating [13] and demonstrating its feasibility with specific projects [14]. A council promoting Smart cities (SCs) or public service providers in general, must comply to ensure seamless delivery of public services.

INCOSE Systems Engineering's Guidelines also introduce the importance of reliability along the systems lifecycle, from concept to disposal. We can imagine that a failure in a waste treatment plant (a part of the circular economy of the city) may induce a problem for the environment, especially if the failure affects the plant's environmental safety measures like barriers or other essential parts. This can be extended to the metering systems providing alerts near populations or areas of vulnerable biodiversity. Also, security failures can happen, for example, after a cyberattack affecting electronic systems.

Recent research reveals that, at least for popular urban energy systems in Smart Cities, the nexus between circular economy and environmental safety is not a current trend in research [15]. One solution for avoiding missing known risks and emerging opportunities regarding this nexus and interoperability management could be promoting collaborative public-private projects designed in open innovation environments for reliability, safety, and security [16]. As emerging opportunities, we can consider those related to renewable energy in cities e. g. solar thermal, PV, biomass, and waste heat use.

3.3. Environmental Measurement and Verification System (EMVS) Definition

To promote open innovation about any environmental system, it is crucial to start using shared terminology. Table 2 contains a generic and non-normalized approach for EMVS:

Figure 1 offers a diagram presenting a generic EMVS where the main categories are represented with rounded rectangles, the potential influence relationships with arrows, and the logical and physical system levels with rectangles.

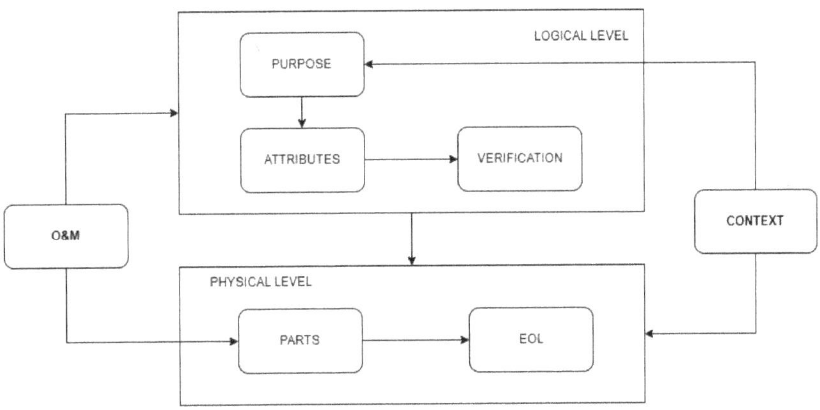

Figure 1. An EMVS conceptual model.

Table 2. Generic EMVS terminology.

Element Name	Element	Definition
CONTEXT	Context of the measurement	Type of impact or damage to detect Regulations, ISO standards, etc. Potentially affected collectives Measurement site
O&M	Operation and maintenance of the measurement and verification system	Measurement quality Models' quality maintenance Forecasting Operator
PURPOSE	Measurement system purpose	Emissions forecasting Urban mapping
PARTS	Measurement parts	Hardware Software Certifications Communications
ATTRIBUTES	Measurement system attributes	Quality Reliability Availability Cybersecurity Recovery plan
VERIFICATION	Verification activities	Models' calibration Compliance with limits and regulations
EOL	End-of-life (EOL) condition	Recyclability Environmental footprints of products and services

4. Motivational Example: Urban Odor Measurement and Verification

At this point, it is important to quantify the opportunity and the cost for a motivational case of odor management where new devices and subsystems can be added after Citizen Science (CS) and public metering, through identifying the lower integral cost of an EMVS.

A population of circa 100,000 citizens was considered, taken from a real case in a district affected by odors and air pollution. The authors prefer not to mention the name of the district, which is a lovely place to live.

4.1. The Opportunity to Use Existing Urbanization

The selected district is currently fully urbanized (with lighted streets), and includes approximately 5000 new houses, a 20% increase. This should provide no less than EUR 200/house of tax income for the City Council, according to the authors' assumptions and public tax rules [17,18]. The main assumption is that future use of such urbanized spaces depends on how and to what extent people will be protected against odors (in what follows, the coverage ratio). Figure 2 shows an aerial view taken from Google Earth.

Figure 2 shows an orange arc corresponding to a 4 km radius from the odor sources from waste management activities. In the same figure, individual housing urbanization areas are depicted in yellow: those filled in yellow are existing, and areas where housing is not present are not filled in. Existing buildings are filled in green, and unfilled green outlines indicate future buildings. Notice that new urbanization is growing in the northeast, inside the same space limited by the orange arc.

A simplified economic model was developed to quantify the opportunity and costs. The following sections describe the insights of this cost and profit model.

Figure 2. Urbanization and buildings in 2024 within a 4 km radius of the odor source. North is vertical upwards. Individual housing urbanization and other existing buildings are filled in yellow and green, respectively. Unfilled areas represent future buildings.

4.2. The Property Loss Risk or Re-Ubanizing Cost

The consequence of not measuring and verifying the risk is property value loss or depreciation for the homeowners, as detailed in recent research, e.g., [19]. This risk is enough to drive regional and local governments to take odor management actions by contracting environmental advisors to conduct odor mapping using CS, measurement, calibration, and modelling techniques.

In the motivational examples, the yearly property loss risk for homeowners leaving the district due to odor problems is estimated as 5% of an average flat sale price of approx. EUR 200,000. The probability of the value loss was estimated as 3% of the proportion of buildings without representative measurements within the district (1−X). Near the condition of X = 100% buildings covered with representative measurement, the persistence effect was modelled in an integral way with another (1−X) factor. The probability used was, at maximum, that of the yearly wind frequency from the closer odor source based on the available mesoscale wind atlas from an ERA-NET project [20]. The property loss was considered equivalent to the re-urbanizing cost (preparing streets for energy, transport, water, and communications, among other things), estimated at EUR 500 per sq. meters.

4.3. The Upfront Costs of the Measurement System

The CS app (official and based on standards) was assumed to be free for any inhabitant of the district. The major upfront cost is the cost of acquiring and installing a high-quality station, for example, EUR 50,000 € per block. This high-quality station is useful for calibrating low-cost IoT devices and sharing wireless or radio communications.

Combining high-quality stations with low-cost devices defines two measurement types: professional system or PRO (100% of the high-quality cost) and cheap but professional or MIX (50% of the high-quality cost) cost scenarios. It is possible to estimate an amortization cost for the upfront cost by assuming the measurement system lasts 10 years.

4.4. The Maintenance Cost of the Measurement System

The yearly maintenance cost of the measurement system is estimated at 20% of the yearly amortization cost of the measurement system. This estimation included the financing cost of the system (finance interest).

4.5. The Cost of Centralizing Professional Measurement and Verification Services

The cost of the measurement and verification center was estimated by adding the cost of renting 500 sq. meters to the cost of hiring and managing a professional staff, in total, circa EUR 200,000 per year. To this total cost, 15% of the amortization and maintenance cost of the measurement system was also added.

4.6. Results and Discussion of the Motivational Economic Model

For a qualitative comparison of the PRO and MIX cost scenarios, the integrated cost was normalized by the minimum cost of the PRO scenario. The opportunity is explained in 2.1. When the opportunity is greater that the integrated cost (PRO or MIX), the Council has an incentive to promote an EMVS to measure odors. This is theoretical, of course, but it motivates thinking about innovation.

Figure 3 represents the normalized integrated cost and the normalized opportunity cost. The measurement coverage (% of dwellings with representative measurements) starts at 50%, which corresponds to the theoretical case in which 50% of the population are not concerned with odors, but this is something to discover with a survey or continuously via a Social Science app (a CS application).

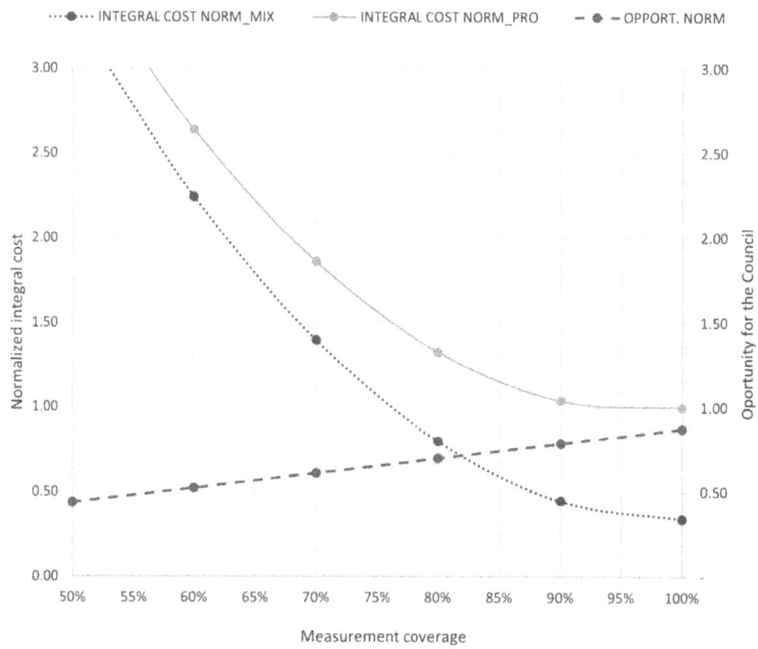

Figure 3. Normalized integrated cost (PRO and MIX scenarios) and normalized opportunity (dashed line).

As a first tentative result for the example, the maximum potential savings through choosing a MIX system instead of a PRO could be, as expected, 0.6 times the maximum cost, and this could make the difference for a generic public purchaser attending to certain feasibility criteria.

The second tentative result is that the environmental system is affordable. It may cost EUR 25/home/year, which is not negligible for a medium-income family, but as an example, it is in the range of the cost of a standard cable-TV subscription, and, as mentioned, it can be reduced via the MIX approach.

The third tentative result is that if the opportunity or the cost is affected by uncertainties, the feasibility is compromised; thus, there is a need for more opportunities (income or scale economics).

5. Cost Optimization and New Opportunities

5.1. Extending Measurement and Verification to Environmental Safety

One can observe that the system can start with a PRO configuration and evolve to a MIX configuration because it is cheaper. This could be a valid (and conservative) strategy for total cost optimization and to obtain profit. This approach requires regular verification actions that are quite automatable using systems engineering enabling software. It would be interesting to include the uncertainty of using only a PRO system to make this optimization more robust. This analysis would require accurate data and could be the object of technological innovation projects (pilots). It would be possible to optimize some costs, as suggested in Figure 4.

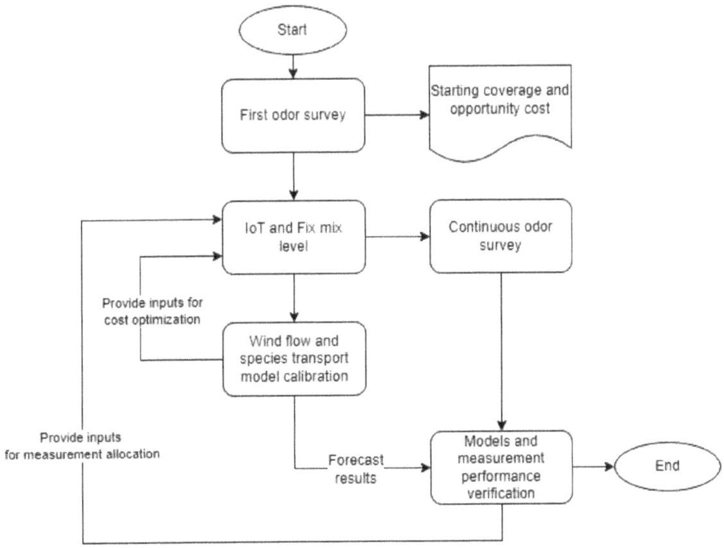

Figure 4. Suggested processes for EMVN system's cost optimization.

Once the convenience of having a measurement and verification center with forecasting capability has been justified, one can think of more opportunities to use such capability for energy management.

5.2. Trends and Opportunities in Energy Management

Choosing the right energy model for a city is not trivial. Urban planners must enable the new energy model's developments and the broader administration must control the related activities in terms of permitting (access to public assets), market trading freedom, and competitiveness. At the same time, the city purchases energy too. In this situation, urban planners and technologists must pay attention to interoperability and engineering trends and collaborate with professionals for sustainability [15]. The current energy context for cities is also conditioned by the European policy of achieving secure, clean, and affordable energy [21]. In this context, energy efficiency and renewable energy generation are key, but new market agents like the energy communities also enable other key activities like demand aggregation and collective self-consumption, making the urban energy users (private or public) act not as mere consumers, but as prosumers [22].

5.2.1. Energy Communities and Innovation

Energy communities (EC) are the objects of new regulation in Spain by order of a Law Project [23]. The agreed definitions currently include the renewable energies community (in Spanish: CER) and the citizen energy community (in Spanish: CCE) [24]. Energy communities (ECs) may try new activities for the community's benefit. For example, an EC can provide air quality services and earn money for investing more or better maintaining its renewable energy assets, and it can commercialize its renewable energy surplus or its contribution to demand management services, like distributed storage services (thermal, electric, or even hybrid) or demand response services, all using IoT to promote a smarter grid. There is nothing that forbids an EC from providing innovative services initialized for example under Public Purchase of Innovation (PPI) [25] schemes, for example, by certificating contribution to resilience in analogy to the Spanish Energy Saving Certificate (in Spanish: CAE) [26].

5.2.2. Heat Pumps & Solar Energy Systems Integration

Electric heat pumps in buildings are mostly used to provide heat (as electric heaters) or to evacuate heat (as electric chillers) from/to low thermal energy sources like the atmosphere (autothermic/aerothermic), aquifers, or the ground (autothermic/geothermic), near district heating. Of course, this can be the case for existing or retrofitted solar thermal storage subsystems allocated in buildings. It is rarer, but also possible, to use heat pumps to downgrade and amplify heat; nowadays, this is not available for acclimatization of dwellings.

Heat pump-based systems, in practice, can be allocated in the façade or specific spaces within buildings or in the city, with different environmental impact (noise generation, heating island effect contribution), risks (eventual falling objects, property value degradation, etc.) and opportunities (electric efficiency by recovery of waste heat, local work generation including maintenance, thermal energy lifecycle price optimization, raw materials consumption, circular economy competitiveness).

Cooling is becoming a growing demand in the southern cities of Europe. It is estimated that the risk of dying due to extreme heat in a city like the region of Madrid was 194 deaths/100,000 h in 2023, according to Instituto de Salud Carlos III MoMo statistics [27]. Schools should also be prepared for this important challenge and can be converted into safe spaces for citizens on the hottest days of the year, since not everybody has access to cooling systems or can afford to pay expensive electricity bills.

It is well known that the electric power required for feeding a heat pump is several times less than the heat it can generate for a higher heat sink. This relationship is called the coefficient of performance (COP) and differs from stational and nominal definitions and examples [28]. The lower the temperature difference between the source and the sink, the lower electricity a heat pump demands (higher COP), and this is interesting for future urban consumption for electric mobility at a minimum electric infrastructure cost or the available one. Electricity in summertime (for cooling) may come from local photovoltaic (PV) generation, saving imported power to the city, although PV does not contribute greatly in wintertime (for heating).

Regarding the motivational example, due to the year of construction, around 5000 m^2 of solar thermal flat plate collectors were installed to cover between 60% to 70% of the heat demand for sanitary hot water after the Spanish Construction Code change in 2006 [29]. The real estate bubble in Spain started in 1997 and burst in 2007 and the quality of installations declined afterwards (see an example in Figure 5). According to the Spanish Association of Solar Thermal Manufacturers (ASIT), 30% of solar energy systems are not well maintained and might not be working properly [30,31]. This problem could be solved with compulsory maintenance supervision, as is starting to happen with fire safety in Spain [32]. The surveillance and maintenance of solar installations could be performed by organizations of engineers and technologists in collaboration with energy agencies and supervised by certification agencies, and it could even be extended to attend to the same potential problem at the European level. Supervising the maintenance of existing and unattended urban thermal networks could be a natural evolution following new standards [33].

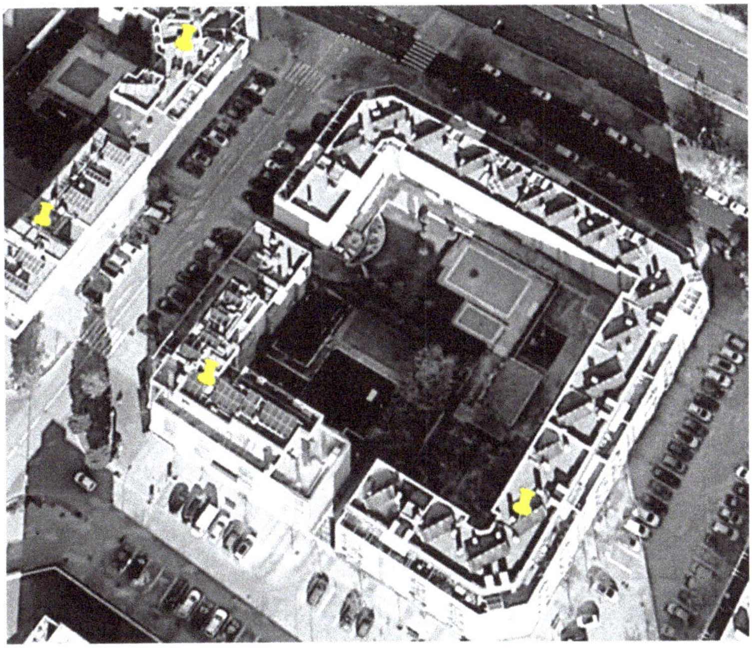

Figure 5. An example of solar thermal systems installed within the district, with 4 different solar thermal systems. The yellow pin represents an existing solar-thermal system.

Introducing a reversible high-temperature water-X heat pump that takes the solar thermal storage energy for a central (existing or new) heating and cooling system would demand minimal electrical power, which is good for future urban mobility. An additional aerothermal heat pump could complement this solution, being necessary to increase the size of the heat storage tanks. The solution can also use low thermal (condensation) boilers during the energy transition and heating and air conditioning (HVAC) installations like community hydronic networks and fan coils in homes, following national and international guidelines [34,35]. For this to be possible, the building management system (BMS) should be interoperable with the existing controls for the solar thermal system and the environmental measurement system suggested in the motivational case.

According to the authors' estimations for the case studied, using solar thermal systems in buildings could contribute to GHG savings from 15% (for central hot water systems) to 30% (for a central heating and cooling system for the same building). Of course, there are additional GHG savings to be made in heating and cooling centralization, allowing up to 70% of GHG savings in total. Properly maintaining (and if possible, extending the lifetime of) the existing solar thermal systems is smart, and so too could be the use of hybrid photovoltaic–thermal technologies (PVTs) for those roofs without solar panels yet, according to international assessments [36]. Notice also that the thermal energy storage can be heated directly using the electricity surplus, contributing to new business models for demand management in the city.

5.2.3. Digitalization and Engineering Quality for Small Projects

Engineering works are present in many activities: in the electronic systems configuration used for the demand response subsystem project the energy community (EC) will use, in the heating, ventilation, and air conditioning (HVAC) retrofit project the building owners or an EC will invest in, or in the accounting of the energy saving certificates (CAEs) being acquired by energy companies for hundreds of ECs in the future. To conduct these works, engineering uses computer-aided design (CAD), and building information modeling (BIM) is part of engineering digitalization, along with Excel® and Word® files, all containing information; exchanging this information between tools and other digital tools (for HVAC, electronics, etc.) is part of the design stage.

The information used in engineering requires specific management to guarantee certain quality, but also to be affordable for small customers. This is part of the mission that professional engineers societies (PESs), especially the industrial engineering ones, have faced for decades in Spain, according to the law [37] enabling specific insurance access to engineers.

Specific information technologies (ITs) are being used in large engineering projects to enable quality and best technical management processes like knowledge management, interoperability management, validation, verification, and risk and opportunity management, including issues such as, for instance, attending to failure cause and effects [38]. These practices are described in systems engineering guidelines [12]. To be able to transfer and use these or other enablers to small projects is a big innovation challenge for both engineering and technology professionals and the professional engineers' societies (PESs), and for this, it is necessary to gain experience through pilots in the use of new ITs and the best practices available.

After reviewing 73 recent research publications about energy systems included in the European taxonomy [39], the authors found evidence that the enablers mentioned above are at least understood by technologists, except for interoperability management. Interoperability management is necessary for automating engineering information to make small projects more affordable for customers and freer from avoidable mistakes and for alerting early engineers. PESs can attend to this need by using specific information technologies (ITs), checking engineering documents, and providing training to at least 18% of the engineers who are dedicated to engineering projects as their main activity, according to the latest report about engineering and technology professionals in Spain [40], which stated that 22% of the engineers chose to be trained in ITs.

To acquire many software licenses could be very expensive for small engineering firms. Cloud computing could reduce this cost. It makes sense to share the related infrastructure or service contracts for the environmental measurement and verification system with other benefits like the cybersecurity the Cloud can afford [41] and it becomes mandatory for essential services [42].

There is an opportunity for a measurement and verification center to be expanded to provide homeowners with energy measurement and verification services standings, based on the fact that the same dataloggers, meteorologic sensors, communications, and in-field services required by the odor metering system can be shared.

5.2.4. Circular Economy & Industry

The more savings the City Council achieves, the more interesting the business can be because it can activate public purchase of an innovation tendering process using legal instruments [43]. Of course, conducting pilots beforehand to demonstrate the feasibility of the full investment is necessary. It could be possible to extend the lifespan of the energy system or to plan a local circular economy [44]. This would also be attractive for the emerging industrial solar sector. Altogether, this is an example of the nexus between environmental safety and circular economy but also an opportunity for the application of Industry 4.0 (I4.0) where a flexible reconfiguration of a recycling plant replaces the specific characteristics of diverse types of solar thermal collectors (I4.0, decentralization design principle), aggregating such demand in a decentralized way (I4.0, service-oriented design principle) [45].

5.3. Validation of Interest and Originality of the re-ISSUES Concept

To validate the interest in the idea of linking odor measurement with solar thermal activities in cities, the authors created an online and anonymous rough questionnaire asking about the type of the respondent organization or person, allowing free answers to the questions described in Table 3, which includes the results:

Table 3. Questions about Energy Management Opportunities for the Spanish Solar Thermal sector.

Energy Management Opportunities	Question	Results	Comments
Energy Communities and Innovation	Would the solar thermal value chain be receptive to attending environmental measurements in cities to increase sales?	Yes 100%	Relevant agents: engineering firm. Not adding other motivations.
Heat Pumps and Solar Energy Systems Integration	Is the sector considering the integration of heat pumps with the existing solar thermal systems?	Yes 100%	Relevant agents: association and engineering firm. As expected.
Digitalization and Engineering Quality For Small Projects	Is the sector concerned about the interoperability challenges in engineering?	Yes 100%	The respondents did not ask about definitions; there are many types of interoperability.
Circular Economy and Industry	Is the sector considering circular economy opportunities around solar thermal installations in cities?	Yes 100%	Relevant agents: association without a specific work group on circular economy.

The number of respondents was only a few (three) and we can offer a qualitative analysis; the Association represents almost all the sector players in Spain and its answers are quite representative. A preliminary answer to the general interest in linking odor measurement with solar thermal activities in cities is that the concept is interesting for the solar thermal sector. This work can help to imagine opportunities for the sector and future EMVM sustainability systems.

Regarding the originality of connecting urban renewables to odor management in cities, a search in the European Commission Database CORDIS was conducted and the results are included in Table 4, none specifically covering this link:

Table 4. Search criteria for European Research and Development projects.

Domain of Application	Keywords	Results
Health OR Society OR Energy OR Digital Economy OR	('odour' OR 'odor') AND ('renewable' OR 'renewables')	Program: H2020. Grant Agreement 788359.

The Scalings project defined a policy roadmap in collaboration with companies, universities, policymakers, and citizens enabling or supporting innovation processes for responsible co-creation projects. The project carried out a comparative study in 10 countries about living labs, public procurement of innovation, and co-creation facilities, across a range of technical domains; nevertheless, the project did not use ontologies to promote semantic interoperability.

6. The re-ISSUES Model Purpose

The authors propose a model for renewable energy-linked interoperable smart and sustainable urban environmental systems, from now on referred to as re-ISSUES.

The re-ISSUES model is intended to be the following:

- An URBAN ENVIRONMENTAL SYSTEM enabling environmental management that combines diverse types of data, devices, agents, measurement and verification processes, and methods;
- SMART, because it allows calibrating chemical species dispersion models to provide technical alerts for industrial activities to operate safely or to take further investment actions in the Smart City (SC) context;
- SUSTAINABLE, because the EMVM system cost can be optimized by reusing knowledge and collaborating with local renewable energy businesses;
- INTEROPERABLE, because it introduces interoperability management at a semantic level, through the concerned professionals, and at a technical level, allowing different IoT devices and data sources to exchange and use valuable data.

A first conceptual model, based on Section 3.3, initiates an ontology to discover research trends and potential for improvements to the model itself, enabling semantic interoperability and knowledge reuse using the same methodology already explained in [15].

7. Research Questions

The following questions may uncover trends in the SC knowledge domain:

- (Q1) Is interoperability being considered in environmental management in cities?
- (Q2) Are risk-centered or risk-based approaches being considered in environmental management or environmental engineering?
- (Q3) Is CS being used for odor management in the SC?

The questions were converted into textual alerts in documents obtained from search portals. The research topic was not an urban energy system (UES), like in [15], but the environmental management in a city, although the workflow represented in Figure 6 is the same.

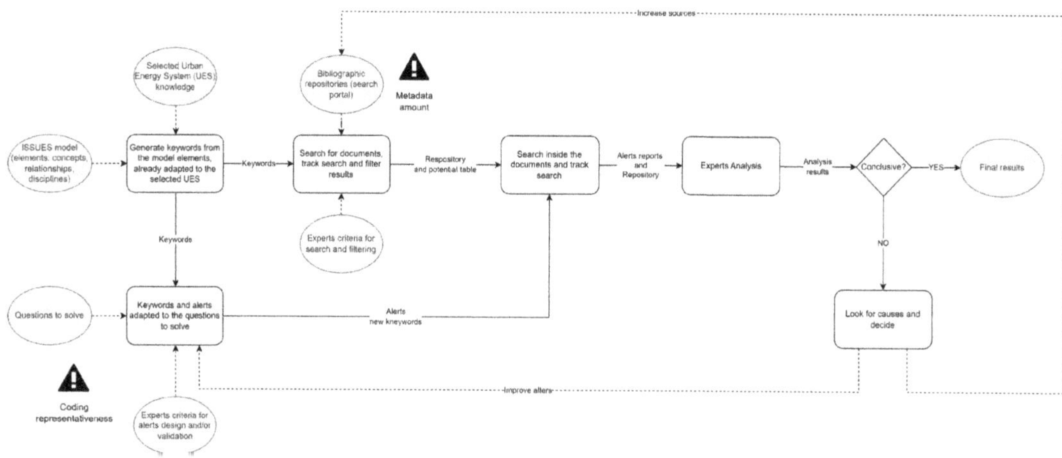

Figure 6. Search methodology representation [15].

8. Ontology for Using the re-ISSUES Model

Table 5 summarizes the ontology of the first version of the re-ISSUES model, extending the ontology initiated in Section 3.3 and described at a high level in Section 6.

Table 5. re-ISSUES Model Ontology.

Concept	Relationship with the Environmental System	Metadata and Operators (for the Search Portal)	Keywords (for the Search Portal)	<Cluster> (for Alerts)
CONTEXT	Context of the measurement	(urban OR city) AND environmental management	Measurement site AND damage AND context AND regulation	Impact, damage, detect, regulation, ISO, affected, affection, collective, measurement, context.
PURPOSE	Measurement system purpose		Purpose AND emission AND measurement system AND (map OR forecasting)	Emission, forecasting, mapping, map, purpose, measurement system.
ATTRIBUTES	Measurement system attributes		Quality AND availability AND recovery	Quality, Reliability, Availability, Cybersecurity, Recovery.
VERIFICATION	Verification activities		Verification AND calibration AND threshold	Calibration, Compliance, comply, limit, threshold.
PARTS	Measurement parts	smart city AND odor	Hardware AND software AND communication	Hardware, Software, Certification, Communication.
O&M	Operation and maintenance of the measurement and verification systems		O&M	Quality, Models' quality Forecasting, Operator, O&M.
EOL	End-of-life condition		Disposal	Recyclability, Footprint, Disposal.

9. Textual Alerts Definition

At this point, the questions were converted to textual alerts and are presented in Table 6. The strategy followed was to use semantic clusters corresponding to principles for triggering a textual alert and to close the alert detection with the more suitable semantic clusters. The reverse composition is also possible.

Table 6. Textual Alerts.

Question	Filter Cluster	Context Cluster	[Pattern 1]	[Pattern 2]
(Q1) Is interoperability being considered in environmental management in cities?	Interoperability, interoperate, interoperable	<CONTEXT>	<Filter>...<Context 1>	The reverse of pattern 1
(Q2) Are risk-centered OR risk-based approaches being considered in environmental management or environmental engineering?	Risk, risk-centered, risk-based	<ATTRIBUTES> <PARTS> <VERIFICATION> <O&M> <EOL>	<Filter>...<Context 1> OR <Context 2>... <Context 5>	The reverse of pattern 1
(Q3) Is CS being used for odors in the SC?	Citizen, science, community, civic	<PARTS> <VERIFICATION>	<Filter>...<Context 1> OR <Context 2>	The reverse of pattern 1

10. Results and Discussion

Using Table 5 search criteria, 247 readings were found in the Science Direct repository. After checking the abstracts, 31 available readings were found within the research scope, but after conducting the textual alerts search, only 9 provided alerts' results were found interesting or relevant in relation to question Q2, 6 for question Q3, and none for Q1. The most promising readings were the following readings, gathered in Table 7:

Table 7. Questions and readings for utility validation (R = relevant; I = interesting; IR = irrelevant).

ID	Bibliography Index (DOI) and Title	Verification by Experts		
		Q1	Q2	Q3
1	https://doi.org/10.1016/j.seps.2024.101834. Psycho-social conditions of urban communities in the complexity of waste management: Are awareness and waste banks the main solution?		IR	
2	https://doi.org/10.1016/j.techfore.2020.120190. Comparative analysis of urban ecological management models incorporating low-carbon transformation			R
3	https://doi.org/10.1016/j.envpol.2024.123385. Assessment of environmental risk areas based on airborne pollen patterns as a response to land use and land cover distribution.		R	
4	https://doi.org/10.1016/j.jenvman.2022.115941. Seemingly bounded knowledge, trust, and public acceptance: How does citizen's environmental knowledge affect facility siting?		R	
5	https://doi.org/10.1016/j.jhazmat.2020.123943. An emerged challenge of air pollution and ever-increasing particulate matter in Pakistan; A critical review		I	I
6	https://doi.org/10.1016/j.psep.2023.04.014. Municipal solid waste landfills in lower- and middle-income countries: Environmental impacts, challenges and sustainable management practices		I	R
7	https://doi.org/10.1016/j.jclepro.2022.135460. Supporting sustainability projects at the neighbourhood scale: Green visions for the San Salvario district in Turin guided by a combined assessment framework		IR	
8	https://doi.org/10.1016/j.atmosenv.2020.117343. Evaluating the impact of PM2.5 atmospheric pollution on population mortality in an urbanized valley in the American tropics		I	I
9	https://doi.org/10.1016/j.envsci.2021.12.022. Barriers and opportunities to incorporate scientific evidence into air quality management in Mexico: A stakeholders' perspective		I	I
10	https://doi.org/10.1016/j.wroa.2024.100212. Low-cost monitoring systems for urban water management: Lessons from the field		R	

With this short list of results, providing statistics is inappropriate. Notice that the nine references are concentrated in the period 2022 to 2024, from a 2020 to 2024 selection.

Q1's answers are somehow disturbing because interoperability management could be a missing piece of environmental management in cities. Of course, the textual alert for Q1 must be improved to secure this result. Regarding Q2 and Q3, Table 8 provides some feedback on the re-ISSUES model including conclusions from the most relevant readings:

Table 8. Potential improvements for the model.

ID	Article Title	Evidence for Potential Improvements	Model Element
2	Comparative analysis of urban ecological management models incorporating low-carbon transformation [46]	A new and comprehensive management framework incorporating urban planning, industrial transformation, organizational model, environmental protection, and institutional systems was proposed.	CONTEXT PURPOSE

Table 8. *Cont.*

ID	Article Title	Evidence for Potential Improvements	Model Element
3	Assessment of environmental risk areas based on airborne pollen patterns as a response to land use and land cover distribution [47]	Spatial regionalisation in environmental risk assessment is a common practice among institutions responsible for managing public health systems. It favors the application of management plans and assessment monitoring measurements and allows the optimization of resources and funds.	CONTEXT PURPOSE
		The risk areas proposed in the Madrid region by scientific criteria can be adjusted to other operational criteria, e.g., demographic, administrative, etc., and an equivalent approach can be applied in other similar monitoring networks.	CONTEXT PURPOSE
4	Seemingly bounded knowledge, trust, and public acceptance: How does citizens' environmental knowledge affect facility siting? [48]	Results showed robust evidence that citizens' acceptance of environmental goods provision was negatively related to their perceived environmental knowledge of pollution and risks.	VERIFICATION
		This study extends the literature on citizens' trust in public service providers. This study adds to the literature by demonstrating the critical role of trust in both the government and facility operators and revealing that environmental knowledge boundedness hurts public trust and further affects public acceptance.	CONTEXT PURPOSE
6	Municipal solid waste landfills in lower- and middle-income countries: Environmental impacts, challenges and sustainable management practices [49]	Environmentally friendly, cost-effective solutions which have community acceptance are needed and are the key to sustainable solutions. Engagement from local governments, generators, NGOs, and community involvement is also required to attain and address the various initiatives taken by the government.	CONTEXT PURPOSE
10	Low-cost monitoring systems for urban water management: Lessons from the field [50]	Low-cost monitoring systems require advanced technology to ensure scientific-grade quality of data for a given research or monitoring objective. To address these limitations, we call for better documentation of the system's design process and performance for the community of practice to learn effectively from each other.	PARTS O&M
		The economic benefits of low-cost systems are promising. The environmental costs of such systems are poorly understood.	EOL
		Socio–technological challenges associated with low-cost monitoring technology—e.g., data management, communication, and cybersecurity—are highlighted in this article.	ATTRIBUTES

11. Conclusions and Further Steps

Although CS and IoT are not new, their combined use with the related methodologies and technologies seems to be promising for improving the air quality in cities, which could be especially important in districts near industrial polygons and/or environmental activities attending the city needs.

The authors provide a good motivational example to connect the solar energy integration challenges of maintenance and circular economy with the opportunity to solve air quality problems more sustainably. This example is not the only case that can be found in cities in southern Europe. The knowledge that pilot projects at the example site may generate could contribute to the sustainability of larger or industrial solar energy scales and more digital and local industry opportunities following several 4.0 principles.

The first conclusion of this research work is that there is a certain lack of knowledge about interoperability and interoperability management in recent research in the open literature. This is not a good signal, because air quality assurance is a public issue and the interoperability governance requirements are growing. Pilot projects should promote semantic and technical interoperability early in the project design, supported within an international system engineering framework. The re-ISSUES model proposed should be improved to detect more textual evidence to assure this conclusion, of course, increasing the literature with more repositories and textual alerts at the same time, as stated by the used methodology [15].

The second conclusion of this work is that risk-based approaches and CS are considered in research in environmental engineering and cities, respectively, allowing improvement of the re-ISSUES model with nine relevant inputs pointing to seven model elements. This is favorable because it should motivate the research community to provide expert support to public administration for public procurement of innovation processes and address the potential connection between allergies and odors with experts in this new topic.

Further steps in this work include estimating the effect of cost uncertainties on the economic model and using this sensitivity analysis to motivate specific studies and pilots attending major uncertainties and cost optimization actions. After verifying that the uncertainties do not compromise the contribution of the re-ISSSUES model to EMVS sustainability, it would be possible to extend the re-ISSUES model in terms of Sustainable Development Goals and impact analysis. Another step is to define a case study for an industrial polygon with manufacturing activities to enrich the types of assessments with an example of a recent industrial and urban decarbonization process [51]. For this, it will be necessary to assess first the convenience of professionally supervising the maintenance of thermal networks to set the basis for future maintenance of PVT installations, and for unattended thermal energy networks in Spain and other southern countries in Europe. Of course, the re-ISSUES model will be exported to interoperability formalization standard OWL2, the Web ontology language for the Semantic Web, and RDF, the standard model for data exchange on the Web, to promote semantic interoperability between different urban and industrial businesses and urban planners, and also provide a technical interoperability basis, thereby enabling the contribution of the re-ISSUES model to future EMVS sustainability.

Author Contributions: Conceptualization, R.P.; methodology, A.L. and A.F.; investigation, R.P., A.F. and A.L.; writing—original draft preparation, R.P.; writing—review and editing, R.P., A.L. and A.F. All authors have read and agreed to the published version of the manuscript.

Funding: This research received no external funding.

Acknowledgments: The authors would like to thank The Reuse Company for assisting us in the use of SES ENGINEERING Studio's RISK&ALERTS and KM-KNOWLEDGE manager capabilities, AMBIENTE ET ODORA, INERCO, ENVIRA, AZENTÚA, and ANDT for the information exchange about EMVS. Also, to Colegio Oficial de Ingenieros Industriales de Madrid (COIIM) [Miriam Rodríguez, Daniel Cerrato, Santiago Calvo] and ASIT/SOLPLAT for validating re-ISSUES interest.

Conflicts of Interest: The authors declare no conflicts of interest.

References

1. European Commission. Publication of new EU Environmental Standards for the Waste Incineration Sector. 18 December 2019. Available online: https://commission.europa.eu/news/publication-new-eu-environmental-standards-waste-incineration-sector-2019-12-18_en (accessed on 10 April 2024).
2. Burbano, L. How and Why Are Some Cities Challenging Ghost Kitchens. 11 January 2024. Available online: https://www.tomorrow.city/how-and-why-are-some-cities-challenging-ghost-kitchens/ (accessed on 10 April 2024).
3. UNE. UNE-EN 13725:2022. 26 October 2022. Available online: https://www.une.org/encuentra-tu-norma/busca-tu-norma/norma?c=N0070408 (accessed on 10 April 2024).
4. UNE. UNE-EN 16841-1:2017. 22 November 2017. Available online: https://www.en.une.org/encuentra-tu-norma/busca-tu-norma/norma?c=N0059294 (accessed on 10 April 2024).
5. UNE. UNE 77270:2023. 26 October 2022. Available online: https://www.en.une.org/encuentra-tu-norma/busca-tu-norma/norma?c=N0070408 (accessed on 8 April 2024).
6. ECSA. ECSA 10 Principles of Citizen Science. 26 October 2021. Available online: https://eu-citizen.science/resource/88 (accessed on 10 April 2024).
7. Science for Change. OdourCollect, Improving Odour Pollution Management through Citizen Science. Available online: https://scienceforchange.eu/en/project/odourcollect/ (accessed on 10 April 2024).

8. CORDIS. Scaling Up Co-Creation: Avenues and Limits for Integrating Society in Science and Innovation. 31 July 2021. Available online: https://cordis.europa.eu/article/id/435881-think-local-act-global-can-community-led-actions-tackle-large-scale-societal-issues (accessed on 11 April 2024).
9. Valdivia, E.; Raúl, P. La ingeniería ambiental y la ciudad inteligente. *TESLA* **2019**, *23*, 60–63.
10. Fàbrega, C.; Fernández, L.; Monereo, O.; Pons-Balagué, A.; Xuriguera, E.; Casals, O.; Waag, A.; Prades, J.D. Highly Specific and Wide Range NO_2 Sensor with Color Readout. *ACS Sens.* **2017**, *2*, 1612–1618. [CrossRef] [PubMed]
11. del Campo, G.; Saavedra, E.; Piovano, L.; Luque, F.A. Virtual Reality and Internet of Things Based Digital Twin for Smart City Cross-Domain Interoperability. *Appl. Sci.* **2024**, *14*, 2747. [CrossRef]
12. INCOSE. *Systems Engineering Handbook*; John Wiley & Sons: Hoboken, NJ, USA, 2015.
13. European Commission. New Interoperable Europe Act to Deliver More Efficient Public Services through Improved Cooperation between National Administrations on Data Exchanges and IT Solutions. 21 November 2022. Available online: https://ec.europa.eu/commission/presscorner/detail/%20en/ip_22_6907 (accessed on 16 March 2024).
14. ELRA. IMOLA III Project. 13 December 2023. Available online: https://www.elra.eu/imola-iii/ (accessed on 8 April 2024).
15. Pastor, R.; Fraga, A.; López-Cózar, L. Interoperable, Smart, and Sustainable Urban Energy Systems. *Sustainability* **2023**, *15*, 13491. [CrossRef]
16. Larrañeta, J.J. Reflexiones Sobre Una Nueva Gobernanza y Gestión de Riesgos Para la Seguridad. 2022. Available online: https://www.seguritecnia.es/tecnologias-y-servicios/seguridad-corporativa-integral/reflexiones-sobre-una-nueva-gobernanza-y-gestion-de-riesgos-para-la-seguridad_20220624.html (accessed on 8 April 2024).
17. Madrid. Calculadora Estimativa de la Plusvalía Municipal del Ayuntamiento de Madrid. Available online: https://agenciatributaria.madrid.es/portales/contribuyente/es/Informacion-y-documentacion/Utilidades/Simuladores-y-programas-de-ayuda/Nueva-calculadora-de-la-plusvalia-municipal-del-Ayuntamiento-de-Madrid/?vgnextfmt=default&vgnextoid=5f69e038103bd710VgnVCM1000001d4a900aRCRD&vgnextchannel=f113e5bcc9c78710VgnVCM1000008a4a900aRCRD (accessed on 26 August 2024).
18. INE. Cuánto ha variado el IPC desde? Available online: https://www.ine.es/varipc/verVariaciones.do?idmesini=2&anyoini=2006&idmesfin=2&anyofin=2024&ntipo=1&enviar=Calcular (accessed on 8 April 2024).
19. Diaz, C.N.; Izquierdo, C.; Antón, A.; Kavanagh, R.; Marmol, I. Economic Impact of the Lack of Odour Management. 22 July 2022. Available online: https://www.olores.org/en/techniques/dispersion-modelling/1225-economic-impact-of-the-lack-of-odour-management (accessed on 10 April 2024).
20. NEWA. NEWA Project. 2013. Available online: https://www.neweuropeanwindatlas.eu/ (accessed on 2 April 2024).
21. EEA. Secure, Affordable, and Clean Energy in Times of Crisis? 28 November 2022. Available online: https://www.eea.europa.eu/signals-archived/signals-2022/articles/secure-affordable-and-clean-energy (accessed on 21 April 2024).
22. EEA. Energy Prosumers in EuropeCitizen Participation in the Energy Transition. 1 September 2022. Available online: https://www.eea.europa.eu/publications/the-role-of-prosumers-of/at_download/file (accessed on 21 April 2024).
23. MITECO. Miteco Energía. 20 April 2023. Available online: https://www.miteco.gob.es/content/dam/miteco/es/energia/files-1/_layouts/15/Proyecto%20de%20Real%20Decreto-61313.pdf (accessed on 21 April 2024).
24. IDAE. Comunidades Energéticas. 2024. Available online: https://www.idae.es/ayudas-y-financiacion/comunidades-energeticas (accessed on 21 April 2024).
25. European Commission. Guidance on Public Procurement of Innovation. 3 October 2017. Available online: https://ec.europa.eu/docsroom/documents/25724/attachments/1/translations/en/renditions/native (accessed on 16 March 2024).
26. MITECO. Sistema de Certificados de Ahorro Energético (CAE). 2023. Available online: https://www.miteco.gob.es/es/energia/eficiencia/cae.html (accessed on 24 April 2024).
27. ISCIII. MoMo. Available online: https://cnecovid.isciii.es/momo.html (accessed on 26 August 2024).
28. IDAE. La Bomba de Calor en la Rehabilitación Energética de Edificios. September 2023. Available online: https://www.idae.es/sites/default/files/documentos/publicaciones_idae/Guias_IDAE_La_Bomba_de_calor_2023_V11.pdf (accessed on 21 April 2024).
29. Ministerio de la Vivienda y Agenda Urbana. ¿Qué es el CTE? 2024. Available online: https://www.codigotecnico.org/QueEsCTE/Historia.html (accessed on 8 April 2024).
30. ASIT. El Economista. 11 April 2016. Available online: https://www.eleconomista.es/energia/noticias/7481784/04/16/El-30-de-las-placas-solares-termicas-no-funciona-por-mal-mantenimiento.html (accessed on 8 April 2024).
31. Fundación Laboral de la Construcción de Madrid. Por Qué el 30% de las Instalaciones de Energía Térmica no Funcionan? 1 February 2017. Available online: https://aragon.fundacionlaboral.org/en/current-affairs/news/reportages/paneles-solares-en-el-olvido (accessed on 8 April 2024).
32. UNE. UNE 192005-2:2021. 6 October 2021. Available online: https://www.en.une.org/encuentra-tu-norma/busca-tu-norma/norma?c=N0066785 (accessed on 22 April 2024).
33. UNE. UNE-EN 13941-2:2019+A1:2021 (Ratificada). 1 February 2022. Available online: https://www.en.une.org/encuentra-tu-norma/busca-tu-norma/norma?c=N0068059 (accessed on 22 April 2024).
34. IDAE. Guía Técnica de Instalaciones de Climatización Por Agua. June 2012. Available online: https://www.idae.es/sites/default/files/publications/documents/documentos_18_Guia_tecnica_instalaciones_de_climatizacion_por_agua_ed78f988.pdf (accessed on 21 April 2024).

35. ASHRAE. Guideline 36-2021: What's New and Why It's Important. 14 September 2021. Available online: https://www.ashrae.org/news/ashraejournal/guideline-36-2021-what-s-new-and-why-it-s-important (accessed on 21 April 2024).
36. Lämmle, M.; Herrando, M.; Ryan, G. Basic Concepts of PVT Collector Technologies, Applications and Markets. 1 May 2020. Available online: https://www.iea-shc.org/Data/Sites/1/publications/IEA-SHC-Task60-D5-Basic-Concepts-of-PVT-Technologies.pdf (accessed on 21 April 2024).
37. BOE. Real Decreto 1332/2000, de 7 de Julio, Por el Que se Aprueban los Estatutos Generales de los Colegios Oficiales de Ingenieros Industriales y de su Consejo General. 22 July 2000. Available online: https://www.boe.es/buscar/doc.php?id=BOE-A-2000-13866 (accessed on 21 April 2024).
38. CIMDATA. Systems Engineering Rigor Needs an Interoperability Framework. 7 December 2022. Available online: https://www.cimdata.com/en/resources/complimentary-reports-research/commentaries/item/20310-systems-engineering-rigor-needs-an-interoperability-framework-the-reuse-company-s-ses-engineering-studio-is-such-a-framework-commentary (accessed on 21 April 2024).
39. European Commission. EU Taxonomy for Sustainable Activities. Finance. 22 June 2020. Available online: https://finance.ec.europa.eu/sustainable-finance/tools-and-standards/eu-taxonomy-sustainable-activities_en (accessed on 4 December 2022).
40. Observatorio de la Ingeniería de España. Observatorio de la Ingeniería de España. November 2022. Available online: http://www.observatorioingenieria.es/uploads/1/2/5/9/125943317/estudio_observatorio_de_la_ingenieri%CC%81a_de_espan%CC%83a__2_.pdf (accessed on 3 January 2022).
41. Schulz, Y. Engineering Companies are Prime Targets for Cybercrime. 15 August 2023. Available online: https://www.engineering.com/story/engineering-companies-are-prime-targets-for-cybercrime (accessed on 21 April 2024).
42. BOE. Real Decreto 311/2022, de 3 de Mayo, Por el Que se Regula el Esquema Nacional de Seguridad. 4 May 2022. Available online: https://www.boe.es/buscar/act.php?id=BOE-A-2022-7191 (accessed on 21 April 2024).
43. European Commission. Innovation Procurement. 2024. Available online: https://commission.europa.eu/funding-tenders/tools-public-buyers/innovation-procurement_en (accessed on 16 March 2024).
44. Milousi, M.; Souliotis, M. A circular economy approach to residential solar thermal systems. *Renew. Energy* **2023**, *207*, 242–252. [CrossRef]
45. Halla, R.; Schumachera, S.; Bildste, A. Systematic Analysis of Industrie 4.0 Design Principles. In Proceedings of the 55th CIRP Conference on Manufacturing Systems, Lugano, Switzerland, 29 June–1 July 2022.
46. Wang, C.; Zhan, J.; Xin, Z. Comparative analysis of urban ecological management models incorporating low-carbon transformation. *Technol. Forecast. Soc. Chang.* **2020**, *159*, 120190. [CrossRef]
47. Rojo, J.; Cervigón, P.; Ferencova, Z.; Cascón, Á.; Díaz, J.G.; Romero-Morte, J.; Sabariego, S.; Torres, M.; Gutiérrez-Bustillo, A.M. Assessment of environmental risk areas based on airborne pollen patterns as a response to land use and land cover distribution. *Environ. Pollut.* **2024**, *344*, 123385. [CrossRef] [PubMed]
48. Zhang, H.; Zhou, L.; Liu, N.; Zhang, L. Seemingly bounded knowledge, trust, and public acceptance: How does citizen's environmental knowledge affect facility siting? *J. Environ. Manag.* **2022**, *330*, 115941. [CrossRef] [PubMed]
49. Mor, S.; Ravindra, K. Municipal solid waste landfills in lower- and middle-income countries: Environmental impacts, challenges and sustainable management practices. *Process Saf. Environ. Prot.* **2023**, *174*, 510–530. [CrossRef]
50. Hamel, P.; Ding, N.; Cherqui, F.; Zhu, Q.; Walcker, N.; Bertrand-Krajewski, J.-L.; Champrasert, P.; Fletcher, T.D.; McCarthy, D.T.; Navratil, O. Low-cost monitoring systems for urban water management: Lessons from the field. *Water Res. X* **2024**, *22*, 100212. [CrossRef] [PubMed]
51. Pastor, R.; Fraga, A.; Larrañeta, J.J. i-ISSUES—Industrial-Interoperable Safe and Secure Urban Energy Systems. *Appl. Sci.* **2024**, *14*, 3188. [CrossRef]

MDPI AG
Grosspeteranlage 5
4052 Basel
Switzerland
Tel.: +41 61 683 77 34

Processes Editorial Office
E-mail: processes@mdpi.com
www.mdpi.com/journal/processes

www.ingramcontent.com/pod-product-compliance
Lightning Source LLC
Chambersburg PA
CBHW061458210926
51850CB00008B/1750
* 9 7 8 3 7 2 5 8 6 2 4 6 7 *